Conducting Polymer Composites

Related Titles

Biopolymers Based Advanced Materials
ISBN: 978-0-6482205-4-1 (e-book)
ISBN: 978-0-6482205-5-8 (hardcover)

Functional Polymer Blends and Nanocomposites
ISBN: 978-0-6482205-6-5 (e-book)
ISBN: 978-0-6482205-7-2 (hardcover)

Functional Nanomaterials and Nanotechnologies: Applications for Energy & Environment
ISBN: 978-0-6482205-2-7 (e-book)
ISBN: 978-0-6482205-3-4 (softcover)

Advances in Polymer Technology: Material Development, Properties and Performance Evaluation
ISBN: 978-1-925823-00-4 (e-book)
ISBN: 978-1-925823-01-1 (hardcover)

Polymer Nanomaterials for Specialty Applications
ISBN: 978-1-925823-03-5 (e-book)
ISBN: 978-1-925823-04-2 (hardcover)

Advanced Materials
ISBN: 978-1-925823-05-9 (e-book)
ISBN: 978-1-925823-06-6 (hardcover)

Biofuels
ISBN: 978-1-925823-12-7 (e-book)
ISBN: 978-1-925823-13-4 (hardcover)

Liquid Crystalline Polymers
ISBN: 978-1-925823-16-5 (e-book)
ISBN: 978-1-925823-17-2 (hardcover)

Polymer Nanocomposites: Emerging Applications
ISBN: 978-1-925823-14-1 (e-book)
ISBN: 978-1-925823-15-8 (hardcover)

Conducting Polymer Composites

Dr. Vikas Mittal
Editor

CWP
Central West Publishing

This edition has been published by Central West Publishing, Australia

For more information about the books published by Central West Publishing, please visit https://centralwestpublishing.com

Disclaimer
Every effort has been made by the publisher, editor and authors while preparing this book, however, no warranties are made regarding the accuracy and completeness of the content. The publisher, editor and authors disclaim without any limitation all warranties as well as any implied warranties about sales, along with fitness of the content for a particular purpose. Citation of any website and other information sources does not mean any endorsement from the publisher and authors. For ascertaining the suitability of the contents contained herein for a particular lab or commercial use, consultation with the subject expert is needed. In addition, while using the information and methods contained herein, the practitioners and researchers need to be mindful for their own safety, along with the safety of others, including the professional parties and premises for whom they have professional responsibility. To the fullest extent of law, the publisher, editor and authors are not liable in all circumstances (special, incidental, and consequential) for any injury and/or damage to persons and property, along with any potential loss of profit and other commercial damages due to the use of any methods, products, guidelines, procedures contained in the material herein.

NATIONAL LIBRARY OF AUSTRALIA

A catalogue record for this book is available from the National Library of Australia

ISBN (print): 978-1-925823-22-6

Contents

Preface

Nanocomposites generated using conducting polymers as well as conducting fillers have emerged as materials of high relevance owing to superior electrical, mechanical and thermal properties, which enable their use for a large variety of commercial applications. Recent advancements towards the expansion of the family of conducting polymer systems as well as functional conducting fillers have further revolutionized the interest in conducting composites. In this respect, the main aim of the book is to summarize the fundamental aspects related to synthesis and properties of various conducting composites, along with focusing on the wide spectrum of their applications.

Chapter 1 summarizes the recent progress in graphene-polymer nanocomposite materials for polymer solar cell applications. Owing to their 2D structure, exceptional electrical and thermal conductivity, transparency, flexibility, high mechanical strength and very large specific surface area, graphene and its derivatives have emerged as ideal candidates for solar cell applications. In Chapter 2, conducting polymer composites have been emphasized as progressive electrode materials for the development of electrochemical sensors. The structural diversity, flexibility, porous morphology, strong sorption and electrochromic properties of the conducting polymers, along with large surface area, enhanced catalytic activity and large electron transfer rate of the incorporated materials are observed to ensure an overall improvement in the properties of the fabricated electrodes. Development of anti-corrosion coatings with incorporation of polyaniline adsorbed graphene nanoparticle pigments in poly(vinyl butyral) was successfully demonstrated in Chapter 3. Chapter 4 reviews the development of conductive polymer composites for electromagnetic interference shielding applications. In Chapter 5, an attempt has been made to comprehensively summarize the recent studies reporting the development of conducting polymer composite-based electrode materials for the electrochemical energy storage systems, with especial emphasis on their importance in lithium-ion batteries, lithium sulfur batteries and supercapacitors. Chapter 6 reviews recent literature studies on the preparation, morphology, mechanical and thermal properties, as well as applications of electrically conductive rubber blends prepared based on the combination of natural rubber (NR), styrene butadiene rubber (SBR), nitrile rubber (NBR), ethylene-vinyl acetate (EVA) and ethylene propylene diene monomer rubber (EPDM) with conducting polymers. In Chapter 7, research studies specifically reporting the development of conducting composites using natural fibers as reinforcement have been summarized. On the other hand, Chapter 8 provides a comprehensive review of the investigations conducted over the last decade for the development of conducting composites for biomedical applications. Chapter 9 presents conducting nanocomposites of poly(butylene succinate) with graphene nanosheets as filler, prepared by melt mixing method.

The book would not have been successfully accomplished without the support of chapter contributors. The book is dedicated to my family for unswerving support and constant motivation.

Vikas MITTAL

Chapter 1

Conductive Polymer/Graphene Composites for Photovoltaic Applications

Ana M. Díez-Pascual,[1*] José A. Luceño Sánchez,[1] Rafael Peña Capilla[2] and Pilar García Díaz[2]

[1]*Department of Analytical Chemistry, Physical Chemistry and Chemical Engineering, Faculty of Sciences, Alcalá University, 28871 Madrid, Spain*
[2]*Department of Signal Theory and Communication, Polytechnic High School, Alcalá University, 28871 Madrid, Spain*
**Corresponding author*: am.diez@uah.es

1.1 Introduction

The development of organic solar cells (OSCs) in the last decades has received specific attention of industry and academia, due to the significant features as compared to silicon-based cells, i.e. environmental friendliness, flexibility, lightweight structure and low processing costs [1,2]. The main reason for low processing costs is the generation of an active layer by solution-based methods (i.e. printing or spin-coating), which reduces device cost as it allows low temperature-based large-scale fabrication of devices.

OSCs comprise of two kind of materials that allow energy production: an electron donor and an electron acceptor. The electron donor material absorbs solar radiation photons, and these captured photons lead to the formation of excited states called excitons. The excitons are electron-hole pairs bounded by electrostatic forces, which can be split into free electrons-hole pairs if electric fields are applied. These electrons-hole pairs are later dissociated, and the resulting electrons are acquired by the electron acceptor material.

OSCs can be classified in two groups, according to the chemical structures of the components of the active layers (i.e. the p-type semiconductor as electron donor): small molecule solar cells (SMSCs) and polymer solar cells (PSCs).

PSCs are a kind of OSCs that can be transparent, a characteristic that allows them to be employed in walls, flexible electronics,

Conducting Polymer Composites, edited by Vikas Mittal

windows, etc. However, PSCs have a low power conversion efficiency (PCE) and experience photochemical degradation [3]. PSCs typically have the following structure, which is also depicted in Figure 1.1:

- Conductive glass of indium tin oxide (ITO)
- Hole transporting layer of PEDOT:PSS
- Active layer (PBDTTT and PCBM)
- Electron transport layer (i.e. LiF)
- Metal electrode with a low work function metal (i.e. Al)

On the other hand, in an inverted device structure, it is common to have ITO with a buffer layer (i.e. TiOx and ZnO) as the cathode electrode and a high work function metal (i.e. Ag or Au with a hole transporter layer, such as PEDOT:PSS) as the anode electrode [4].

The PCE and performance of PSC devices are strongly affected by the order and nature of the different layers and electrodes [5]. Bulk heterojunction (BHJ) structure can be considered the most successful PSC architecture, which is made by blending an electron-deficient fullerene as acceptor and an electron-rich conjugated polymer as donor with a bi-continuous nanoscale interpenetrating network [6].

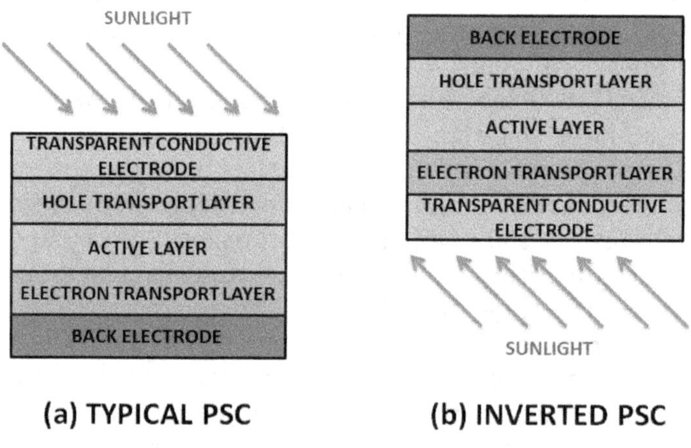

Figure 1.1 Schematic representation of a typical (a) and an inverted (b) PSC.

The characterization of a solar cell is normally carried out based on three parameters:

a) The open circuit voltage (V_{oc}), which is the maximum voltage available from a solar cell when the current through the device is zero.

b) The short circuit current density (J_{SC}), i.e. the maximum current through the solar cell at zero voltage.

c) The fill factor (FF), which can be calculated using equation 1 [7]:

$$FF = \frac{J_m \cdot V_m}{J_{SC} \cdot V_{OC}} \tag{1}$$

where J_m and V_m are the current density and voltage at the maximum power point. It shows how close the current-voltage performances of the PSCs are to ideal values; the maximum theoretical value of FF is 1. Generally, FF values ≥ 0.75 are regarded to be very high. A typical current density-voltage curve and the photovoltaic parameters are represented in Figure 1.2.

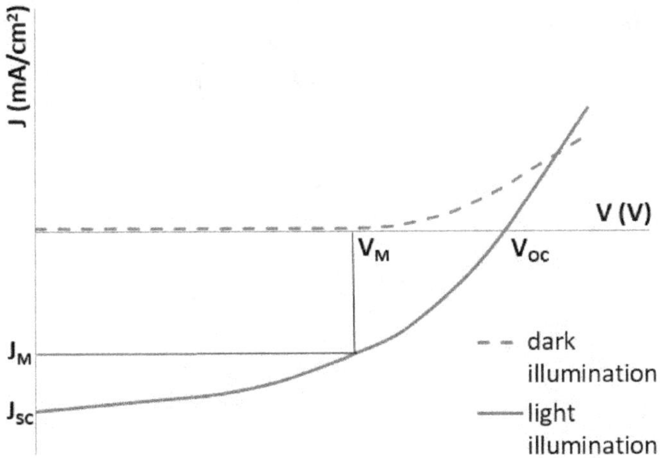

Figure 1.2 Typical current density-voltage curve of a PSC under dark (- - -) and under illumination (—).

The overall performance or power conversion efficiency (PCE) can be calculated according to equation 2:

$$PCE(\%) = \frac{V_{OC} \cdot J_{SC} \cdot FF}{P_{in}} \cdot 100 \tag{2}$$

where P_{in} is the incident solar power. The value of V_{oc} is limited by the difference between the value of the highest occupied molecular orbital (HOMO) of the polymer donor and the value of the lowest free molecular orbital (LUMO) of the fullerene acceptor [8]. The value of J_{SC} is related to the product of the photo-induced charge carrier density and the mobility within the organic semiconductors.

Recently, the synthesis of conjugated polymers and fullerene derivatives for BHJ cells has been the focus of a large number of investigations [2,6]. These compounds have the electronic properties of typical semiconductors and conductors, flexibility and lightweight features of plastics, and the possibility to be produced by a solution process. Figure 1.3 demonstrates the chemical structure of the most widely applied conjugated polymers in BHJ PCSs. As can be observed, the C=C double bonds appear in all of these compounds, which are the origin of their low-energy optical transitions, electronic properties and high-electron affinities. However, these polymers in a neutral state have insulating behavior, and these need to be doped (oxidation for p-doping or reduction for n-doping) in order to become conductive materials.

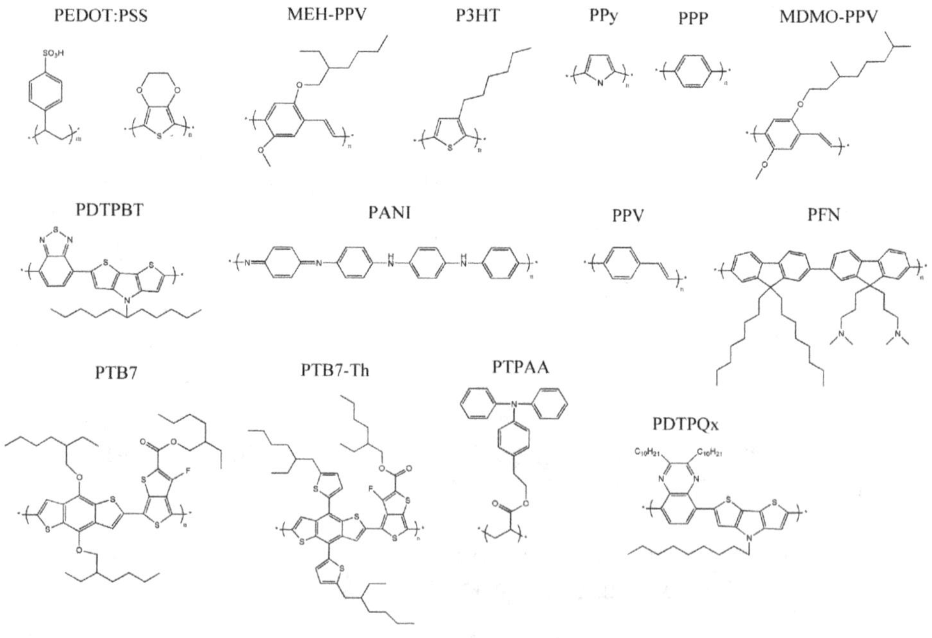

Figure 1.3 Chemical structure of the conjugated polymers widely used in PSCs.

The monomer ethylenedioxythiophene (EDOT) can be polymerized to obtain poly(3,4-ethylenedioxythiophene) (PEDOT), which exhibits efficient performance and good electrochemical properties compared to other polythiophenes [9]. However, it presents an unstable behavior in neutral state, since it rapidly oxidizes with air, and it is not soluble in common solvents. The most common approach to improve its processability is the addition of a polyelectrolyte, poly(styrenesulfonate) (PSS), leading to an aqueous PEDOT:PSS dispersion, in which PEDOT is oxidized. PSS role is to act as a counter ion and keep the PEDOT chains dispersed. Nonetheless, PEDOT:PSS has other limitations like hygroscopic behavior, high acidity and heterogeneous electrical properties, resulting in a low durability [10].

One of the most commonly employed compounds in photovoltaics is poly(3-hexylthiophene) (P3HT) due to its electronic properties, solubility in organic solvents and self-assembling behavior. Poly(p-phenylene vinylene) (PPV) has limitations due to its solubility, but it can be significantly improved if side-groups are added to the conjugated backbone, resulting in compounds such as poly[2-methoxy-5-(3′,7′-dimethyloctyloxy)-1,4-phenylene vinylene] (MDMO-PPV) or poly[2-methoxy-5- (2'-ethylhexyloxy)-p-phenylene vinylene] (MEH-PPV). Nevertheless, the absorption of PPV and its derivatives is low, and these also suffer from photo-degradation [3]. Polypyrrole (PPy), other widely employed polymer, is insoluble in the neutral state owing to the crosslinking of the polymeric segments. However, PPy can be swollen by several solvents, and the swollen polymer can be dissolved in solvents such as dimethyl sulfoxide (DMSO), chloroform, tetrahydrofuran (THF) and N-methyl-2-pyrrolidone (NMP). Further, it can be doped both in basic and acid media. Polyaniline (PANI) is extensively used owing to its interesting characteristics such as stability, ease of synthesis and low cost. The electrical properties of PANI can be tailored by modifying its protonation and oxidation state, and it shows an acid/base doping response [11]. Poly(p-phenylene) (PPP) is highly stable, particularly when undoped, hence, is broadly applied as electrode, although it is just only soluble in organic solvents.

Originally, BHJ-based PSCs employed MEH-PPV as electron donor material and a fullerene C60 derivative material as electron acceptor. These cells were commonly manufactured using spin-coating. Nevertheless, the cell efficiency was quite low (around 7.7%) [12]. In recent years, the performance of this type of cells has been enhanced, reaching PCE values >9% for single-junction cells [13] and >11% for tandem solar cells [14]. These improved PCE values have been possible due to the development of new polymeric donor materials like polythieno[3,4-b]-thiophene-co-benzodithiophene] (PTB7). This semiconducting polymer presents a low bandgap, a low HOMO level and absorbs in the NIR region, which allow the attainment of the highest efficiencies reported to date for BHJ-based cells [13,15]. Also, PTB7 displays high solubility in a lot of solvents, making it possible to process at room temperature. Also, it can be coated using numerous techniques such as spray and blade coating, ink-jet printing, etc. For example, a highly efficient PSC has been developed using an inverted device structure, where ITO was substituted by poly[(9,9-bis(3′-(N,N-dimethylamino)propyl)- 2,7-fluorene)-alt-2,7-(9,9–dioctylfluorene)] (PFN), a conjugated polymer soluble in water and alcohol, and the photoactive layer was formed by a mixture of PTB7 and [6,6]-phenyl C71-butyric acid methyl ester (PC71BM). The efficiency can be additionally enhanced by using poly[4,8-bis(5-(2-ethylhexyl)thiophen-2-yl)benzo[1,2-b:4,5-b′]dithiophene-co-3-fluorothieno [3,4-b]thiophene-2-carboxylate-2-6-diyl)], known as (PTB7-Th) or PBDTTT-EFT. The HOMO/LUMO levels decrease and the performance is improved as compared to PTB7, along with enhanced stability. As a result of its increased photo-response approaching 800 nm, the PTB7-Th/PC71BM blend has been employed as active layer in single junction PSCs [16,17]. However, due to the difficulty of scaling up the production process, the PCE of this kind of cells cannot be further improved, and the mechanical fragility together with ion diffusion within the polymer layers reduce their practical applications. Consequently, the development of new, cheaper and more stable materials for use as hole transport or electron transport layers in PSCs is of paramount interest.

During the last years, graphene (G) has emerged as an effective candidate to prepare polymer nanocomposites with excellent mechanical, electrical and thermal properties, suitable to be used in enhanced photovoltaic devices [18]. The addition of functional groups to G changes its properties and behavior in different solvents (i.e. it is possible to avoid G aggregation when hydrophobic or hydrophilic groups are added) [18]. Owing to the versatility and unique properties, G has evolved as a suitable nanomaterial for most components of the photovoltaic devices, including the transparent conductive electrodes (TCEs) and active (ALs) as well as interfacial (IFLs) layers.

The present chapter summarizes the relevant advances in the preparation of G/polymer nanocomposites for PSC applications. The synthesis of G and its derivatives (graphene oxide (GO) and reduced graphene oxide (rGO)), synthesis methods of G/polymer nanocomposites and the properties of the resulting PSCs are discussed. Even though quite a few reviews dealing with PSCs have been published [1-6], to the best of our knowledge, literature reviews about solar cells comprising polymers and G-based materials are scarce [19,20], and none of them concentrates on the nanocomposites.

1.2 Graphene and Its Derivatives: Properties and Synthesis

1.2.1 Properties and Synthesis of Graphene

G consists of a flat atomic-thick single layer of sp^2 carbon atoms creating a 2D structure similar to a honeycomb (Figure 1.4). Also, this atomic distribution exhibits an electronic band structure with a linear profile of energy on momentum close to the Fermi level [21]. G presents exceptional thermal, mechanical and electronic properties, some of them are listed below:

- Thermal conductivity of 3000-6000 W/(m K) [22], significantly higher than that of Cu (400 W/m K)
- Electronic properties: electron mobility of around 15,000 $cm^2/(V\ s)$ [23] and electrical conductivity of 9.6×10^7 S/m [24], higher than that of metals like Ag (6.3×10^7 S/m) or Cu (6.0×10^7 S/m)
- Mechanical properties: elastic modulus near to 1 TPa [25], ultimate strength value of 130 GPa and ~40 N/m of breaking strength [26]
- Surface area of 2630 m^2/g [23], much higher than amorphous carbon (900 m^2/g) or carbon nanotubes (100-1000 m^2/g)
- Optical absorption of visible light close to 2.3% [27]

Owing to the abovementioned excellent properties, G is a versatile material for photovoltaic cell technologies. Furthermore, G sheets have good flexibility and are largely unreactive, which facilitates their application as electrode or protective layers. However, some characteristics should be improved to extend its uses in the field of solar cells. In particular, its sheet resistance is considerably higher than that of other materials used as electrodes (i.e. ITO), therefore, it is essential to generate a good balance between properties and improve the electronic transport without decreasing its optical transparency.

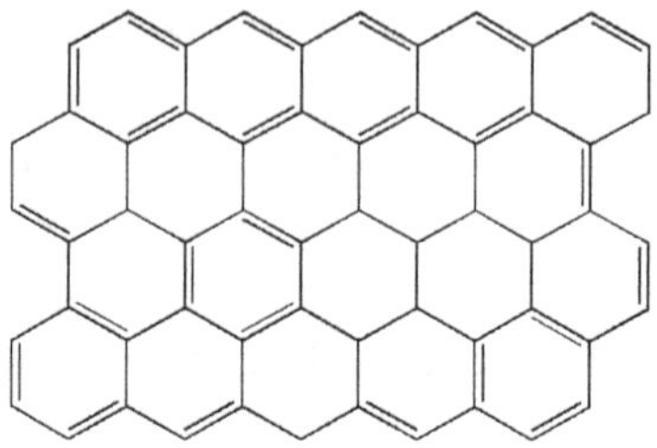

Graphene

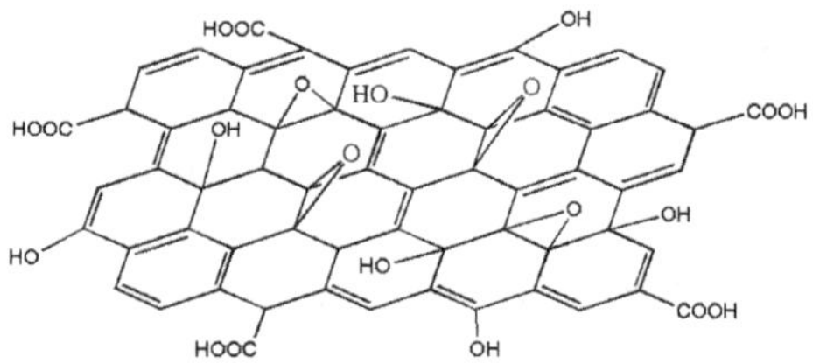

Graphene Oxide

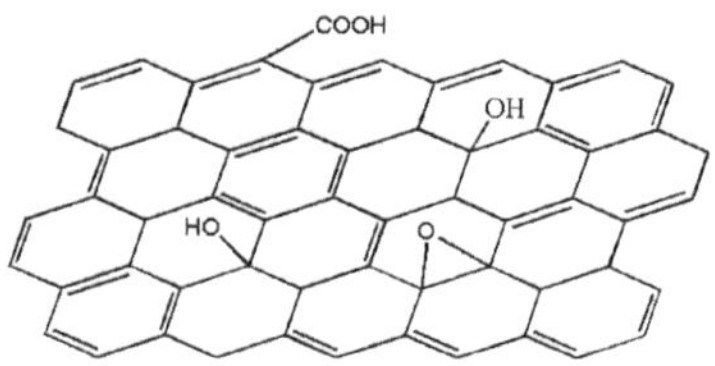

Reduced Graphene Oxide

Figure 1.4 Chemical structures of graphene (G), graphene oxide (GO) and reduced graphene oxide (rGO).

G can be obtained by a number of methods [28]. However, in this chapter, the techniques that could be applied for photovoltaic device manufacturing are elaborated:

- Chemical vapor deposition (CVD)
- Epitaxial growth
- Exfoliation
- Chemical and electrochemical techniques

CVD procedures are the most extended ones to produce G at a large-scale, due to their simplicity and low cost. The first step is the growth of G on a substrate (typically a transition metal such as Ni, Cu,

Pd, Ru, Ir or Pt) at high temperature (500-1200 °C), which is obtained by saturating the surface of the substrate with a hydrocarbon gas (i.e. CH_4, CH_3OH) [23]. The second step consists of the cooling of the substrate, which reduces the solubility of carbon and precipitates it in the form of sheets (mono- or multilayer). Graphene synthesized by this technique is highly appropriate for transparent electrodes applications, however, the main issues of CVD are the expensive substrates and the complex control on the film thickness.

Epitaxial growth is other substrate-based procedure, in which G is grown by graphitization on silicon carbide (SiC) under vacuum and at temperatures close to 1300 °C. In this case, it is possible to adjust the sheet thickness by controlling the temperature and time. Also, an annealing process can increase the homogeneity of the thickness across the layer [29]. Nevertheless, it is not recommended for photovoltaic purposes owing to its high cost and solid adhesion of G layers to the substrate.

Exfoliation, or mechanical exfoliation, was the first technique employed to obtain G in 2004 by Geim and Novoselov [30]. This technique consists of peeling it off from graphite flakes with a scotch tape [30]. Though G sheets isolated by means of this method have good properties, however, it is not viable as large-scale production procedure.

Chemical solution processes are low-cost large-scale production methods, and supply G layers that can be further treated to develop transparent electrodes or dopant additives for other materials [31]. The major weakness of these methods is the poor dispersibility of G. Nonetheless, upon reaction with strong oxidizing agents, oxygenated functional groups are generated on the edges (e.g. carboxylic groups (-COOH)) and basal plane (e.g. hydroxyl, carbonyl and epoxy moieties (-OH, -C=O, -O-)) of the G sheets. This affords graphene oxide (GO) layers which are soluble in various solvents (i.e. water, chloroform, dimethyl-formamide (DMF) and tetrahydrofuran (THF) [32]) (Figure 1.3). Owing to its improved dispersibility in organic solvents, GO is more widely used as nanofiller in polymer-based composites.

The electrochemical exfoliation (a kind of electro-chemical method) method to synthesize G is carried out via infiltration of graphite by ions from the solution under an applied potential. G synthesized by this approach can be used to produce thin films for solar cells [33], since it can be readily dissolved in organic solvents (i.e. DMF). Furthermore, it can be modified under vacuum or by using HNO_3 treatment, in order to be used for transparent electrodes.

1.2.2 Properties and Synthesis of Graphene Oxide

GO is an oxidized form of G that comprises oxygenated functional groups on the basal plane and edges (Figure 1.4). Hence, GO has different properties as compared to G, as mentioned below:

- Easy dispersability in water and other organic solvents, as well as in different matrixes, which is a significant advantage for large-scale applications.
- Amphipathic behavior, surface functionalization potential
- Biocompatibility, capability to interact with tissues and cells [34]
- Large interlayer spacing, capacity to intercalate compounds between its layers
- Sheet resistance of 10^{12} Ω/sq or higher due to the presence of functional groups that act as defects on the structure, which obstruct electrical transport and decrease electron mobility.

Several methods have been reported to synthesize GO: Staudenmaier, Hofmann, Brodie and Hummers [29,34]. These methods involve oxidation of graphite in the presence of strong acids and oxidants. The Hummers' method is the most utilized, which consists of the oxidation of graphite using potassium permanganate and sulfuric acid. The degree of oxidation can be controlled by the oxidizing agents, synthesis conditions and precursor graphite. The active species in the oxidation process is dimanganese heptoxide (Mn_2O_7). A large number of modifications of the Hummers' approach have been published over the last decades, in order to improve the performance and/or decrease costs [29,34]. As an example, the use of $NaNO_3$ is avoided due to the generation of toxic gasses (NO_2/N_2O_4) and the delivery of Na^+ and NO^{3-} ions in water [35].

Alternatively, GO can be obtained from graphite oxide by means of stirring, ultrasonication or a mixture of both. Ultrasonication is a faster technique, even though graphene layers can be fragmented and damaged. Mechanical stirring takes longer, although graphene layers are less damaged.

1.2.3 Properties and Synthesis of Reduced Graphene Oxide

Reduced graphene oxide (rGO) is another graphene-based material that can be obtained upon reaction of GO with a reducing agent.

Consequently, its structure comprises less oxygenated functional groups (Figure 1.4). The idea is to attain a material similar to mechanically exfoliated G or graphite. Nevertheless, residual oxygenated groups and defects alter the structure of the carbon plane, hence, the properties of rGO are different from those of G [36]:

- Dispersible at low concentrations in organic solvents such as DMSO, NMP and DMF
- Surface area of 500 m^2/g
- Electrical conductivity of 670 S/m

The reduction of GO can be achieved via three different ways: thermal, chemical, or electrochemical approaches. The thermal reduction is a heat-based treatment, which involves a very fast heating of GO (>2000 °C/min [37]), thus, inducing the exfoliation of the G sheets due to the expansion of CO_x gases (CO and CO_2) between the GO sheets. This technique is not recommended for solar cell technologies, due to the low melting-point of certain substrates (i.e. polymers or glasses). Thermally reducing GO at >1000 °C produces a high surface area material similar to pristine G. However, the structure of the GO flakes is damaged by the heating process, resulting in significant mass loss, and potentially impacting the mechanical strength of the rGO material. Other alternative thermal treatment consists of using microwave irradiation or photo-reduction [38], where GO is heated using a laser or a flash lamp. This method leads to a higher degree of GO reduction due to the large amount of energy provided by the laser/lamp, thus, resulting in highly conductive rGO films [39], which can be directly used to fabricate electronic devices.

The chemical reduction approaches use chemical reagents in order to reduce GO and can be performed at ambient temperature or under moderate heating. The most widely employed chemical reagent is hydrazine monohydrate ($N_2H_4 \cdot H_2O$) [40], which increases the hydrophobicity and induces the agglomeration of the rGO nanosheets when added to an aqueous GO dispersion. Nevertheless, the toxicity of this reagent is an important hazard for large-scale synthesis, thus, the use of other reducing reagents (i.e. urea, hydroiodic acid (HI) or $NaBH_4$ [41]) can be a more suitable solution. Further, rGO produced by these methods has often sub-standard electrical conductivity and surface area.

Electrochemical production of rGO has the potential to be the preferred commercial method in the future, since it produces a material almost identical in structure to pristine G. The process takes place at room temperature in an electrochemical cell. Substrates like ITO are coated with a thin layer of GO. Electrodes are attached to each end of the substrate, and linear sweep voltammetry is subsequently performed in a sodium phosphate buffer, without adding any chemical reagent. This method is highly suitable for solar cell applications.

1.3 Preparation of Polymer Nanocomposites

The preparation approaches of polymer nanocomposites can be classified in two groups: covalent and non-covalent methods [26,42].

1.3.1 Covalent Methods

These are based on the attachment of the polymeric chains on the filler sheets as well as the wrapping of chains around them to prevent aggregation. Such anchoring can be performed via "grafting from" or "grafting to" approaches [43]:

- The "grafting from" approach is based on atom transfer radical polymerization (ATRP) in order to grow the polymer chains from the surfaces of the filler sheets. The high reactivity of monomers makes this strategy efficient and controllable, thus, resulting in nanocomposites with high degree of grafting. However, this approach needs accurate control on the amount of each reactant and polymerization conditions.
- The "grafting to" consists of the reaction of functional groups of the polymer chains with G or its derivatives to form chemical bonds. In the case of GO, the carboxylic groups can react with a polymer that has amine or hydroxyl groups by amidation or esterification reactions [44]. A disadvantage of this method is that the grafted polymer content is restricted due to the low reactivity and high steric hindrance of the polymer chains.

1.3.2 Non-covalent Approaches

These include three types of strategies: *in-situ* polymerization, solution mixing and melt-blending.

- The *in-situ* polymerization has two steps: (1) the swelling of filler sheets within the liquid monomer and (2) the addition of the initiator and subsequent heating in order to initiate the polymerization. This approach, considering an electrochemical polymerization, can be applied to produce nanocomposites with conductive polymers [45]. The electrochemical polymerization produces composite films with high mechanical stability, which can be applied as electrodes for energy storage devices. It is also possible to perform polymerization of aniline to generate PANI, which has high stability and applicability in PSCs under controlled acidic medium, although it causes agglomeration of the G sheets [18].

- The solution mixing is based on the stable dispersion of both filler and polymer matrix in a common solvent. After the dispersion of filler sheets, the polymer chains are adsorbed onto delaminated sheets. The solvent is subsequently removed, leading to sandwich-structured nanocomposites [44]. Few literature studies are available about solution mixing as compared to *in-situ* polymerization due to the poor solubility of conducting polymers in regular solvents [18], however, the solubility can be enhanced by chemical modification.
- The melt-blending consists of the mixing of G or its derivatives into molten polymer matrices while maintaining strong shearing. Although it is an eco-friendly process (there is no consumption of solvents) and is easily applicable in industry, its use can lead to the damage to modified graphene surface due to high temperature and shear [46]. However, the use of lower temperatures or thermally stable modifications can prevent the degradation process [46].

1.4 Polymer Nanocomposites in Solar Cells

The applications of G and its derivatives in PSCs are numerous and cover different layers: conductive electrodes (CE) or TCE, IFLs and ALs. Furthermore, these can also be applied in dye-sensitized solar cells (DSSCs). As the literature on this subject is wide-ranging, only representative examples of polymer nanocomposites with G or its derivatives have been discussed in this review. The results have been divided into four sections: (1) G or derivatives based polymer nanocomposites as transparent conductive electrodes, (2) G or derivatives based polymer nanocomposites as active layers, (3) G or derivatives based polymer nanocomposites as interfacial layers, and (4) G or derivatives based polymer nanocomposites in dye-sensitized solar cells. The most important properties are also presented in Table 1.1.

Table 1.1 Summary and performance of PSCs based on G or derivates based polymer nanocomposites

Nanocomposite	Device Structure	Jsc (mA cm^{-2})	V_{oc} (V)	FF	PCE (%)	Ref
PMMA-G	PET/PMMA-G/MoO_3/PEDOT:PSS /PCDTBT:PC_{70}BM/Ca/Al	8.88	0.83	0.45	3.3	47
T/rGO	PET/rGO/PE-DOT:PSS/P3HT:PCBM/TiO_2/Al	4.39	0.56	NA	0.78	48
poly(ethylene 2,6-naphthalate) (PEN)/G	PEN/G/ZnO/PTB7:PC_{71}BM/MoO_3/Ag	14.9	0.71	0.58	6.2	51
Au/G/PEDOT:PSS	ITO/ZnO/P3HT:PCBM/PE-DOT:PSS/Au/G	10.58	0.59	0.43	2.7	56
G/Au/PEDOT:PSS	G/Au/PE-DOT:PSS/P3HT:PCBM/ZnO/Ag/PI	10.6	0.597	0.50	3.17	57
EG/PEDOT:PSS	EG/PEDOT:PSS/PTB7:PC71BM/Ba /Al	6.53	0.707	0.56	4.23	64
P3OT:G	ITO/ PEDOT:PSS/P3OT:G/LiF/Al	4.2	0.92	0.37	1.4	65
P3HT:G	ITO/ PEDOT:PSS/P3HT:G/LiF/Al	4.0	0.72	0.38	1.1	66
P3OT:PCBM–G	ITO/PEDOT:PSS/P3OT:PCBM–G/LiF/Al	4.6	0.67	0.37	1.14	68
P3HT-grafted GO	ITO/PEDOT:PSS/G-P3HT:C60/Al	3.5	0.43	0.41	0.61	69
C60-rGO:P3HT	ITO/PEDOT:PSS/C60-rGO:P3HT/Al	4.45	0.56	0.49	1.22	70
P3HT-f-MWCNTs-SPFG	ITO/PEDOT:PSS/P3HT-f-MWCNTs-SPFG/LiF/Al	4.7	0.67	0.32	1.05	71
PTM21-GO	ITO/PEDOT:PSS/P3HT:PTM21-GO:PCBM/LiF/Al	8.11	0.59	0.30	1.43	72
P3HT:GQDs	ITO/PEDOT:PSS/P3HT:GQDs/Al	6.33	0.67	0.30	1.28	73
P3HT/ANI-GQDs	ITO/PEDOT:PSS/P3HT/ANI-GQDs/Al	3.51	0.61	0.53	1.14	74
P3HT:PCBM:GQDs	ITO/PE-DOT:PSS/P3HT:PCBM:GQDs/LiF/Al	26.46	0.60	0.33	5.24	75
GOQDs:PTB7: PC_{71}BM	ITO/PEDOT:PSS/GOQDs:PTB7: PC_{71}BM/Al	15.2	0.74	0.68	7.6	76
N-dopedG:P3HT: PCBM	ITO/PEDOT:PSS/N-doped G:P3HT:PCBM/Al	14.95	0.59	0.51	4.48	77
GO/PEDOT:PSS	ITO/GO/PEDOT:PSS/PCDTBT: PC_{71}BM/Al	10.44	0.82	0.50	4.28	78
Au@PEG–GO	ITO/PEDOT:PSS(Au@PEG-GO)/PBDTTT-CT	16.1	0.75	0.60	7.26	79

PRGO-PTB7-th	NA	14.78	0.76	0.64	7.24	80
ZnO@G:EC	ITO/ZnO@G:EC/P3HT:PC61BM/MoO_3/Ag	7.73	0.63	0.73	3.9	81
GO/P3HT:PCBM	ITO/GO/P3HT:PCBM/Al	11.4	0.57	0.54	3.5	82
G:PEDOT:PSS	ITO/FLGs-PE-DOT:PSS/PCBM/P3HT/Ca/Al	9.44	0.58	0.55	3.7	83

1.4.1 Polymer Nanocomposites with G or Derivatives as Transparent Conductive Electrodes

ITO electrodes, the most extensive TCEs in PSC structures, can be substituted by G or G-based derivatives, such as GO or rGO. Flexible polymers have been widely studied as substrate materials for TCEs, although two polymers are more commonly employed: polyethylene terephthalate (PET) and PEDOT:PSS.

PET-based Electrodes

G was mixed with poly(methyl methacrylate) (PMMA) to obtain PMMA/G nanocomposites, which can work as anode in PSCs [47]. An *et al.* [47] spin coated PMMA onto homogeneous CVD-synthesized G in order to attain a bilayer composite that was placed onto a PET substrate. The G-substrate adhesion was increased by PMMA, avoiding contamination and crack development. The resistance of the G sheets decreased around 50%, while their transmittance was retained. The device structure was PET/PMMA-G/MoO_3/PEDOT:PSS/Poly[N-9'-heptadecanyl-2,7-carbazole-alt-5,5-(4',7'-di-2-thienyl-2',1',3'-benzothiadiazole)] (PCDTBT):PC70BM/Ca/Al and displayed V_{OC} of 0.83 V, J_{SC} of 8.88 mA/cm^2, FF of 0.45 and PCE of 3.3%, which was about twice that of the reference cell without G.

It is also feasible to fabricate photovoltaic devices using rGO films onto PET to obtain rGO/PET hydrophilic surfaces, leading to a device with a structure of PET/rGO/PEDOT:PSS/PCBM/TiO_2/Al [48]. The combination reached the best results when the rGO film thickness was 16 nm, and the transmittance was 65%, thus, leading to PCE of 0.78%, J_{SC} of 4.39 mA/cm^2 and V_{oc} of 0.56 V. Furthermore, the performance of the device through bending cycles was preserved much better than that of ITO-based devices (at least until 1200 cycles), given that the fragile nature of ITO is observed to cause degradation and cracking.

Other suitable electrode was manufactured by placing a rGO micromesh onto PET, resulting in a PCE value of 3.05%, which is comparable with that of ITO-based devices (3.82%) [49]. Similar to the previous example, the device had a good bending stability. Due to the use of a rGO micromesh instead of pristine rGO, it exhibited lower sheet resistance (around 565 Ω/sq) and better transparency (59%) [49]. However, the efficiency of these electrodes is limited since these usually have a high concentration of structural defects.

PEDOT-based Electrodes

The surface of G layers can be modified with different compounds, such as PEDOT-block-poly(ethylene glycol) (PEDOT:PEG) copolymer doped with perchlorate (PC) (Figure 1.5) [50], forming a device with a structure of PET/GO:PEDOT:PSS/F8T2/C_{60}/Al. In this case, a low pressure CVD procedure (LPCVD) was applied to obtain three monolayer-G films, which exhibited both good transparency and conductivity (92% transmittance and 300 Ω/sq electrical resistivity). Besides, it was possible to use these as conventional PSC anode (PCE of

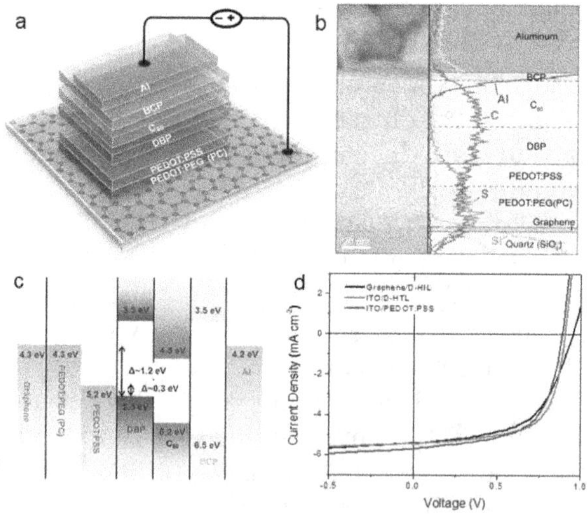

Figure 1.5 (A) Schematic diagram of the G/PEDOT:PEG(PC)/PEDOT:PSS/DBP/C60/BCP/Al PSC with G anode; (B) Representative TEM micrograph (left) of the device described in (A), with an EDX scan of the device cross-section (right). (C) Energy level diagram of the PSC; (D) J-V characteristics of the G-based device compared to ITO reference cells. Reproduced from Reference 50 with permission from Springer Nature.

6.1%) and as inverted PSC cathode (PCE of 7.1%) [51]. The device, with G/PEDOT:PEG(PC)/PEDOT:PSS/DBP/C60/BCP/Al structure, reached the following values: FF of 57.6%, V_{OC} of 0.71 V and J_{SC} of 14.8 mA/cm^2. Furthermore, these devices exhibited a good resistance against flexing cycles, as there was hardly any performance loss after 100 cycles and a good resistance against mechanical deformation was achieved. LPCVD G has also been applied to manufacture PSCs in which it acted as both anode and cathode [52]; a maximum PCE of 3.7% was attained by using G as the bottom film and depositing the top G film onto flexible poly(ethylene 2,6-naphthalate) (PEN).

La Notte *et al.* [53] employed spray coating method to develop a cathode for inverted PSCs using doped 4-layered CVD G modified with PEDOT:PSS. The active layer had phenyl-C71-butyric acid methyl ester (PC70BM) as acceptor and poly[(5,6-difluoro -2,1,3-benzothiadiazol-4,7-diyl)-alt-(3,3‴-di(2-octyldodecyl)2,2′;5′,2″;5″,2‴-quaterthiophen-5,5‴-diyl)] (PffBT4T-2OD) as donor. The resulting cell exhibited PCE of 2.8%, FF of 0.37, V_{OC} of 0.72 V and a J_{SC} of 10.5 mA/cm^2, which are acceptable results compared with those of an ITO-based solar cell. Furthermore, this type of nanocomposites can be prepared via a vibration-assisted ultrasonic spray coating, which is a suitable industrial solution since it is a single-step, rapid and scalable procedure [54,55]. The fabricated films exhibited a maximum electrical conductivity of 298 S·cm^{-1}, which was ten-times higher than that of pristine PEDOT:PSS films, and had a transparency similar to that of ITO-coated glasses. The observed property enhancement was attributed to the increase in carrier concentration and mobility, given that there are strong π-π interactions between G sheets and PEDOT:PSS rings that behave as high-motion channels. Also, the π-π interactions decrease the amount of defects in PEDOT:PSS. In addition, the properties can be modified by adjusting the energy level of G-PEDOT:PSS films, which improves the stability and mechanical properties (hardness, wear resistance, etc.).

Another option is to synthesize single-layer G films via a CVD process, add Au nanoparticles as dopant material and combine them with P3HT:phenyl-C61-butyric acid methyl ester (PC61BM) to obtain a semitransparent material. This nanocomposite was applied as top electrode in PSCs with ITO as bottom electrode, resulting in the following structure: ITO/ZnO/P3HT:PCBM/PEDOT:PSS/Au/G (Figure 1.6a) [56]. The doped G-based electrode exhibited a conductivity value about 4 times higher than that of pristine G. The device exhibited a maximum PCE of 2.7% (Figure 1.6b) and higher external quantum efficiency (EQE), which is the ratio of the number of charge carriers collected by the solar cell to the number of photons of a certain energy shining on the solar cell from outside, when the illumination takes place from the G side (Figure 1.6c), due to the superior transmittance of the G electrode. It is envisaged that if the synthesis method and the G quality are improved, the efficiency of these PSCs will be enhanced.

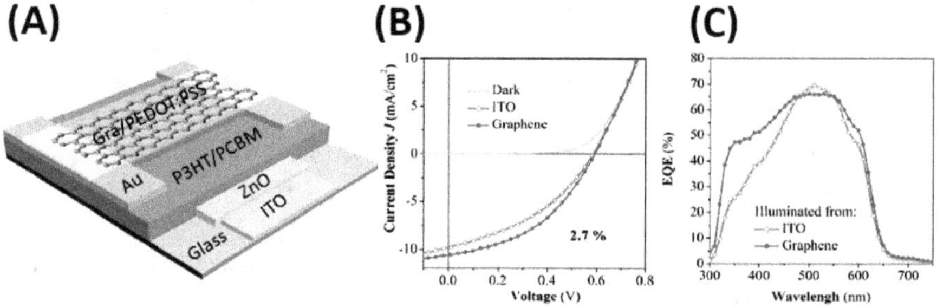

Figure 1.6 (A) Schematic representation and band structure of a PSC with the structure Glass/ITO/ZnO/P3HT:PCBM/Au/PEDOT:PSS/G; (B) J-V characteristics and (C) external quantum efficiency (EQE) measured from two sides of the PSC with G top electrode and different active layer thicknesses. Reproduced from Reference 56 with permission from American Chemical Society.

It is also feasible to manufacture a TCE using CVD multi-layer G films doped with Au nanoparticles and PEDOT:PSS (instead of single-layer G films) on a polyimide (PI) substrate. In a recent study, such a TCE was applied as top electrode and P3HT:PC61BM as active layers [57]. The resulting device, G/Au/PEDOT:PSS/P3HT:PCBM/ZnO/Ag/PI, displayed a PCE of 3.2%, along with significant stability and flexibility. Further, the PCE was largely unaffected after 1000 bending cycles. More outstandingly, the higher number of G layers avoided the diffusion of air across G, resulting in a strong layer-packaging effect on flexible solar cells. Multilayer G can behave as an environmental barrier, thus, preserving the PSC from air contamination, along with decreasing the related costs.

TCEs based on functionalized G such as sulfonated G (SG)/PEDOT nanocomposites obtained by *in-situ* polymerization have also been described [58]. SG was obtained from GO via reduction with $NaBH_4$ and subsequent sulfonation with the aryl diazonium salt of sulfanilic acid. Subsequently, the monomer EDOT and $Fe_2(SO4)_3$ were added to an aqueous dispersion of SG, and the mixture was heated at 50 °C and stirred for 2 days. The G composites exhibited good processability in aqueous and organic media, high transparency, elevated thermal stability and conductivity, etc. Nanocomposites with a thickness of a few nanometers reached transmittance greater than 80% (400-1800 nm wavelength) and conductivity of 0.2 S/cm, which is considerably higher than that of PEDOT:PSS-based devices (10^{-6}-10^{-5} S/cm).

More devices have been developed using functionalized G. Chang *et al.* [59] fabricated a nanocomposite electrode comprising surfactant-functionalized G and PEDOT:PSS by spin coating. The graphene sheets were observed to be homogeneously dispersed in the PEDOT:PSS. The TCE showed high electrical and mechanical stability, conductivity and transparency comparable to ITO electrodes. The resistance after bending (5% in 1000 cycles) also remained largely unchanged. Jo *et al.* [60] avoided the use of surfactants for the synthesis of rGO/PEDOT composite, which is more desirable from an environmental view-point. The authors synthesized a filler suspension stabilized by the GO reduction in the presence of the polymer. The resulting nanocomposite was stabilized via π-π interactions between the rGO sheets and PEDOT backbone, as well as electrostatic repulsions between the negatively charged PSS and rGO sheets. The film displayed transmittance and conductivity of 80% and 2.3 kΩ/sq respectively, which were almost preserved even after 100 bending cycles.

It is also suitable to use GO in TCEs, as revealed by Lima *et al.* [61], who developed a GO/PEDOT:PSS nanocomposite by a large-scale compatible fabrication process. The cell structure was PET/GO:PEDOT:PSS/F8T2/C_{60}/Al, and it reached a PCE of 1.10%, along with exhibiting flexibility and no changes in conductivity after 1000 bending cycles. The nanocomposite acted as a hole transport material, and it was fabricated without the need of surfactants or dopants. It was also possible to modify the sheet resistance and optical transmittance of the nanocomposite films by varying the GO/PEDOT:PSS ratio. Furthermore, the Fenton's reaction has been be applied to produce GO/PEDOT nanocomposites suitable as TCEs [62]. The synthesis involves two steps: (1) intercalating GO with $FeCl_3$ and H_2O_2 and (2) *in-situ* polymerization of EDOT on GO sheets (Figure 1.7). The nanocomposites exhibited enhanced optical properties and electro-activity.

Other Electrodes

ITO electrodes could also be modified with a G/ZnS/PPy ternary

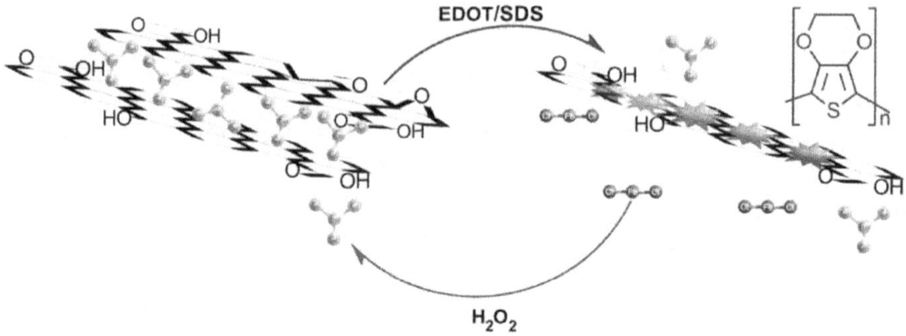

Figure 1.7 Schematic of the composite synthesis. Reproduced from Reference 62 with permission from Elsevier.

nanocomposite synthesized by layer-by-layer electrophoretic deposition [63]. The device, with ITO/G/ZnS/PPy/electrolyte/Pt structure, achieved PCE of 0.92%, V_{OC} of 0.48 V, J_{SC} of 0.42 mA/cm^2 and FF of 0.23, which were higher than the values for ZnS-based or PPy-based electrodes. The nanocomposite enhanced the efficiency due to the synergistic combination of components: ZnS nanoparticles behaved as bridges, PPy acted as hole acceptor and sensitizer, and G behaved as conductive transporter and conductive collector.

Another technique to prepare G electrodes is electro-chemical exfoliation of graphite with the intercalation of sulfate ions (Figure 1.8) [64]. The exfoliated graphite (EG) was employed as anode in PSCs after being deposited onto flexible PEN. The active layer of the device with structure EG/PEDOT:PSS/PTB7:PC71BM/Ba/Al was PTB7:PC71BM, and it reached a PCE of 4.23% (which was preserved after 150 bending cycles). Additionally, the device exhibited sheet resistance of 0.52 kΩ/sq and transparency of 70%.

1.4.2 Polymer Nanocomposites with G or Derivatives as Active Layers

Originally, OSCs had similar structure to conventional solar cells: two electrodes separated by a heterojunction. However, this architecture had a poor performance because of the limited interface area, where the exciton dissociation takes place into free electron-hole pairs. This structure implied that only excitons formed near the interface (a few nm range) could be dissociated. An alternative approach to reduce the thickness of the heterojunction was employed, however, it resulted in low J_{SC} and poor light absorption.

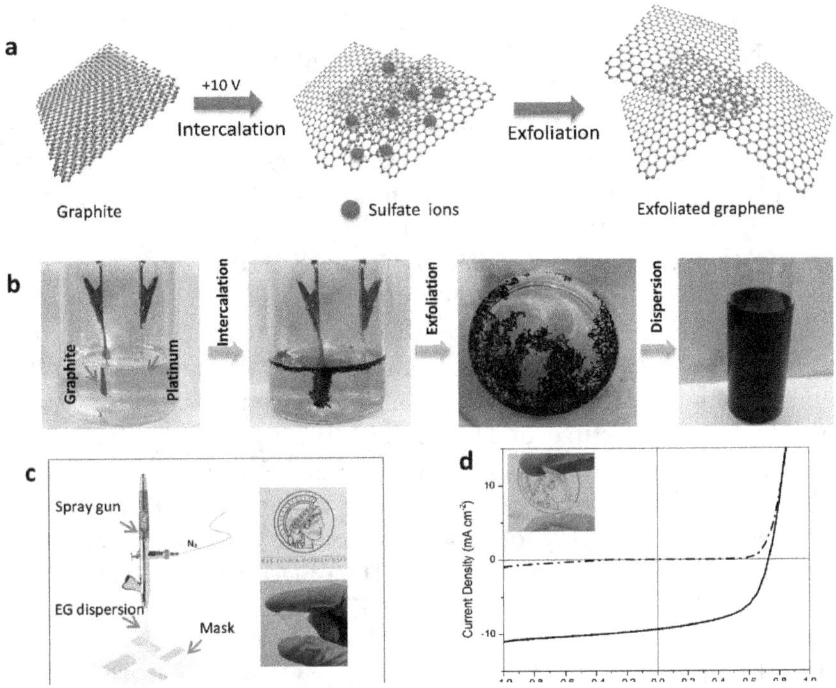

Figure 1.8 (a) Scheme of the electrochemical exfoliation of graphite; (b) optical images of the exfoliation process; (c) schematic representation of spray deposition of EG dispersion onto PEN; and (d) J-V characteristics of the cell under light (solid line) and dark conditions (dashed line). Reproduced from Reference 64 with permission from American Chemical Society.

Subsequently, OSCs were generated using the BHJ approach [6]. The BHJ consists of a bulk mixture between both phases, with nanoscale separation of phases in the active layer, usually formed by a fullerene acceptor or derivatives and a polymer donor. This architecture improved the PCE compared to previous heterojunction cells, which can be attributed to the increased interfacial surface, which enabled better exciton dissociation and improved charge carrier collection. Therefore, the properties of G (2D structure and large specific surface area) can benefit the exciton dissociation.

G can act as an acceptor in PSCs upon functionalization with phenyl isocyanate [65]. The functionalized G was combined with poly(3-octylthiophene) (P3OT), which acted as donor material, to develop a cell with structure ITO/PEDOT:PSS/P3OT:G/LiF/Al. The device exhibited a PCE of 1.4% at 5 wt% G concentration, which was attained by an annealing process at 160 °C for 20 min. The same functionalized G at a higher concentration (10 wt%) was applied as an acceptor in a similar PSC with P3HT donor instead of P3OT [66], which led to PCE of 1.1%, FF of 0.38, V_{OC} of 0.72 V and J_{SC} of 4.0 mA/cm^2 after annealing for 10 min at 160 °C. The enhanced performance of the device was due to the improved exciton dissociation area, which was explicated by the enhanced electron transport through G. However, when the annealing temperature was increased up to 210 °C, the PCE decreased to 0.57%. The P3HT:functionalized G nanocomposites are soluble in common organic solvents, which facilitates the structure/property characterization and the device fabrication via solution processing. Another BHJ photovoltaic device based on functionalized G with a structure of ITO/ZnO/P3HT:F-graphene/MoO_3/Ag was recently developed [67], showing PCE of 1.9%, FF of 0.52, V_{OC} of 0.61 V and J_{SC} of 5.95 mA/cm^2 (Figure 9D). Wang *et al.* [68] prepared a similar PSC with a P3OT/solution-processable G nanocomposite as active layer, with a structure of ITO/PEDOT:PSS/P3OT:PCBM-G/LiF/Al, yielding PCE of 1.14%, FF of 0.37, V_{OC} of 0.67 V and J_{SC} of 4.6 mA/cm^2.

CH_2OH-terminated P3HT was also anchored on the COOH groups of GO surface by esterification reaction [69]. The resulting compound was suitable for solution processing as P3HT-g-G was soluble in several organic solvents. The PSC structure was ITO/PEDOT:PSS/G-P3HT:C_{60}/Al (Figure 1.9), where P3HT-g-GO/C_{60} acted as active layer, and showed a PCE of 6.01%, which is twice that of the P3HT/C_{60}-based device, attributed to the improved light absorption and electron delocalization by the chemical grafting of P3HT on GO. The authors also applied a lithiation reaction to graft C_{60} onto the surface of rGO, leading to a hybrid material useful in BHJ cells as electron acceptor [70]. The proposed structure device was ITO/PEDOT:PSS/C_{60}-rGO:P3HT/Al, and it reached PCE of 1.22%, V_{OC} of 0.56 V and J_{SC} of 4.45 mA/cm^2. T observed performance was better than PSCs without fullerenes as electron acceptors, due to the improved electron transport of the rGO-C_{60} hybrid.

Liu *et al.* [71] synthesized a functionalized multi-walled carbon nanotube (f-MWCNT)/P3HT/solution processable G nanocomposite for application in PSCs. The device structure was ITO/PEDOT:PSS/P3HT-f-MWCNTs-SPFGraphene/LiF/Al, where G acted as percolation path for the electrons and electron acceptor, P3HT as an electron donor and f-MWCNTs as percolation paths of holes. However, the performance of the device was not optimal: PCE of 1.05%,

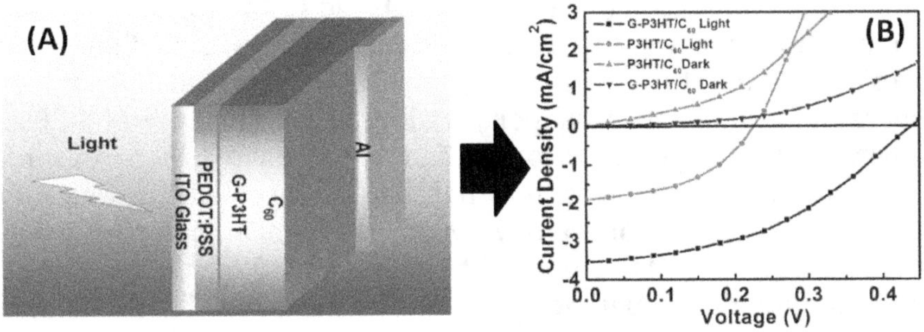

Figure 1.9 (A) Schematic representation of ITO/PEDOT:PSS/G-P3HT:C_{60}/Al PSC and (B) J-V characteristics using P3HT:C_{60} or G-P3HT:C_{60} as active layer. Reproduced from Reference 69 with permission from American Chemical Society.

FF of 0.32, J_{SC} of 4.7 mA/cm^2 and V_{OC} of 0.67 V. However, the low PCE of these G-based PSCs could be enhanced by optimizing the processing conditions and G content, according to the theoretical efficiency limit of 12% for this type of PSCs.

A number of PSCs with carbon nanotubes (CNTs) have been reported, and these exhibit poorer performances compared to those incorporating G. Lee *et al.* [72] studied the behavior of P3HT/PCBM blends with maleimide-thiophene copolymer-functionalized CNTs (PTM21-CNTs) and GO (PTM21-GO) for application as active layer in PSCs. The device, with the structure ITO/PEDOT:PSS/P3HT:PTM21-GO:PCBM/LiF/Al, showed a maximum PCE of 1.7% at 0.3 wt% of PTM21-GO. The property enhancement in the device was achieved due to the larger contact area provided by the functionalized GO sheets, along with better charge-transporting capacity and improved dispersion capability compared to the CNTs. Furthermore, GO is amphipathic, hence, can be used in aqueous solutions, whilst CNTs are hydrophobic and insoluble in most of the common solvents.

An alternative approach is the use of G quantum dots (GQDs) [73-77], which exhibit wider band gap than G. Moreover, the GQDs are rich in oxygenated groups and soluble in aqueous and organic solvents, hence, allowing additional functionalization. Li *et al.* [73] developed a device based on electrochemically synthesized GQDs (3-5 nm size), which acted as electron acceptors. Surface groups such as hydroxyl, carbonyl and carboxylic acid facilitated the dispersion in common organic solvents and subsequent mixing with polymers leading to the structure ITO/PEDOT:PSS/P3HT-GQD/Al. The value of the LUMO level (4.2-4.4 eV) of the GQDs led to the formation of an electron transport cascade in the system P3HT-GQDs-Al. Compared to reference P3HT-based devices, the GQDs increased the exciton separation and carrier transport, leading to an efficiency of 1.28%. However, the efficiency was lower than that of devices prepared with fullerenes as electron acceptors because of lower electron affinity and non-optimized morphology, which resulted in lower FF. Similarly, Gupta *et al.* [74] compared the effect of GQDs synthesized by acidic and hydrothermal methods and G sheets, both functionalized with aniline, as electron acceptors in a device with a structure of ITO/PEDOT:PSS/P3HT:ANI-GQD/LiF/Al. P3HT was combined with increasing amounts of GQDs in order to optimize the devices. The best results were obtained with 1 wt% GQDs, reaching a PCE of 1.14%. GQDs resulted in higher J_{SC} values than G sheets owing to their homogenous and uniform distribution within P3HT, enhanced exciton separation and transport towards electrodes, which also led to higher FF.

Li *et al.* [75] prepared GQDs from double-walled CNTs by a solution-based process, which were applied onto a P3HT:PCBM active layer. The device, with a structure of ITO/PEDOT:PSS/P3HT:PCBM:GQDs/LiF/Al (Figure 1.10), reached a PCE of 5.24%. The reported process represented a new approach to enhance the efficiency of PSCs. Kim *et al.* [76] compared the effects of GQDs with different oxidation degrees in PSCs. The GQDS were prepared using the Hummers' method, oxidized and subsequently hydrothermally reduced for 5 or 10 h prior to addition of the PTB7:PC71BM active layer, resulting in a device with structure of ITO/PEDOT:PSS/GOQDs:PTB7:PC71BM/Al. During the reduction process, the oxygen-related functional groups were gradually removed while the size remained unchanged below 1 nm. In addition, the reduction had a negative effect on the light absorption, although it enhanced the electrical conductivity. Once the GQDs concentration was optimized, the positive effect of the dots was observed to depend on their reduction time, as J_{SC} increased for devices comprising the oxidized dots whereas FF increased for those with the dots reduced for 5 h. This was consistent with the morphological observations which showed that the functional groups, richer in oxidized GQDs, played a positive role in light harvesting, while sp^2 carbon-richer reduced samples made a beneficial contribution to charge conductivity, decreasing the leakage current and enhancing shunt resistance as well as FF. A maximum PCE of 7.6% was achieved with 5 h-reduced GQDs.

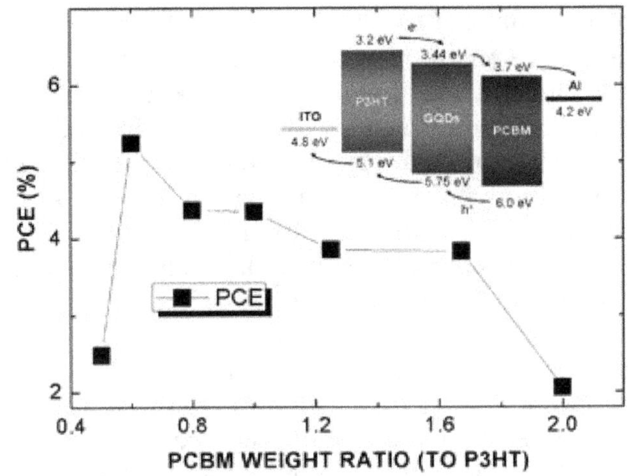

Figure 1.10 Schematic representation of ITO/PE-DOT:PSS/P3HT:PCBM:GQDs/LiF/Al device and PCE variation profile with P3HT content. Reproduced from Reference 75 with permission from Springer Nature.

An alternative approach to improve G band gap is the use of heteroatoms as doping materials [77]. The doping of rGO with nitrogen, and its application in P3HT:PCBM active layers, presented opportunities to develop a device with a structure ITO/PEDOT:PSS/N-doped graphene:P3HT:PCBM/Al, which reached a PCE of 4.5%. The PCE was about 40% higher than that of a device without G. However, the performance of these PSCs was in general below par than that of C_{60} or its derivatives based conventional cells [2-4].

Further research is necessary in order to tailor the structure and properties of G and its derivatives to optimize the fabrication of PSCs with higher PCEs.

1.4.3 Polymer Nanocomposites with G or Derivatives as Interfacial Layers

The IFLs, which involve electron transporting layer (ETL) and hole transporting layer (HTL), are located between both active layer-cathode and active layer-anode and affect considerably the PSC performance. G and its derivatives are frequently employed to enhance the electrical contacts, charge collection and charge transport. Furthermore, an adequate IFL can enhance light absorption and distribution as well as stability. IFLs are an important element in PSCs since these can allow or obstruct the passage of certain carriers through them, reducing the charge recombination in both electrodes. This is particularly positive in BHJ PSCs, due to the fact that the random distribution of acceptor and donor semiconductors can induce a direct contact between both semiconductors as well as the anode and cathode, which increases the chances of recombination. The contact resistance at the electrodes can be reduced with an IFL design that avoids the requirement to be located at the active layer. Furthermore, IFLs can reduce the reflection of light at interfaces, thus, enhancing the light absorption and distribution.

G and its derivatives can be applied as IFLs (ETL or HTL) owing to their energy band structures. Furthermore, these alternatives are less corrosive than ITO electrodes and lead to efficient charge transport [19]. Rafique *et al.* [78] developed a GO/PEDOT:PSS composite, which was applied as HTL in a BHJ PSC based on PCDTBT:PC71BM (Figure 1.11). The device, with a structure of ITO/GO/PEDOT:PSS/PCDTBT: PC71BM/Al had a PCE of 4.28%, which was higher than the PCEs attained with PEDOT:PSS or GO (3.57% and 2.77%). The device also displayed improved stability and better reproducibility. The observed performance can be attributed to the well matched work function of GO and PEDOT:PSS that decreased resistance and improved charge carrier mobilities. Also, the shunt resistance could increase as the electrons could be blocked by GO, which has a band-gap of ~3.6 eV.

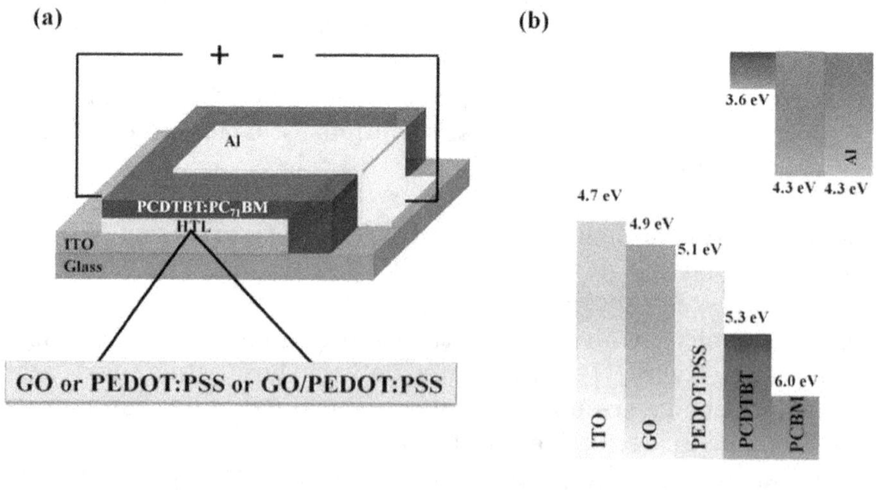

Figure 1.11 (A) Schematic device structure. (B) Energy band diagram. Reproduced from Reference 78 with permission from Springer Nature.

Other alternative has been the use of a PEG-modified Au nanoparticles/GO nanocomposite as HTL in a PSC with a structure of ITO/PEDOT:PSS(Au@PEG-GO)/PBDTTT-CT [79]. The nanocomposite could be dispersed in organic solvents and water, due to the beneficial effects of PEG on the solubility. The PCE reached 7.26%, and different spectral enhancement regions were detected depending on the position of the nanocomposite, which was useful for enhancing the absorption of solar radiation due to the different dielectric behavior around the nanoparticles.

Another option is the use of a nanocomposite based on polyacrylonitrile-grafted rGO (PRGO) as HTL in PTB7-Th-based PSCs [80]. The nanocomposite was prepared by an *in-situ* reduction with radiation, followed by a grafting polymerization with styryl-functionalized GO and acrylonitrile. It exhibited a good stability against weathering, work function of 4.87 eV, electrical conductivity of 0.87 S/cm and homogeneous thin-film morphology. These properties confirmed the potential of the developed material for PSC interfacial layers. The developed device reached PCE of 7.24%, J_{SC} of 14.78 mA/cm^2, V_{OC} of 0.76 V and FF of 0.64, which were observed to be close to PEDOT:PSS-based PSC (PCE of 7.17%, J_{SC} of 13.91 mA/cm^2, V_{OC} of 0.78 V and FF of 0.66), although with better durability.

Regarding ETL, novel nanocomposites based on ZnO nanocrystals dispersed in a G matrix stabilized with ethyl cellulose (EC) have been developed [81]. The resulting ZnO@G:EC nanocomposites exhibited a smooth morphology (Figure 1.12) and high conductivity for all G concentrations, which could be explained considering that the G structure was preserved. The authors developed a device with a structure of ITO/ZnO@G:EC/P3HT:PC61BM/MoO_3/Ag, which exhibited a PCE of 3.9%, better than bare ZnO nanocrystals (around 20% lower). As P3HT:PC61BM was substituted by PTB7:PC71BM, the maximum PCE was observed to be 8.4%, which demonstrated the possibility of achieving highly conductive ETL for PSCs.

In order reduce the recombination of holes and electrons as well as avoiding the leakage currents, Li *et al.* [82] prepared GO thin films deposited between ITO and an active layer of P3HT:PCBM by spin coating. The thicker GO ETL films exhibited lower performance, due to the increase in the resistance of the HTL. The best results were attained for 2 nm thick films, which led to the maximum PCE of 3.5%, FF of 0.54, V_{OC} of 0.57 V and J_{SC} of 11.4 mA/cm^2. However, it is not straightforward to obtain the GO thin films via spin coating. To overcome this issue, composites based on single-walled carbon nanotubes

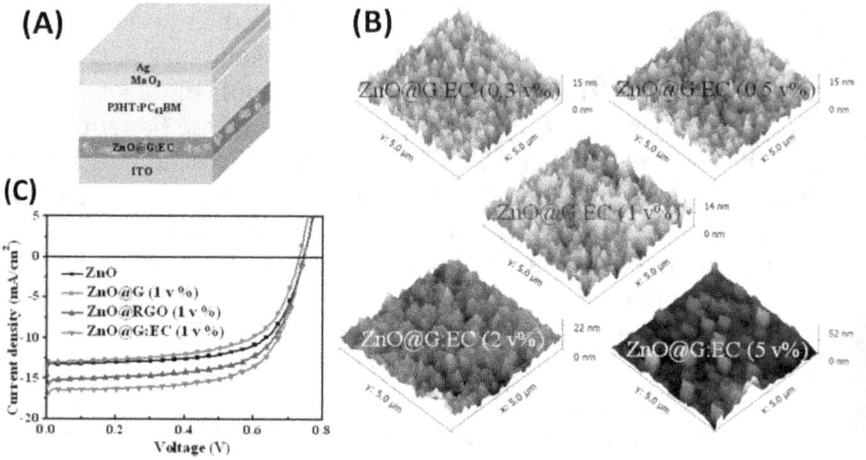

Figure 1.12 (A) Schematic representation of ITO/ZnO@G:EC/P3HT:PC61BM/MoO_3/Ag device. (B) AFM images of ZnO@G:EC nanocomposites with different G contents. (C) J-V curves of PSCs with the different nanocomposites. Reproduced from Reference 81 with permission from American Chemical Society.

(SWCNTs) were developed. The combination of GO and SWCNTs improved the nanocomposite conductivity. The PSC with a mass ratio of 1:0.2 (GO:SWCNT) exhibited PCE of 4.1%, FF of 0.63, V_{OC} of 0.6 V and J_{SC} of 10.8 mA/cm^2, which were better than PSC with raw GO-based HTL. The PCE of a conventional PEDOT:PSS device could be enhanced from 3.1 to 3.7% when PEDOT:PSS was replaced by a nanocomposite based on PE-DOT:PSS and few-layer G nanosheets [83]. The PCE increased from 0.42 to 0.74%, in the case of butylamine-modified G/PE-DOT:PSS nanocomposite. Also, when the HTL was based on a GO/PE-DOT:PSS nanocomposite, the obtained PCE was 1.43%, which was higher than the reference PSC without GO due to the improved charge carrier transport and conductivity [84]. On the whole, the addition of G or G derivatives as dopants or additives to PEDOT:PSS enhances the PSC efficiency, though the reason for this improvement is not completely understood to date.

1.4.4 Polymer Nanocomposites with G or Derivatives in Dye-Sensitized Solar Cells

Dye-sensitized solar cells (DSSCs) are composed of a working electrode of mesoporous dye-sensitized TiO_2 nanocrystals, an electrolyte incorporating a redox pair (typically I_2/I_{3-}) and a Pt-based counter electrode [85]. DSSCs are semitransparent, semi-flexible and are easy to manufacture with traditional roll-printing techniques, thus, enabling a range of uses not applicable to rigid photovoltaic systems, along with the development of PCEs comparable to silicon solar cells. Ruthenium sensitizer-based DSSCs show a maximum PCE of 12.3%. However, the high cost of ruthenium-based sensitizers and Pt-based electrodes limits the commercialization of DSSCs, hence, new dye materials are required. G and its nanocomposites have high potential as efficient promoters of electron transport in TiO_2 electrodes and dyes for DSSCs with good photoelectrochemical stability [86]. Hong *et al.* [87] prepared a counter electrode of DSSC by spin coating functionalized graphene/PEDOT:PSS nanocomposite on ITO. The electrode with a 60 nm nanocomposite layer (1.0 wt% G) exhibited high transmittance in the visible wavelength (>80%) and catalytic activity. The PCE of the DSSC with this electrode was 4.5%, slightly lower than the cell with Pt counter electrode (6.3%) under the same conditions.

Nanocomposites of G and conductive polymers have also been explored as flexible counter electrodes of DSSCs. Lee *et al.* [88] spin coated G with a PEDOT layer, and the corresponding DSSC demonstrated a PCE of 5.1%, close to Pt/fluorine-doped tin oxide (FTO)-based DSSC (PCE of 5.9%), albeit its FF was only 0.6 due to the high resistance of the G-based counter electrode. Another counter electrode was developed by spin coating a PEDOT layer on a CVD-grown G-covered PET sheet, and the PCE was increased up to 6.26%. Moreover, the DSSC showed excellent performance even when bent.

Kaniyoor and Ramaprabhu [89] developed a counter electrode by drop casting thermally exfoliated G dispersed in a Nafion:ethanol solution on FTO substrates. The role of the polymer was to prevent the peeling off of the film, thus, ensuring a good substrate adhesion. The DSSC showed I_{SC} of 7.7 mA/cm^2, V_{OC} of 0.68 V, FF of 0.54 and PCE of 2.8%, comparable to Pt-counter electrode based DSSC with an efficiency of about 3.4%. Another strategy was wrapping G by PVP, leading to a uniform G thin film on FTO [90]. The DSSC with G/PVP counter electrode exhibited PCE of 3.0%, I_{SC} of 11.2 mA/cm^2, V_{OC} of 0.73 V and FF of 0.37.

Wang *et al.* [91] prepared PANI/G nanocomposites by *in-situ* polymerization of aniline monomer in a G dispersion. A uniform dispersion of PANI nanoparticles on the G surface was attained (Figure 1.13), which improved both the electroactive region of PANI and the conductivity of the nanocomposite, thus, enhancing its electrocatalytic activity as a counter electrode. The corresponding DSSC displayed I_{SC} of 13.3 mA/cm^2, V_{OC} of 0.69 V, FF of 0.67 and PCE of 6.09%, only slightly lower than that the cell with a Pt counter electrode.

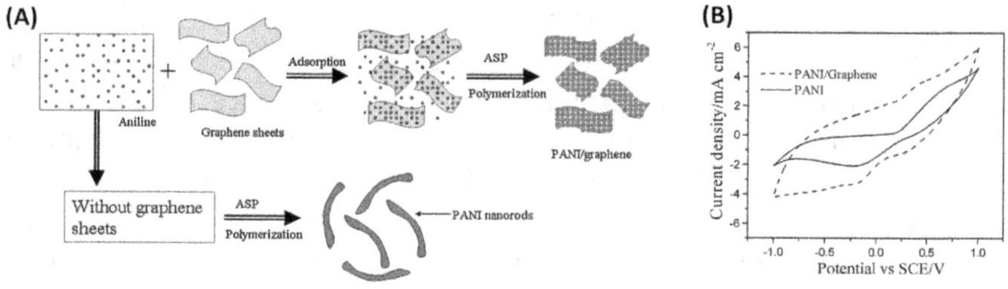

Figure 1.13 (A) Schematic representation of the synthesis procedure of PANI/G counter electrode. (B) J-V curves of DSSCs with PANI and PANI/G as counter electrode. Reproduced from Reference 91 with permission from Elsevier.

Gong *et al.* [92] incorporated GO in PPy matrix and reduced the nanocomposite to form rGO/PPy counter electrode. The nanocomposite exhibited outstanding catalytic performance for the reduction of triiodide and led to a DSSC with I_{SC} of 15.8 mA/cm^2, V_{OC} of 0.73 V, FF of 0.71 and PCE of 8.14%, parameters comparable to the cell with a Pt cathode. Kaniyoor and Ramaprabhu [93] found that the addition of a cationic polyelectrolyte such as polydiallyldime-thylammonium chloride (PDDA) to rGO improved the performance of the counter electrode, while the addition of an anionic polyelectrolyte such as PSS reduced it, owing to charge repulsion between negatively charged polymer and negatively charged triiodide. Thus, in the latter case, a DSSC with poor performance was attained: I_{SC} of 8.7 mA/cm^2, V_{OC} of 0.69 V, FF of 0.35 and PCE of 2.1%. Overall, to date, no polymer/G nanocomposite has been able to surpass the performance of Pt cathodes for the I_2/I_{3-} mediator.

1.5 Conclusions and Outlook

In this chapter, the recent progress in G/polymer nanocomposite materials for PSC applications has been summarized. Owing to their 2D structure, exceptional electrical and thermal conductivity, transparency, flexibility, high mechanical strength and very large specific surface area, G and its derivatives, GO and rGO, have emerged as ideal candidates for solar cell applications. Further, the blending with polymers improves their processability, chemical and electrochemical properties, thereby, extending their range of applications. Consequently, G/polymer nanocomposites have high potential to be used in several device components, such as TCEs, ALs and IFLs. Even so, whilst certain enhancements have already been achieved in replacing conventional materials (i.e. ITO, PEDOT:PSS), a number of issues need to be overcome before using G/polymer nanocomposites as the main components in commercial PSCs.

- New methods to synthesize high-quality G thin films with controllable morphology and electronic properties are necessary, given that the performance of PSCs comprising G is related to the nanomaterial characteristics: purity, quality, band gap, structural morphology, etc. In particular, for TCE applications, solution processable films with an optimal balance between high transparency and low sheet resistance are needed. Additionally, inexpensive techniques for synthesizing G at an industrial level with reproducible properties, no defects and high stability need to be developed. Despite a lot of progress in this direction, the present methods are strongly restricted by their low efficiencies, which has to be addressed for commercial applications.
- The electrical conductivity of chemically modified G is significantly lower than that of ideal G. Accordingly, G/polymer nanocomposites display poor electrical properties and do not fulfill the requirements for use in TCEs and counter electrodes of solar cells. Thus, the functionalization and ultrasonication processes applied to G before and during the fabrication of polymer nanocomposites need to be optimized in order to minimize the reduction in electrical conductivity. Novel doping or functionalization approaches compatible with the fabrication process of PSCs should be examined to reach high charge carrier densities, improved stability and tailorable energy levels in G-based materials.
- The actual specific surface area of G and its derivatives is considerably smaller than the theoretical one due to the strong agglomeration of the G layers by means of van der Waals forces and π-π stacking interactions. Further, blending with polymers makes the issue even worse. Consequently, novel strategies to successfully exfoliate the G sheets and retain a

high specific surface area when blended with polymers are essential. Though significant achievements have been attained through addition of stabilizers such as surfactants, these occasionally have negative effects on the final device performance. Hence, innovative synthetic routes are pursued. Another key feature is the possibility to produce G on different substrates to eliminate the problems that arise during the transfer of G from one material to another.

- The performance of G/polymer nanocomposites in PSCs is well below than described for conventional materials. Hence, the composition and morphology of the nanocomposites as well as the structure of the corresponding PSCs require additional optimization. In this regard, the use of different types of G-based materials like GQDs is an interesting choice. Owing to quantum confinement and edge effect, the GQDs can display superior properties than neat G, which could allow for the fabrication of PSCs with better PCE.
- The toxicity of nanostructured materials is a point of concern. G-based materials may induce cytotoxicity to humans [94], and this subject should be investigated in detail and clarified prior to their commercial application.

Overall, the research on this topic is still in its early infancy, and the mechanisms used to elucidate the observed effects are not fully clear yet. Nevertheless, it is envisaged that constant innovative efforts towards the development of G/polymer nanocomposites and device manufacturing processes for high-performance PSCs could immensely benefit the solar energy industry.

Acknowledgement

Dr. A. M. Diez-Pascual would like to thank the Spanish Ministry of Economy and Competitivity for a "Ramón y Cajal" postdoctoral fellowship co-financed by the EU.

References

1. Brabec, C. J., Sariciftci, N. S., and Hummelen, J. C. (2001) Plastic solar cells. *Advanced Functional Materials*, **11**(1), 15-26.

2. Cheng, Y.-J., Yang, S.-H., and Hsu, C.-S. (2009) Synthesis of conjugated polymers for organic solar cell applications. *Chemical Reviews*, **109**(11), 5868-5923.
3. Jorgensen, M., Norrman, K., and Krebs, F. C. (2008) Stability/degradation of polymer solar cells. *Solar Energy Materials and Solar Cells*, **92**(7), 686-714.
4. Thompson, B. C., and Frechet, J. M. (2007) Polymer-fullerene composite solar cells. *Angewandte Chemie, International Edition*, **47**(1), 58-77.
5. Chen, H.-Y., Hou, J., Zhang, S., Liang, Y., Yang, G., Yang, Y., Yu, .L, Wu, Y., and Li, G. (2009) Polymer solar cells with enhanced open-circuit voltage and efficiency. *Nature Photonics*, **3**, 649-653.
6. Dennler, G., Scharber, M. C., and Brabec, C. J. (2009) Polymer-fullerene bulk-heterojuntion solar cells. *Advanced Materials*, **21**(13), 1323-1338.
7. Green, M. A. (1981) Solar cell fill factors: General graph and empirical expressions. *Solid State Electronics*, **24**(8), 788-789.
8. Luo, J., Wu, H., He, C., Li, A., Yang, W., and Cao, Y. (2009) Enhanced open-circuit voltage in polymer solar cells. *Applied Physics Letters*, **95**, 043301.
9. De Kok, M. M., Buechel, M., Vulto, S. I. E., van de Weijer, P., Meulenkamp, E. A., de Winter, S. H. P. M., Mank, A. J. G., Vorstenbosch, H. J. M., Weijtens, C. H. L., van Elsbergen, V. (2004) Modification of PEDOT:PSS as hole injection layer in polymer LEDs. Physica Status Solidi A, **201**(6), 1342-1359.
10. Yun, J.-M., Yeo, J.-S., Kim, J., Jeong, H.-G., Kim, D.-Y., Noh, Y.-J., Kim, S.-S., Ku, B.-C., and Na, S.-I. (2011) Solution-processable reduced graphene oxide as a novel alternative to PEDOT:PSS hole transport layers for highly efficient and stable polymer solar cells. *Advanced Materials*, **23**(42), 4923-4928.
11. Taleghani, H. G., Aleahmad, M., and Eisazadeh, H. (2009) Preparation and characterization of polyaniline nanoparticles using various solutions. *World Applied Sciences Journal*, **6**, 1607-1611.
12. Walker, B., Kim, C., and Nguyen, T. Q. (2011) Small molecule solution-processed bulk heterojunction solar cells. *Chemistry of Materials*, **23**(3), 470-482.
13. Das, S., Keum, J. K., Browning, J. F., Gu, G., Yang, B., Dyck, O., Chen, W., Chen, J., Ivanov, I. N., Hong, K., Rondinone, A. J., Joshi, P. C., Geohegan, D. B., Duscher, G., and Xiao, K. (2015) Correlating high power conversion efficiency of PTB7:PC71BM inverted organic solar cells with nanoscale structures. *Nanoscale*, **14**(38), 15576-15583.
14. You, J., Dou, L., Yoshimura, K., Kato, T., Ohya, K., Moriarty, T., Emery, K., Chen, C.-C., Gao, J., Li, G., and Yang, Y. (2013) A polymer tandem solar cell with 10.6% power conversion efficiency. *Nature Communications*, **4**, 1-10.

15. Xue, Z., Wang, S., Yang, J., Zhong, Y., Qian, M., Li, C., Zhang, Z., Xing, G., Huettner, S., Tao, Y., Li, Y., and Huang, W. (2018) Enhanced power conversion efficiency in iridium complex-based terpolymers for polymer solar cells. *npj Flexible Electronics*, **2**, 1.
16. Wan, Q., Guo, X., Wang, Z., Li, W., Guo, B., Ma, W., Zhang, M., and Li, Y. (2016) 10.8% efficiency polymer solar cells based on PTB-Th and PC71BM via binary solvent additive treatment. *Advanced Functional Materials*, **26**(36), 6635-6640.
17. Chen, J.-D., Cui, C., Li, Y.-Q., Zhou, L., Ou, Q.-D., Li, C., Li, Y., amd Tang, J.-X. (2014) Single junction polymer solar cells exceeding 10% power conversion efficiency. *Advanced Materials*, **27**(6), 1035-1041.
18. Singh, K., Ohlan, A., and Dhawan, S. K. (2012) Polymer-graphene nanocomposites: Preparation, characterization, properties, and applications. In: *Nanocomposites – New Trends and Developments*, Ebrahimi, E. (ed.), InTech, UK, pp. 37-71.
19. Lin, X.-F., Zhang, Z.-Y., Yuan, Z.-K., Li, J., Xiao, X.-F., Hong, W., Chen, X.-D., and Yu, D.-S. (2016) Graphene-based materials for polymer solar cells. *Chinese Chemical Letters*, **27**(8), 1259-1270.
20. Sun, Y., Zhang, W., Chi, H., Liu, Y., Hou, C .L., and Fang, D. (2015) Recent development of graphene materials applied in polymer solar cell. *Renewable and Sustainable Energy Reviews*, **43**, 973-980.
21. Wallace, P. R. (1947) The band theory of graphite. *Physical Review*, **71**, 622-634.
22. Chang, H., and Wu, H. (2013) Graphene-based nanomaterials: Synthesis, properties, and optical and optoelectronic applications. *Advanced Functional Materials*, **23**(16), 1984-1997.
23. Weiss, N. O., Zhou, H., Liao, L., Liu, Y., Jiang, S., Huang, Y., and Duan, X. (2012) Graphene: An emerging electronic material. *Advanced Materials*, **24**(43), 5782-5825.
24. Castro Neto, A. H., Guinea, F., Peres, N. M. R., Novoselov, K. S., and Geim, A. K. (2009) The electronic properties of graphene. *Reviews of Modern Physics*, **81**, 109-162.
25. Lee, C., Wei, X., Kysar, J. W., and Hone, J. (2008) Measurement of the elastic properties and intrinsic strength of monolayer graphene. *Science*, **321**(5887), 385-388.
26. Diez-Pascual, A. M., Gomez-Fatou, M. A., Ania, F., and Flores, A. (2015) Nanoindentation in polymer nanocomposites. *Progress in Materials Science*, **67**, 1-94.
27. Nair, R. R., Grigorenko, A. N., Novoselov, K. S., Booth, T. J., Stauber, T., Peres, N. M. R., and Geim, A. K. (2008) Fine structure constant defines visual transparency of graphene. *Science*, **320**(5881), 1308.
28. Bhuyan, M. S. A., Uddin, M. N., Islam, M. M., and Bipasha, F. A. (2016) Synthesis of graphene. *International Nano Letters*, **6**(2), 65-83.
29. Zheng, Q., and Kim, J. K. (2015) *Graphene for Transparent Conducto-*

rs. Synthesis, Properties and Applications, Springer, USA.
30. Novoselov, K. S., Geim, A. K., Morozov, S. V., Jiang, D., Zhang, Y., Dubonos, S. V., Grigorieva, I. V., and Firsov, A. A. (2004) Electric field effect in atomically thin carbon films. *Science*, **306**(5696), 666-669.
31. Wu, J., Becerril, H. A., Bao, Z., Liu, Z., Chen, Y., and Peumans, P. (2008) Organic solar cells with solution processed graphene transparent electrodes. *Applied Physics Letters*, **92**, 263302.
32. Diez-Pascual, A. M., and Diez-Vicente A. L. (2016) Poly(propylene fumarate)/polyethylene glycol-modified graphene oxide biocomposites for tissue engineering. *ACS Applied Materials & Interfaces*, **8**(28), 17902-17914.
33. Su, C.-Y., Lu, A.-Y., Xu, Y., Chen, F.-R., Khlobystov, A. N., and Li, L.-J. (2011) High-quality thin graphene films from fast electrochemical exfoliation. *ACS Nano*, **5**(3), 2332-2339.
34. Dreyer, D. R., Park, S., Bielawski, C. W., and Ruoff, R. S. (2010) The chemistry of graphene oxide. *Chemical Society Reviews*, **39**(1), 228-240.
35. Chen, J., Yao, B., Li, C., and Shi, G. (2013) An improved Hummers method for eco-friendly synthesis of graphene oxide. *Carbon*, **64**, 225-229.
36. Eda, G., and Chhowalla, M. (2010) Chemically derived graphene oxide: Towards large-area thin-film electronics and optoelectronics. *Advanced Materials*, **22**(22), 2392-2415.
37. Pei, S., and Chen, H.-M. (2012) The reduction of graphene oxide. *Carbon*, **50**(9), 3210-3228.
38. Cote, L. J., Cruz-Silva, R., and Huang, J. (2009) Flash reduction and patterning of graphene oxide and its polymer composite. *Journal of American Chemical Society*, **131**(31), 11027-11032.
39. Zhang, Y., Guo, L., Wei, S., He, Y., Xia, H., Chen, Q., Sun, H.-B., and Xiao, F.-S. (2010) Direct imprinting of microcircuits on graphene oxides film by femtosecond laser reduction. *Nano Today*, **5**(1), 15-20.
40. Stankovich, S., Dikin, D. A., Dommett, G. H., Kohlhaas, K. M., Zimmey, E. J., Stach, E. A., Piner, R. D., Nquyen, S. T., and Ruoff, R. S. (2006) Graphene-based composite materials. *Nature*, **442**, 282-286.
41. Periasamy, M., and Thirumalaikumar, M. (2000) Methods of enhancement of reactivity and selectivity of sodium borohydride for applications in organic chemistry. *Journal of Organometallic Chemistry*, **609**(1-2), 137-151.
42. Salavagione, H. J., Diez-Pascual, A. M., Lazaro, E., Vera, S., and Gomez-Fatou, M. A. (2014) Chemical sensors based on polymer composites with carbon nanotubes and graphene: The role of the polymer. *Journal of Materials Chemistry A*, **2**(35), 14289-14328.
43. Bai, H., Li, C., and Shi, G. (2011) Functional composite materials based on chemically converted graphene. *Advanced Materials*, **23**(9), 1089-1115.

44. Diez-Pascual, A. M., Sanchez, J. A. L., Capilla, R. P. and Diaz, P. G. (2018) Recent developments in graphene/polymer nanocomposites for application in polymer solar cells. *Polymers*, **10**, 217.
45. Wang, H. L., Hao, Q. L., Xia, X. F., Wang, Z. J., Tian, J., Zhu, J. H., Tang, C., and Wang, X. (2011) In situ fabrication of nanoscale graphene oxide/polyaniline composite and its electrochemical properties. *Advanced Materials Research*, **148-149**, 1547-1550.
46. Fawaz, J., and Mittal, V. (2014) Synthesis of polymer nanocomposites: Review of various techniques. In: *Synthesis Techniques for Polymer Nanocomposites*, Mittal, V. (ed.), Wiley, USA, pp. 1-30.
47. An, C. J., Kim, S. J., Choi, H. O., Kim, D. W., Jang, S. W., Jin, M. L., Park, J.-M., Choi, J. K., and Jung, H.-T. (2014) Ultraclean transfer of CVD-grown graphene and its application to flexible organic photovoltaic cells. *Journal of Materials Chemistry A*, **2**(48), 20474-20480.
48. Yin, Z., Sun, S., Salim, T., Wu, S., Huang, X., He, Q., Lang, Y. M., and Zhan, H. (2010) Organic photovoltaic devices using highly flexible reduced graphene oxide films as transparent electrodes. *ACS Nano*, **4**(9), 5263-5268.
49. Konios, D., Petridis, C., Kakavelakis, G., Sygletou, M., Savva, K., Stratakis, E., and Kymakis, E. (2015) Reduced graphene oxide micromesh electrodes for large area, flexible, organic photovoltaic devices. *Advanced Functional Materials*, **25**(15), 2213-2221.
50. Park, H., Chang, S., Smith, M., Gradecak, S., and Kong, J. (2013) Interface engineering of graphene for universal applications as both cathode and anode in organic photovoltaics. *Scientific Reports*, **3**, 1581.
51. Park, H., Chang, S., Zhou, X., Kong, J., Palacios, T., and Gradecak, S. (2014) Flexible graphene electrode-based organic photovoltaics with record-high efficiency. *Nano Letters*, **14**(9), 5148-5154.
52. Song, Y., Chang, S., Gradecak, S., and Kong, J. (2016) Visibly transparent organic solar cells on flexible substrates with all-graphene electrodes. *Advanced Energy Materials*, **6**(20), 1600847.
53. La Notte, L, Bianco, G. V., Palma, A. L., Carlo, A. D., Bruno, G., and Reale, A. (2018) Sprayed organic photovoltaic cells and mini-modules based on chemical vapor deposited graphene as transparent conductive electrode. *Carbon*, **129**, 878-883.
54. Soltani-kordshuli, F., Zabihi, F., and Eslamian, M. (2016) Graphene-doped PEDOT:PSS nanocomposite thin films fabricated by conventional and substrate vibration-assisted spray coating (SVASC). *Engineering Science and Technology, an International Journal*, **19**(3), 1216-1223.
55. Chen, Q., Zabihi, F., and Eslamian, M. (2016) Improved functionality of PEDOT:PSS thin films via graphene doping, fabricated by ultrasonic substrate vibration-assisted spray coating. *Synthetic Metals*, **222**, 309-317.

56. Liu, Z., Li, J., Sun, Z.-H., Tai, G., Lau, S.-P., and Yan, F. (2012) The application of highly doped single-layer graphene as the top electrodes of semitransparent organic solar cells. *ACS Nano*, **6**(1), 810-818.
57. Liu, Z., Li, J., and Yan, F. (2013) Package-free flexible organic solar cells with graphene top electrodes. *Advanced Materials*, **25**(31), 4296-4301.
58. Xu, Y., Wang, Y., Liang, J., Huang, Y., Ma, Y., Wan, X., and Chen, Y. (2009) A hybrid material of graphene and poly (3,4-ethyldioxythiophene) with high conductivity, flexibility, and transparency. *Nano Research*, **2**(4), 343-348.
59. Chang, H., Wang, G., Yang, A., Tao, X., and Liu, X. (2010) A transparent, flexible, low-temperature, and solution-processible graphene composite electrode. *Advanced Functional Materials*, **20**(17), 2893-2902.
60. Jo, K., Lee, T., Choi, H. J., Park, J. H., Lee, D. J., Lee, D. W., and Kim, B.-S. (2011) Stable aqueous dispersion of reduced graphene nanosheets via non-covalent functionalization with conducting polymers and application in transparent electrodes. *Langmuir*, **27**(5), 2014-2018.
61. Lima, L. F., Matos, C. F., Gonçalves, L. C., Salvatierra, R. V., Cava, C. E., Zarbin, A. J. G., and Roman, L. S. (2016) Water based, solution-processable, transparent and flexible graphene oxide composite as electrodes in organic solar cell application. *Journal of Physics D: Applied Physics*, **49**, 105106.
62. Carrasco-Valenzuela, L., Zaragoza-Contreras, E. A., and Vega-Rios, A. (2017) Synthesis of graphene oxide/poly(3,4-ethylenedioxythiophene) composites by Fenton's reagent. *Polymer*, **130**, 124-134.
63. Sookhakian, M., Amin, Y. M., Baradaran, S., Tajabadi, M. T., Golsheikh, A. M., and Basirun, W. J. (2014) A layer-by-layer assembled graphene/zinc sulfide/polypyrrole thin-film electrode via electrophoretic deposi-tion for solar cells. *Thin Solid Films*, **552**, 204-211.
64. Ricciardulli, A. G., Yang, S., Feng, X., and Blom, P. W. M. (2017) Solution-processable high-quality graphene for organic solar cells. *ACS Applied Materials & Interfaces*, **9**(30), 25412-25417.
65. Liu, Z., Liu, Q., Huang, Y., Ma, Y., Yin, S., Zhang, X., Sun, W., and Chen, Y. (2008) Organic photovoltaic devices based on a novel acceptor material: Graphene. *Advanced Materials*, **20**, 3924-3930.
66. Liu, Q., Liu, Z., Zhang, X., Yang, L., Zhang, N., Pan, G., Yin, S., Chen, Y., and Wei, J. (2009) Polymer photovoltaic cells based on solution-processable graphene and P3HT. *Advanced Functional Materials*, **19**(6), 894-904.
67. Ye, J., Li, X., Zhao, J., Mei, X., and Li, Q. (2016) Solution-processed P3HT-functional graphene for efficient heterojunction organic photoelectronics. *Journal of Materials Research*, **31**(14), 1985-1992.

68. Wang, H., He, D., Wang, Y., Liu, Z., Wu, H., and Wang, J. (2011) Organic Photovoltaic Devices Based on graphene as an electron-acceptor ma-terial and P3OT as a donor material. *Physica Status Solidi A*, **208**(10), 2339-2343.
69. Yu, D., Yang, Y., Durstock, M., Baek, J.-B., and Dai, L. (2010) Soluble P3HT-grafted graphene for efficient bilayer-heterojunction photovoltaic devices. *ACS Nano*, **4**(10), 5633-5640.
70. Yu, D., Park, K., Durstock, M., and Dai, L. (2011) Fullerene-grafted graphene for efficient bulk heterojunction polymer photovoltaic devices. *The Journal of Physical Chemistry Letters*, **2**(10), 1113-1118.
71. Liu, Z., He, D., Wang, Y., Wu, H., Wang, J., and Wang, H. (2010) Improving photovoltaic properties by incorporating both SPFGraphene and functionalized multiwalled carbon nanotubes. *Solar Energy Materials and Solar Cells*, **94**(12), 2148-2153.
72. Lee, R.-H., Huang, J.-L., and Chi, C.-H. (2013) Conjugated polymer-functionalized graphite oxide sheets thin films for enhanced photovoltaic properties of polymer solar cells. *Journal of Polymer Science, Part B: Polymer Physics*, **51**(2), 137-148.
73. Li, Y., Hu, Y., Zhao, Y., Shi, G., Deng, L., Hou, Y., and Qu, L. (2011) An electro-chemical avenue to green-luminescent graphene quantum dots as potential electron-acceptors for photovoltaics. *Advanced Materials*, **23**(6), 776-780.
74. Gupta, V., Chaudhary, N., Srivastava, R., Sharma, G. D., Bhardwaj, R., and Chand, S. (2011) Luminescent graphene quantum dots for organic photovoltaic devices. *Journal of American Chemical Society*, **133**(26), 9960-9663.
75. Li, F., Kou, L., Chen, W., Wu, C., and Guo, T. (2013) Enhancing the short-circuit current and power conversion efficiency of polymer solar cells with graphene quantum dots derived from double-walled car-bon nanotubes. *NPG Asia Materials*, **5**, e60.
76. Kim, J. K., Park, M. J., Kim, S. J., Wang, D. H., Cho, S. P., Bae, S., Park, J. H., and Hong, B. H. (2013) Balancing light absorptivity and carrier conductivity of graphene quantum dots for high-efficiency bulk hetero-junction solar cells. *ACS Nano*, **7**(8), 7207-7212.
77. Jun, G. H., Jin, S. H., Lee, B., Kim, B. H., Chae, W.-S., Hong, S. H., and Jeon, S. (2013) Enhanced conduction and charge-selectivity by N-doped graphene flakes in the active layer of bulk-heterojunction organic solar cells. *Energy & Environmental Science*, **6**(10), 3000-3006.
78. Rafique, S., Abdullah, S. M., Shahid, M. M., Ansari, M. O., and Sulaiman, K. (2016) Significantly improved photovoltaic performance in polymer bulk heterojuntion solar cells with graphene oxide/PEDOT:PSS double decked hole transport layer. *Scientific Reports*, **7**, 39555.
79. Chuang, M.-K., and Chen, F.-C. (2015) Synergistic plasmonic effects

of metal nanoparticle-decorated PEGylated graphene oxides in polymer solar cells. *ACS Applied Materials & Interfaces*, **7**(13), 7397-7405.

80. Jung, C.-H., Noh, Y.-J., Bae, J.-H., Yu, J.-H., Hwang, I.-T., Shin, J., Shin, K., Lee, J.-S., Choi, J.-H., Na, S.-I. (2017) Polyacrylonitrile-grafted reduced graphene oxide hybrid: An all-round and efficient hole-extraction material for organic and inorganic-organic hybrid photovoltaics. *Nano Energy*, **31**, 19-27.
81. Hu, A., Wang, Q., Chen, L., Hu, X., Zhang, Y., Wu, Y., and Chen, Y. (2015) In situ formation of ZnO in graphene: A facile way to produce a smooth and highly conductive electron transport layer for polymer solar cells. *ACS Applied Materials & Interfaces*, **7**(29), 16078-16085.
82. Li, S.-S., Tu, K.-H., Lin, C.-C., Chen, C.-W., and Chhowalla, M. (2010) Solution-processable graphene oxide as an efficient hole transport layer in polymer solar cells. *ACS Nano*, **4**(6), 3169-3174.
83. Nguyen, D. D., Tai, N. H., Chueh, S. Y., Chen, Y. J., Kuo, W. S., Chou, T. W., Hsu, C. S., and Chen, L. J. (2011) Synthesis of ethanol-soluble few-layer graphene nanosheets for flexible and transparent conducting compo-site films. *Nanotechnology*, **22**(29), 295606.
84. Hu, X., and Zhou, Q. (2013) Health and ecosystem risks of graphene. *Chemical Reviews*, **113**(5), 3815-3835.
85. Gratzel, M. (2003) Dye-sensitized solar cells. *Journal of Photochemistry and Photobiology C*, **4**(2), 145-153.
86. Yan, X., Cui, X., Li, B., and Li, L.-S. (2010) Large, solution-processable graphene quantum dots as light absorbers for photovoltaics. *Nano Letters*, **10**(5), 1869-1873.
87. Hong, W., Xu, Y., Lu, G., Li, C., and Shi, G. (2008) Transparent graphene/PEDOT-PSS composite films as counter electrodes of dye-sensitized solar cells. *Electrochemistry Communications*, **10**(10), 1555-1558.
88. Lee, K. S., Lee, Y., Lee, J. Y., and Ahn, J.-H. (2012) Flexible and platinum-free dye-sensitized solar cells with conducting-polymer-coated graphene counter electrodes. *ChemSusChem*, **5**(2), 379-382.
89. Kaniyoor, A., Ramaprabhu, S. (2011) Thermally exfoliated graphene based counter electrode for low cost dye sensitized solar cells. *Journal of Applied Physics*, **109**, 124308.
90. Li, Z.-Y., Akhtar, M. S., Kuk, J. H., Kong, B.-S., and Yan, O-B. (2012) Graphene application as a counter electrode material for dye-sensitized solar cell. *Materials Letters*, **86**, 96-99.
91. Wang, G., Xing, W., Zhuo, S. (2012) The production of polyaniline/graphene hybrids for use as a counter electrode in dye-sensitized solar cells. *Electrochimica Acta*, **66**, 151-157.
92. Gong, F., Xu, X., Zhou, G., and Wang, Z.-S. (2013) Enhanced charge transportation in a polypyrrole counter electrode via incorporation

of reduced graphene oxide sheets for dye-sensitized solar cells. *Physical Chemistry Chemical Physics*, **15**(2), 546-552.
93. Kaniyoor, A., and Ramaprabhu, S. (2012) Soft functionalization of graphene for enhanced tri-iodide reduction in dye sensitized solar cells. *Journal of Materials Chemistry*, **22**(17), 8377-8384.
94. Chng, E. L. K., and Pumera, M. (2013) The toxicity of graphene oxides: dependence on the oxidative methods used. *Chemistry: A European Journal*, **19**, 8227-8235.

Chapter 2

Conducting Polymer Composites in Electrochemical Sensors

Shahid Bashir,[a*] S. Ramesh,[a*], K. Ramesh,[a] Arshid Numan,[b] and Javed Iqbal[c,d]

[a]*Centre of Ionics University of Malaya, Department of Physics, Faculty of Science, University of Malaya, Kuala Lumpur 50603, Malaysia*
[b]*Graphene & Advance 2D Materials Research Group (GAMRD), School of Science and Technology, Sunway University, No. 5, Jalan University, Bandar Sunway, 4700 Subang Jaya, Selangor, Malaysia*
[c]*Department of Chemistry, Faculty of Science, University of Malaya, Kuala Lumpur 50603, Malaysia*
[d]*Centre of Nanotechnology, King Abdulaziz University, Jeddah 21589, Saudi Arabia*
**Corresponding authors:* shahidbashirbaig@gmail.com; rameshtsubra@gmail.com

2.1 Introduction

A polymer is a macromolecule comprised of many repeating units called monomers. The term "polymer" was derived by Berzelius in 1833 from ancient Greek words "polus" meaning many and "meros" meaning parts [1]. There are two major types of polymers, such as natural and synthetic polymers. Polymers are essential in everyday life due to their wide range of properties owing to the multiple repeating units forming long chains [2]. Among the wide range of polymers, conducting polymers have received a significant attention from the scientific community due to their remarkable properties. There are three main types of organic compounds which can transport charge: a) ion radical salts, b) charge transfer complexes and c) conjugated organic polymers. Another type of polymers, known as intrinsic conducting polymers, have recently emerged, which exhibit interesting optical and electrical properties [3]. Their development was initiated in the 1970s, when polymers having high electronic conductivity were discovered in the partially oxidized state. This was possible owing to the efforts of Alan G. MacDiarmid, Alan J. Heeger and

Conducting Polymer Composites, edited by Vikas Mittal

Hideki Shirakawa, who got Noble Prize in Chemistry in 2000 for the discovery of electronically conducting polymers. The preparation, characterization and applications of electronically conducting and electrochemically active polymer systems are still of significant research interest [4-7]. Of particular research interest is understanding the behavior of these systems, especially the mechanism of charge transfer and charge transport processes that occur during redox reactions. Moreover, conducting polymers are conjugated polymers having delocalized π-bonds on the backbone of polymers. These π-bonds assist in the migration of electrons throughout the polymeric chain. Thus, these polymers can act as conductors, semiconductors or superconductors. There are many polymers which show conducting properties such as polyaniline (PANI), polypyrrole (PPy), polyacetylene [8], polyparaphenylene (PPPh), polyparaphenylene vinylene (PPV), polythiophenes (PTh), polydiaminonaphthalene, polyazulene, and poly (3,4-ethylenedioxythiophene) (PEDOT). The structures of these polymers are given in Figure 2.1.

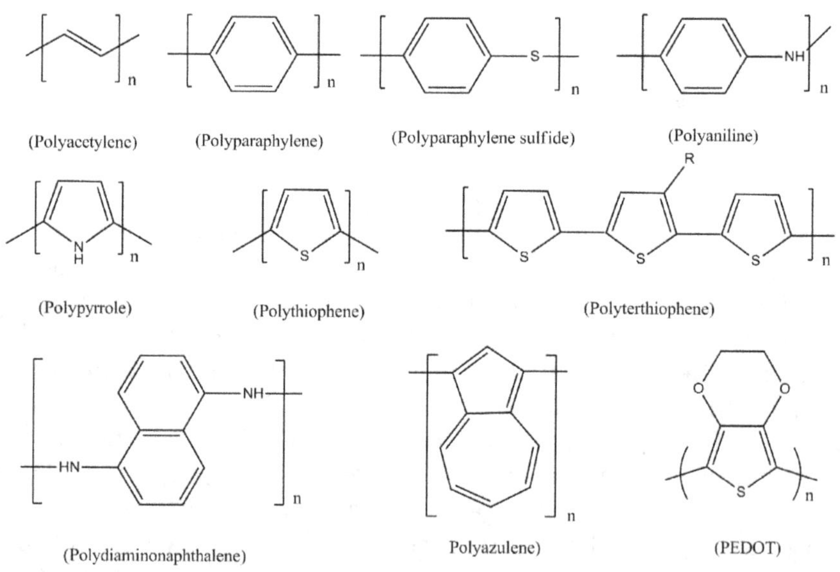

Figure 2.1 Structures of conducting polymers.

Conducting polymers exhibit inherent features such as light-weight, electrochromic and photochromic nature, low cost, variable conductivity, corrosion resistant behavior, etc. Thus, these polymers are useful for a broad range of potential applications in fields such as energy storage, energy conversion, electrocatalysis, electroanalysis, microsystem technologies, microwave screening, electronic devices, corrosion protection and sensors. However, conducting polymers have poor solubility, stability and processability. In the last two decades, scientists have applied different novel approaches to transform the properties of the conducting polymers and to overcome these drawbacks. The numerous approaches proposed are blending, copolymerization, interpenetrating networks formation and composite synthesis.

Composites are materials that comprise of two or more physically or chemically different phases (continuous and dispersed phases) separated by a distinct interface. The different phases can be combined judiciously in an appropriate ratio to obtain new system with enhanced functional or structural properties, not attainable using the components individually. Composites are now a day's essential part of materials applications due to the advantages such as low weight, high fatigue strength, corrosion resistance, high mechanical strength, good energy storage capacity and excellent sensing properties. In polymer composites, the polymer acts as the matrix phase, while carbonaceous, metal, metal oxide and clay materials constitute the dispersed phase. The incorporation of these reinforcements in the polymer matrices enhances the mechanical strength and electron transfer rate at the modified surface interface. Especially, conducting polymer composites exhibit distinctive physico-chemical and electro-optical characteristics. These composites can be synthesized through several methods such as melt intercalation, *in-situ* polymerization, sol-gel, direct mixing of polymer and filler and solvent method. The properties of the composites depend on the nature of the components and methods of preparation [9,10]. Compared to conducting polymers, the conducting polymer composites have even higher potential of applications in energy storage devices (batteries, supercapacitors), solar cells, fuel cells, catalysis, anticorrosive coatings, electrochromic devices, artificial muscles, biomedicine and sensors. The overview of the applications of conducting polymer composites is presented in Figure 2.2. For the last two decades, conducting polymer composites have been especially considered to be effective materials for the fabrication of electrochemical sensors. The significant advantage of the conducting polymer composites for sensors is the enhanced response characteristics and sensitivity to small perturbation. Conducting polymer composites improve the sensitivity due to the enhanced charge transport properties [11,12].

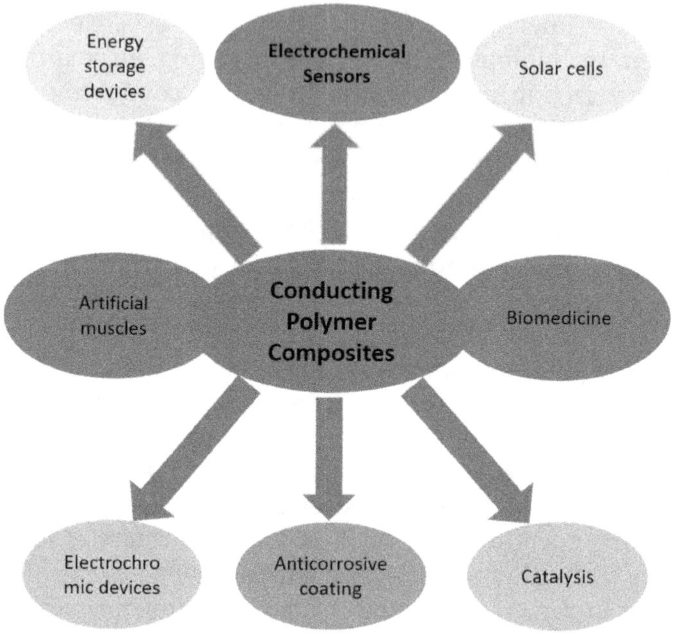

Figure 2.2 Application of conducting polymer composites in different fields.

2.2 Basic Principle of Electrochemical Sensors

The exceptional interest in the exploitation and development of electrochemical devices in the detection, quantification and observation of specific chemical species directed towards the development of electrochemical sensors. Analytical chemistry suggests that the electrochemical sensors are the rapidly developing type of chemical sensors. An electrochemical sensor can be defined as "a device which delivers uninterrupted information about its surrounding environment". More specifically, the electrochemical sensors provide the quantitative response that directly signifies the quantity of the chemical species. Electrochemical sensors usually consist of a transducer and sensing element or catalyst. The transducer (which is more commonly known as electrode) converts the chemical energy into electrical energy and the sensor generates the electrical signals of the detected analyte by displaying on the instrument for further analysis [13]. The schematic diagram of an electrochemical sensor is represented in Figure 2.3. Electrochemical sensors are classified into three main types such as amperometric, voltammetric/potentiometric, and impedimetric sensors, based on measurable current (amperometric detection), measurable potential (potentiometric detection) and measurable conductivity change of the medium (impedimetric), respectively, between the electrodes. The electrochemical sensor device consists of three electrodes such as working electrode, reference electrode and auxiliary or counter electrode. The most important electrode is the working electrode which is made of platinum, fluorine doped tin oxide (FTO), indium doped tin oxide (ITO), carbon or gold. Silver-silver chloride (Ag/AgCl) is usually used as a reference electrode. Furthermore, saturated calomel electrode (SCE) or reverse hydrogen electrode (RHE) can also be used as reference electrodes. The auxiliary or counter electrode is made of platinum foil, platinum wire, platinum mesh or graphite rod [14,15].

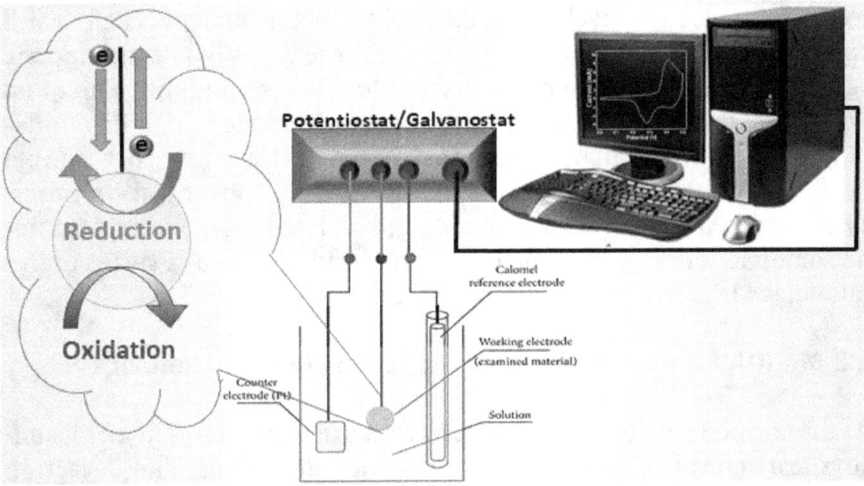

Figure 2.3 Schematic diagram of the electrochemical sensor.

Voltammetric sensor usually applies a potential between the working and reference electrodes that causes oxidation or reduction of the target analyte which generates an electrical current. Basically, the generated current is the result of the occurrence of an electrochemical reaction at the electrode surface and electrode/electrolyte surface interface. Various useful voltammetric methods are cyclic voltammetry, square-wave voltammetry, differential-pulse voltammetry, linear sweep voltammetry and hydrodynamic voltammetry. Voltammetric sensors have low detection limit and broad dynamic range. On the other hand, a constant potential (at which oxidation or reduction of analyte occurs) is applied to the working electrode in amperometric sensors. The oxidation or reduction potential is first determined by voltammetric technique. The generated current is recorded with respect to time in this sensor. One can conclude that the main difference between the voltammetric and amperometric sensors is the current step in amperometric sensors, rather than potential sweep. Furthermore, the generated current is directly proportional to the concentration of the target analyte. These sensors are more selective and possess higher sensitivity which might be due to the oxidation and reduction potential of the target analyte [16-18]. In potentiometric sensors, a potential difference is measured between the working and reference electrodes. This potential difference provides the information about the target analyte. Potentiometric sensors can also be divided into three types based on the type of electrode: ion-selective electrode (IES), field effect transistors (FETS) and coated wire electrodes (CWES). Among these types of potentiometric sensors, ion-selective electrodes have low detection limit. The most widely used ion-selective electrode is pH electrode which has undeniable advantages such as non-destructive nature, low cost, simple design, easy handling and applicability to wide concentration range. The third type of electrochemical sensor is conductometric or impedometric sensor which measures the conductivity at a series of frequencies [19,20].

2.3 Modification of Electrodes in Electrochemical Sensors

As mentioned earlier, the three electrodes in the electrochemical sensors are reference, auxiliary and working electrodes. The detection properties of the electrochemical sensors mainly depend on the working electrode. The chemical reaction at the electrode surface is the basis of the electrochemical analysis which influences the efficiency of the working electrode. So far, many electrode substrates such as glassy carbon, FTO, ITO, platinum or gold have been used for electrochemical analysis [21,22]. However, the efficiency of the bare working electrode is not good (slow electron transfer rate, high detection limit, poorly resolved peaks, low electrocatalytic activity, small surface area and electrochemical instability). Therefore, catalytically active nanomaterials are used to overcome the drawbacks of the working electrode and to enhance the performance of the electrochemical sensors. The nanomaterials used to improve the efficiency of the electrode involve coating of metals, metal oxides, metal halides, clays, ceramics, carbonaceous materials (such as CNTs and graphene, conducting polymers) and conducting polymer composites. These materials can be deposited on the surface of the bare electrode to improve the electrode efficiency. There are many methods of electrode modification such as drop casting, spin coating, electrodeposition, chemical vapor deposition, metal-organic chemical vapor deposition, covalent attachment and adsorption. However, it is difficult to fabricate the electrode using inorganic, carbonaceous or conducting polymeric materials alone, to fulfill the aforementioned requirements of the electrode for sensors. Therefore, the concept of the modification of electrodes using conducting polymer composites (hybrid materials) have been introduced. The significant advantages of such materials are the availability of a large variety of combinations of inorganic materials with organic matrices. These hybrid materials enhance the properties of the sensor, either by inherent characteristic of the conducting polymer or by improving the conducting polymer matrix through inorganic dopants [23]. Importantly, many types of dopants can be used to improve the properties of conducting polymers. These dopants may be metal nanoparticles (Au, Ag, Pt, Pd, Cu, etc.), inorganic oxides and halides, carbonaceous materials (CNTs, graphene), clays and ceramics. As mentioned earlier, conducting polymers incorporated with such dopants attained substantial attention in the development of electrochemical sensors due to their unique chemical structures and offer a lot of advantages such as low cost, stable composite formation, excellent reproducibility and lower detection limits. Furthermore, conducting polymer composites are helpful in sensing and electron transportation through redox mediation between electrode electrolyte interfaces [12].

Conducting polymer composites modified electrodes can be obtained through spinning, casting or screen-printing. Conducting polymer composites are mostly synthesized in the form of powders or films. The conductive dopant is added during the polymerization of monomers. The resulting powder or film is casted on the electrode through drop casting. Another method of electrode modification using conducting polymer composites is the electropolymerization method. In this method, electrical current is supplied to the cell containing monomers, dopant and solvent through three electrode systems. Electric current tends to deposit the monomers and dopant on the electrode surface and subsequently oxidizes the monomers to form polymer. Thus, films with controlled thickness, porosity and permeation characteristics are achieved [24]. Metal nanoparticles can also be synthesized *in-situ* in the polymers through this method from the metal precursors which are incorporated as dopants during electropolymerization process. Controlled uptake of salts and subsequent reduction of metal precursors such as $PdCl_4^{2-}$, $PdCl_6^{2-}$ or $PtCl_6^{2-}$ play important role in the synthesis of required nanoparticles [25,26]. Similarly, conducting polymer composites containing carbon nanotubes (CNTs) and graphene can also be prepared through chemical and electrochemical polymerization. Moreover, conducting polymer composites containing inorganic compounds (metal oxides) can also be prepared in the form of powders rather than thin films because of ease in preparation. A brief overview of electrode modification is presented in the Figure 2.4.

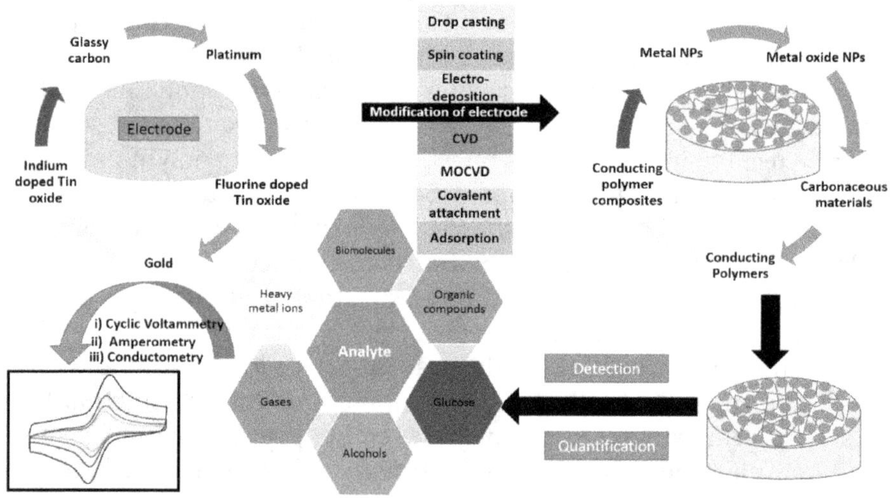

Figure 2.4 Brief overview of electrode fabrication.

2.4 Conducting Polymer Composites in Sensors

As mentioned earlier, conducting polymers can be used to form composites with carbonaceous materials such as carbon nanotubes, graphene, graphene oxide (GO) and reduced graphene oxide, metal and metal oxides nanoparticles and mineral clays. The conducting polymer composites can be prepared through different chemical and electrochemical methods before or during the electrode modification process.

2.4.1 Conducting Polymer Composites with Carbonaceous Fillers

Carbonaceous materials are significant in electrochemical sensors due to the high conductivity, large surface area, functionalization ability and low charge transfer resistance. Carbonaceous materials are mainly divided into two classes such as carbon nanotubes and graphene. CNTs, discovered in 1991, are advantageous to use in electrochemical sensing due to high electrical conductivity and catalytic properties to catalyze redox reaction of target analyte. Secondly, CNTs can be easily immobilized on the electrode surface due to high mechanical strength and small size. CNTs are classified as single-walled carbon nanotubes and multi-walled carbon nanotubes (MWCNTs). CNTs are not usually used alone because of poor sensitivity and selectivity. Minimal selectivity is observed because the response mechanism is based on the electron affinity and density ratio of the target analyte relative to the CNT transducer [27]. Conducting polymers can be combined with CNTs to form composites. These composites have significant advantages over conducting polymers or CNTs alone. The presence of conducting polymers enhances the attachment of the active layer on the substrate whereas the CNTs enhance the electrical conductivity and mechanical strength of the composites. Thus, the synergistic combination of the components results in strong, durable and highly conducting material for sensing applications [28,29]. CNTs-conducting polymer composites have wide range of potential applications for electrochemical sensing such as detection of water contaminants and biological molecules. Ding *et al.* [30] modified glassy carbon electrode (GCE) with polyaniline (PANI)/multi-walled CNTs using *in-situ* chemical oxidation method to investigate the electrocatalytic reduction of bromate present as by-product in drinking water. The composite exhibited effective electrochemical performance. The electrochemical reduction of bromate occurred while PANI transformed from emeraldine form to leucoemeraldine in acidic solution. The results revealed the increase in peak current of bromate reduction linearly [30]. In other study, a carbon ionic liquid electrode (CILE) was modified with polythionine (PTH)/MWCNTs composite and used for the detection of reduced nicotinamide adenine dinucleotide (NADH). The electrode was prepared by electrochemical polymerization of thionine on MWCNTs in neutral medium. Cyclic voltammetric analysis indicated that the electrode could mediate the oxidation of NADH at an overpotential as low as 0.03 V. Amperometric experiments demonstrated that a sensitive and stable response towards NADH was obtained within 5 s. The linear range for the determination of NADH was from 0.8 μmol L^{-1} to 422 μmol L^{-1} with a detection limit of 0.26 μmol L^{-1} (S/N = 3). The wide linear range, lower detection limit and faster response towards NADH suggested that the new method is potentially useful for developing NAD^{+}-dependent enzyme-based biosensors.

Polypyrrole gained increasing interest as conducting polymer owing to its optical, electrical and sensing applications. A flow injection catechol biosensor based on tyrosinase (Tyr) entrapped in carbon nanotube modified polypyrrole biocomposite film on a glassy carbon surface has been developed [32]. Amperometric response was measured as a function of the concentration of catechol, at a fixed bias voltage of −50 mV and a flow rate of 1 mL/min. The proposed biosensor exhibited impressive analytical performance in terms of linear range (from 3 to 50 μM), response time (10 s), detection limit of 0.671 μM and operational (with a relative standard deviation of 0.54%) and long-term stability (85% remained after 10th week). A comparison of the analytical parameters of the developed biosensor with polypyrrole/tyrosinase film electrode was performed in the study. CNTs showed enhanced electron transfer between the electrode and enzyme and were capable to exhibit higher bioactivity owing to intensified surface area [32].

Recently, one dimensional conducting polymer structures (nanowires) attracted significant attention because of high mechanical strength, low cost, facile synthesis, benign synthesis environment and simple doping mechanism. Hui *et al.* [33] synthesized conducting PANI nanowires on glassy carbon electrode, which were further modified with CNTs. The ordered PANI nanowire arrays and CNTs composites were integrated to form 3-D structures. The synthesized PANI/CNTs composites showed excellent electrical properties due to the unique microstructure and inherent conductivity. The composites exhibited remarkable catalytic activity towards nitrite electrochemical reduction. Moreover, these could be utilized as an electrochemical sensor for the detection of nitrite, with outstanding sensitivity, low detection limit (6.1 μM, (S/N 3)) and fast response time of less than 5 s. In another study, Zhong *et al.* [34] reported non-enzymatic amperometric sensor for hydrogen peroxide (H_2O_2). It was fabricated by electrodeposition of multi-walled carbon nanotubes and polyaniline along with platinum nanoparticles on the surface of a glassy carbon electrode. The modification was probed by scanning electron microscopy and cyclic voltammetry. The sensor exhibited high sensitivity (748.4 μA·mM^{-1}·cm^{-2}), wide linear range (7.0 μM–2.5 mM), low detection limit (2.0 μM) (S/N=3), short response time (>5 s) and long-term stability, and was not interfered by common species. It could be successfully applied to determine H_2O_2 in disinfectants.

Graphene is a carbonaceous material with distinctive characteristics, which include excellent thermal conductivity, high specific surface area, mechanical strength, distinctive electronic configuration and ballistic conductivity. Graphene is a single-atom thick and extensively conjugated two-dimensional carbonaceous nanomaterial with sp^2 hybridized carbon atoms attached in a six-membered arrangement through weak π–π interactions [35]. The surface modification of graphene by attachment of active groups through covalent and non-covalent strategies can effectively inhibit the aggregation of single layer graphene due to the attached hydrophobic and hydrophilic groups. This facilitates the dispersion and homogeneity of graphene as well as maintaining its inherent properties [36]. Raj *et al.* [37] synthesized novel poly(4-amino-3-hydroxy-1-naphthalenesulfonic acid)/graphene screen printed carbon sensor through electrochemical method for the simultaneous determination of dopamine (DA) and 5-hydroxytryptamine (5-HT). The electrochemical studies were performed using cyclic and square wave voltammetry. The results revealed that fabricated sensor facilitated the analysis of dopamine and 5-hydroxytryptamine in the concentration range of 0.05-100 μM and 0.05-150 μM with the detection limit of 2 nM and 3 nM, respectively. DA and 5-HT were analyzed in biological fluids in the presence of metabolites. Furthermore, pharmacological formulations, human urine and blood samples were analyzed for the simultaneous detection of DA and 5-HT through the fabricated sensor.

Heavy metal ions can cause a variety of environmental problems and hazards. For instance, these ions, stored in the human body through water, can cause serious health issues. Therefore, it is critical to minimize the contaminants and their exposure of human beings, which can be ensured through effective monitoring of water quality. Park *et al.* [38] developed durable GO anchored functionalized polyterthiophene composite electrochemical sensor for the detection of heavy metal ions. 3'-(2-aminopyrimidyl)-2,2':5',2"-terthiophene (PATT) monomer was polymerized in the presence of GO through potential cycling method followed by Nafion coating. Interference effect of metal ions having almost comparable redox potentials was observed. The performance of chronocoulometry (CC) without predeposition was compared with the results of square wave anodic stripping voltammetry (SWASV) with predeposition. The dynamic range of CC for the target ions was between 1 ppb and 10 ppm respectively, with the detection limits between 0.05 and 0.20 ppb for the CC method without predeposition and between 0.08 and 0.30 ppb for the SWASV with 300 sec of deposition time (n=3). The reliability of the method was evaluated by continuously analyzing environmental water samples using a single sensor probe in a flow system for 93 days [38].

Conducting polymer PEDOT has excellent electrical conductivity properties. It can be widely exploited in electrochemical applications such as energy storage devices and sensors. However, limited catalytic activity inhibits its use in sensors. Therefore, PEDOT can be combined with other materials having enhanced electrical conductivity, catalytic activity and stability. For this purpose, Au@graphene core-shell nanoparticles were doped with PEDOT to form the composites by facile electrodeposition method [39]. The synthesized composites showed excellent conductivity, catalytic activity and sensitivity towards the redox processes of paracetamol. Furthermore, the composites exhibited high sensitivity to paracetamol over a wide linear range of concentrations from 0.15 μM to 5.88 mM, with a detection limit of 41 nM. Furthermore, another study reported the use of reduced graphene oxide with PANI to form thin film composite on flexible polyethylene terephthalate through *in-situ* chemical oxidative polymerization for electrochemical sensing of ammonia gas [40]. Aniline monomer polymerized in the presence of oxidizing agent ammonium peroxydisulfate. The resulting composite possessed excellent gas sensing and remarkable selectivity properties.

2.4.2 Conducting Polymer Composites Containing Metals

Metals have received significant research attention due to excellent electrical, catalytic and optical properties. Therefore, metals have wide range of potential applications in electronics, energy storage/conversion and sensing devices. In the last two decades, metals at nanoscale have been extensively employed to explore the enhancement in properties, specifically in electrochemical field, as these materials effectively improve electron transfer rate and catalytic properties due to the large surface area. However, metal nanoparticles have aggregation tendency at nanoscale which results in the reduction in catalytic properties Therefore, to avoid aggregation and to overcome the drawbacks of pure metal nanoparticles, these are often incorporated in the polymer matrices. Noble metals such as Ag, Au, Pt, and Pd are commonly used to form composites with conducting polymers. Moreover, Ni and Cu have also been investigated as fillers [41]. Due to the enhanced sensitivity as well as electrical and catalytic properties, metal-conducting polymer composites exhibit high potential for use in sensing devices. Many chemical species of concern, including toxic gases, volatile organic compounds, alcohols and humidity, can be detected using metal-conducting polymer composites in these sensing devices. In the composites, the conducting polymers transduce the electronic response of the nanoparticles as well as directly interact with the analytes. Mostly, thin films and powder forms of metal-conducting polymer composites are applied in the sensing devices. In one of the synthesis methods for these composites, the metal acts as core while the conducting polymer as shell, and the composite can be prepared via chemical or electrochemical polymerization. Subsequently, the synthesized composites are deposited on the electrode substrate. In the second method, the composites can be obtained by chemical reduction of metal salts at the conducting polymer interface. Many conducting polymers, when present in their reduced form, have a sufficiently high reducing power with respect to some metal ions. Thus, metal ions having a relatively high positive redox potential, e.g. gold, silver, platinum and copper, can be reduced, thus, forming clusters or small particles within a porous conducting polymer layer [42].

Many studies have been reported on metal-conducting polymer composites for sensing devices. In one such study, Jiang *et al.* [43] reported Au-PANI nanobowl monolayer nanocomposite for NH_3 sensor. The sensor exhibited very fast response of 5 s and recovery behavior of 7 s, which confirmed its potential for gas detection and control. The sensing mechanism involved protonation/deprotonation phenomena carried by NH_3 gas. Moreover, hybrid Au-PANI nanocomposites chemiresistive sensors have been reported for H_2S detection and quantification [44]. These sensors showed excellent detection limit of 0.1 ppb, dynamic range of 0.1 to 100 ppb, good reproducibility and high selectivity. Furthermore, Pd-PEDOT films fabricated on GCE have been reported for the simultaneous detection of dopamine and uric acid through differential pulse voltammetry [45].

Metal-conducting polymer composite based sensors have been generated for the detection of dopamine as it is crucial catecholamine neurotransmitter molecule and widely distributed in the central nervous system for message transfer [46,47]. Due to their capability to immobilize biomolecules, metal-conducting polymer composites have also been widely investigated as biosensors. In addition, the use of these composites was also reported for the electrocatalytic oxidation of NADH [48,49].

Harish *et al.* [50] reported the incorporation of palladium nanoparticles in the PEDOT matrix. In this method, 3,4-ethylenedioxythiophene (EDOT) and $PdCl_2$ were mixed and redox reaction between these precursors led to the formation of Pd nanoparticles supported on freshly formed PEDOT. The scanning (SEM) and transmission electron microscopy (TEM) confirmed the successful synthesis of Pd-PEDOT composite. The catalytic activity of the composite was investigated for the reduction of 4-nitrophenol to 4-aminophenol. The value of the apparent rate constant, obtained using UV-visible spectroscopy of the reduction of 4-nitrophenol at the Pd–PEDOT composite, was ca. $65.89 \times 10^{-3}\ s^{-1}$, which is comparable to other catalytic systems [50]. Jin *et al.* [51] reported lollipop like gold nanoparticles incorporated with PANI. The composite was synthesized by interfacial polymerization through water/oil phase. The catalytic activity of the synthesized composite was analyzed for the reduction of 4-nitrophenol to 4-aminophenol. Overall, the results revealed a high catalytic activity of the synthesized composite.

The detection of hydrogen peroxide is of emerging relevance using sensors as it is detrimental to biological systems and appears to be involved in the neuropathology of central nervous system. Te-PEDOT nanocables have been reported as peroxidase-like catalyst for the calorimetric detection of H_2O_2 [52]. These nanocables were synthesized through self-assembly redox polymerization. The synergistic effect between the Pd nanowires and electrically conducting PEDOT enhanced the catalytic activity towards the oxidation of the peroxidase substrate 3,3′,5,5′-tetramethylbenzidine in the presence of H_2O_2. The results revealed that detection limit towards H_2O_2 was 4.83 μM, and linear range from 10 to 100 μM was achieved. The peroxidase-like activity of the synthesized Pd@PEDOT coaxial nanocables can be observed from Figure 2.5.

2.4.3 Conducting Polymer Composites Containing Metal Oxides

Metal oxide nanoparticles are significant alternative to metals and carbonaceous materials in electrochemical applications. Metal oxides are widely available in nature, have low cost and are less toxic. The preparation of metal oxide nanoparticles is also straightforward, and

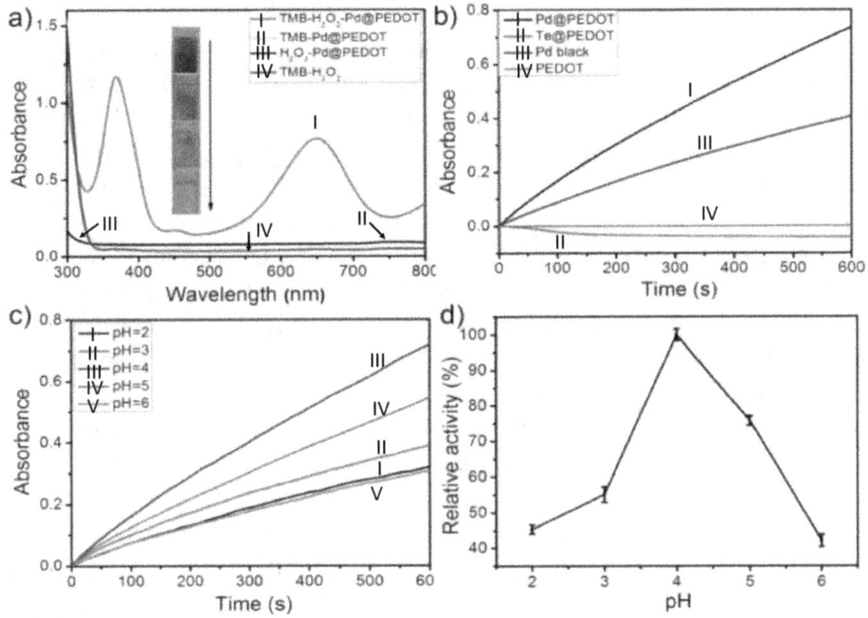

Figure 2.5 a) Typical photographs and UV-vis absorption curves of TMB solutions oxidized by different systems in acetate buffer solution (pH = 4.0) recorded at 10 min of the catalytic reaction, b) time-dependent absorption changes at 651 nm with the addition of different types of catalysts (Pd@PEDOT, Pd black, Te@PEDOT, and pure PEDOT material); the same amounts of Pd were used for Pd@PEDOT and Pd black catalysts, c) dependence of the peroxidase-like activity of the synthesized Pd@PEDOT coaxial nanocables on the pH values and d) the corresponding line chart. Reproduced from Reference 52 with permission from American Chemical Society.

many methods have been reported in the last two decades. These materials have many excellent properties such as large surface-to-volume ratio, high electron transfer rate, strong adsorption ability, high surface reaction activity and small particle size. Due to these properties, metal oxide nanoparticles can be used in sensors. However, aggregation of metal oxide nanoparticles hinders their applications in sensors due to reduction of active area and catalytic properties. Metal oxide-based sensors work at high temperature and consume high energy. On the other hand, conducting polymers have received substantial interest as effective materials to work at room temperature in sensors. Conducting polymers have their own shortcomings such as low response, long response time and low thermal stability. These shortcomings hinder their large- scale exploitation in sensors. However, the combination of metal oxides and conducting polymers prevents the aggregation of nanoparticles, thereby, resulting in the enhancement of active surface area and catalytic properties. The thermal stability and response time of conducting polymers in combination with metal oxides also improve significantly. Such enhanced performance of metal oxide-conducting polymer composites drives their use for different sensor applications. In sensors, these composites are excellent candidates for immobilization and biocatalytic activity of biomolecules such as dopamine, ascorbic acid, NADH, uric acid, hydrazine, epinephrine, etc. [53,54]. Due to catalytic properties of metal oxides, these composites exhibit superior performance for catalytic oxidation and reduction of organic compounds such as 4-nitophenol and bisphenol-A. These composites can detect the organic compounds in the real samples such as meat, human serum and blood [55,56]. Metal oxide-conducting polymer-based sensors have sufficient potential for the removal of toxic organic pollutants from water and air. The heavy metal ions (Pb^{2+}, Ni^{2+}, Fe^{2+}, Zn^{2+}, Hg, Cu^{2+}, etc.) and dyes from drinking water can be potentially removed using these composites due to the very low detection ability at ppb level [57]. The detection and quantification of pharmaceutical formulations is best possible solution using these composites in sensors. Furthermore, the detection of glucose in the food is investigated through non-enzymatic metal oxide-conducting polymer composites modified electrode in electrochemical sensors. Moreover, these composites are excellent for gas sensing (NH_3, H_2S, CO, CO_2, H_2, NO_2 etc.). The gas sensing ability of the composites is due to metal oxide component, as it has the inherited ability of metal oxides to absorb/desorb gases [58].

Tsele *et al.* [54] reported TiO_2-MWCNTs-PANI and RuO_2-MWCNTs-PANI composites modified gold electrode for the catalytic oxidation of epinephrine biomolecules. The cyclic voltammetry technique was applied to detect epinephrine in buffer solutions of pH 1.0-7.0 and real samples of adrenaline injected with epinephrine. The detection limit of epinephrine was observed to be 0.16 μM and 0.18 μM for TiO_2-MWCNTs-PANI and RuO_2- MWCNTs-PANI modified gold electrodes, respectively. Furthermore, Yang *et al.* [53] synthesized Fe_3O_4/polypyrrole/graphene oxide composites for the detection of hydrazine. The composites were synthesized by combining chemical oxidative polymerization and co-precipitation. The synthesized composites were coated on glassy carbon electrode. The resulting sensor, best operated around 0.2 V (vs. SCE), exhibited excellent response to dissolved hydrazine over 5.0 μM to 1.3 mM concentration range, sensitivity of 449.7 μA mM^{-1} cm^{-2} and low detection limit of 1.4 μM (at an S/N ratio of 3).

Ternary metal oxides (TMO) composites-based electrode fabrication in electrochemical sensors is a new approach for the detection of organic compounds. In a recent study, TMO (ZnO, CoO, FeO) composites were synthesized through wet chemical method in basic medium at low temperature. The synthesized composites were applied on glassy carbon electrode using 5% Nafion at room temperature and pressure. The resultant electrode was used for selective detection of bisphenol-A (BPA) using cyclic voltammetry (CV). It was observed that TMO electrode exhibited an excellent sensitivity (3.28 μAμ$M^{-1}$$cm^{-2}$), low detection limit (LOD: 1.2 ± 0.1 nM; S/N = 3), high stability, effective repeatability and reproducibility. This method represents an efficient way of sensor development for the detection of toxic and carcinogenic phenolic compounds [59]. Al-Mokaram *et al.* [60] synthesized polypyrrole-chitosan-TiO_2 nanocomposite thin films and fabricated ITO glass electrode using these thin films. The synergic effect of TiO_2 and conducting polymer on the current responses of the electrode resulted in an effective sensitivity over a linear range of 1-14 mM with a detection limit of 614 μM for glucose. The modified electrode with polypyrrole-chitosan-TiO_2 nanocomposite also exhibited good selectivity, long-term stability with no interference effect and high electron transfer kinetics.

2.4.4 Conducting Polymer Composites Containing Clays

Clay is finely grained natural rock soil material that combines one or more clay minerals. Clays have absorptivity, ability to exchange ions, large surface area, catalytic properties, electron transfer ability, porous morphology, swellability, unique layered structure, intercalated structure and ability to incorporate organic and inorganic materials. These characteristics make them attractive candidates for use in numerous fields. Recently, clays have been the focus of intense research efforts for developing electrochemical applications, especially, sensors. However, poor selectivity and limited adsorption capacity of pristine clays create hindrance for exploitation of the pristine clays in sensors [61]. Usually, clays can be used to form hybrids with polymers, metal oxides, CNTs and graphene for different sensor applications. On the other hand, conventional bare electrodes in sensors suffer extensive surface fouling during electrochemical oxidation.

Therefore, electrode modification more or less inhibits the electrode poisoning and improves reliability [62]. In this respect, clays-conducting polymer composites are one such solution to overcome the fouling drawback. Clay-conducting polymer composites have ability to immobilize the biomolecules. Therefore, these composites can be used for the detection and catalytic oxidation of biomolecules [63]. Shao *et al.* [64] reported hierarchical nanotube structured halloysite-PANI composite via. chemical oxidative polymerization of PANI on halloysite. The authors deposited the composite electrophoretically on ITO. The composite modified ITO (fabricated electrode) was introduced to detect ascorbic acid. The fabricated electrode showed unprecedented sensitivity towards ascorbic acid, as high as 826.53 μA $mM^{-1}cm^{-2}$ within a linear range of 0.005 to 5.5 mM, along with a low detection limit of 0.21 μM. The observed ideal sensing performance could be attributed to the porous structure affording efficient sensing channels for improving electron transport and interactions between PANI and ascorbic acid. Hsu *et al.* [65] modified carbon paste electrode (PCE) with silica/PANI decorated with Au nanoparticles for the detection of ascorbic acid. The modified PCE exhibited high sensitivity, low detection limit, excellent sensing repeatability and broader dynamic range for ascorbic acid detection than bare PCE. Furthermore, modified electrode showed better selectivity in a ternary mixture of ascorbic acid, dopamine, and uric acid.

Clay-conducting polymer composites have been reported for the removal of heavy metal ions from drinking water. De Barros *et al.* [66] synthesized layer-by-layer montmorillonite/PANI and montmorillonite/polyethyleneimine. These films were deposited on ITO substrate and heavy metal ions (Cd^{2+} Pb^{2+}, Cu^{2+}) detection was carried out through square wave anodic stripping voltammetry. In other work, de Barros *et al.* [67] synthesized Langmuir-Blodgett (LB) films of PANI/organophilic montmorillonite composite for the detection of trace levels of cadmium (Cd^{2+}), lead (Pb^{2+}) and copper (Cu^{2+}). Composite films were coated on ITO and detection was carried out with anodic stripping voltammetry. Square wave anodic stripping voltammograms for simultaneous detection of metal ions are demonstrated in Figure 2.6.

2,4-Dichlorophenol (2,4-DCP) compound is widely employed as an important intermediate in the manufacturing of commonly used phenoxy herbicides, pharmaceuticals, fungicides and insecticides. It poses a serious ecological problem as an environmental pollutant due to high toxicity, strong odor emission, persistence in the environment

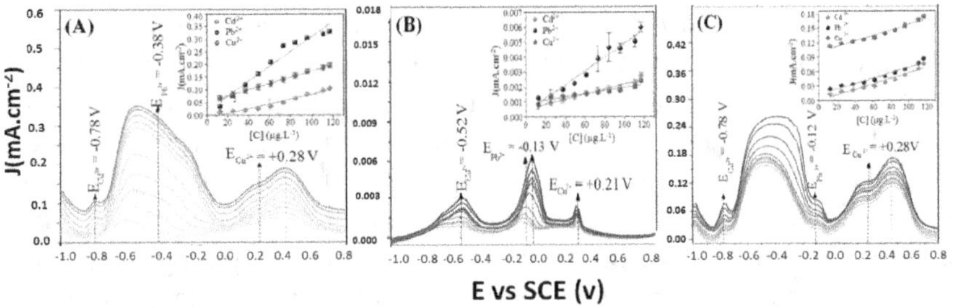

Figure 2.6 Square wave anodic stripping voltammograms for simultaneous detection of Cu^{2+}, Pb^{2+}, and Cd^{2+} in HCl 0.1 mol L^{-1} with a deposition potential of −1.0 V, an accumulation time of 300 s, an amplitude of 50 mV, and a frequency of 15 Hz: (A) PANI-ES LB films, (B) OMt clay LB films, and (C) PANI-ES/OMt nanocomposite LB films. All the films had 20 layers deposited onto ITO plates. Reproduced from Reference 67 with permission from American Chemical Society.

and suspected carcinogenic/mutagenic properties. The highly hazardous and toxic characteristics of 2,4-DCP require the development of simple, fast, sensitive and accurate analytical methods for its detection and quantification. Such as, poly(sodium 4-styrenesulfonate) (PSS) intercalated graphene was prepared via *in-situ* reduction of exfoliated graphite oxide in the presence of PSS, followed by mixing with cetyltrimethylammonium bromide (CTAB) to form nanocomposite through electrostatic-assembly [68]. The synthesized nanocomposite was deposited on GCE and used for the electrocatalytic reduction of 2,4-DCP through linear sweep voltammetry. In another study, Pontes *et al.* [69] synthesized montmorillonite/polypyrrole composite via *in-situ* polymerization with anionic surfactant dodecylbenzene sulfonic acid sodium salt (DBSA). The synthesized composites were tested for the detection of volatile organic compounds such as methanol, ethanol, hexane and benzene.

2.5 Conclusions and Outlook

In this chapter, conducting polymer composites have been emphasized as progressive electrode materials for the development of electrochemical sensors. From the mentioned studies, it can be concluded that conducting polymers alone are not suitable candidates as electrode materials. The major problems encountered with the electrodes fabricated with conducting polymers include low stability, poor cyclic stability and low charge transfer. These problems can be overcome by combining conducting polymers with the materials having superior charge transfer and enhanced mechanical strength, which help to improve the cyclic stability and durability of the electrode. In this respect, such materials include carbonaceous fillers, metal particles, metal oxide nanoparticles and mineral clays. Thus, the structural diversity, flexibility, porous morphology, strong sorption and electrochromic properties of the conducting polymers, while large surface area, enhanced catalytic activity and large electron transfer rate of the incorporated materials ensure an overall improvement in the properties of the fabricated electrodes. The combination of conducting polymers with the nano-reinforcements results in minimization of detection limit and increase in sensitivity. These materials also provide suitable platform for the attachment of different recognition elements due to the presence of diverse functional groups. The ongoing research efforts in this direction are expected to result in significant advancement of the electrochemical sensors in the near future.

Conflict of Interest

The authors declare no conflict of interest.

List of Abbreviations

BPA	bisphenol-A
CNTs	carbon nanotubes
PCE	carbon paste electrode
CC	chronocoulometry
DA	dopamine
2, 4-DCP	2,4-dichlorophenol
FTO	fluorine doped tin oxide
GCE	glassy carbon electrode
GO	graphene oxide
HR-TEM	high resolution-transmission electron microscopy
5-HT	5-hydroxytryptamine
ITO	indium doped tin oxide
MWCNTs	multi-walled carbon nanotubes
NADH	nicotinamide adenine dinucleotide
PANI	polyaniline

PEDOT	poly(3,4-ethylenedioxythiphene)
PPy	polypyrrole
PTH	polythionine
PTh	polythiophene
SEM	scanning electron microscopy
SWCNTs	single-walled carbon nanotubes
SWASV	Square wave anodic stripping voltammetry
TEM	transmission electron microscopy
Tyr	tyrosinase

References

1. Jensen, W. B. (2008) The origin of the polymer concept. *Journal of Chemical Education*, **85**(5), 624.
2. Young, R. J., and Lovell, P. A. (2011) *Introduction to Polymers*, 3rd edition, CRC Press, USA.
3. Skotheim, T. A. (1997) *Handbook of Conducting Polymers*, 2nd edition, CRC Press, USA.
4. Gerard, M., Chaubey, A., and Malhotra, B. (2002) Application of conducting polymers to biosensors. *Biosensors and Bioelectronics*, **17**(5), 345-359.
5. Skotheim, T. A., and Reynolds, J. (2006) *Conjugated Polymers: Processing and Applications*, 3rd edition, CRC Press, USA.
6. Chiang, C. K., Fincher, Jr., C., Park, Y. W., Heeger, A. J., Shirakawa, H., Louis, E. J., Gau, S. C., and MacDiarmid, A. G. (1977) Electrical conductivity in doped polyacetylene. *Physical Review Letters*, **39**(17), 1098.
7. Shirakawa, H., Louis, E. J., MacDiarmid, A. G., Chiang, C. K., and Heeger, A. J. (1977) Synthesis of electrically conducting organic polymers: halogen derivatives of polyacetylene, (CH)x. *Journal of the Chemical Society, Chemical Communications*, (16), 578-580.
8. Rong, M., Mahapatra, N. K., and Maiti, M. (2008) A two warehouse inventory model for a deteriorating item with partially/fully backlogged shortage and fuzzy lead time. *European Journal of Operational Research*, **189**(1), 59-75.
9. Jose, J. P., and Joseph, K. (2012) Advances in polymer composites: Macro- and microcomposites - state of the art, new challenges, and opportunities. In: *Polymer Composites*, Thomas, S., Kuruvilla, J., Malhotra, S. K., Goda, K., and Sreekala, M. S. (eds.), doi:10.1002/9783527645213.ch1.
10. Pang, H., Xu, L., Yan, D.-X., and Li, Z.-M. (2014) Conductive polymer composites with segregated structures. *Progress in Polymer Science*, **39**(11), 1908-1933.

11. Iqbal, S., and Ahmad, S. (2017) Recent development in hybrid conducting polymers: Synthesis, applications and future prospects. *Journal of Industrial and Engineering Chemistry*, **60**, 53-84.
12. Naveen, M. H., Gurudatt, N. G., and Shim, Y.-B. (2017) Applications of conducting polymer composites to electrochemical sensors: A review. *Applied Materials Today*, **9**, 419-433.
13. Stetter, J. R., Penrose, W. R., and Yao, S. (2003) Sensors, chemical sensors, electrochemical sensors, and ECS. *Journal of The Electrochemical Society*, **150**(2), S11-S16.
14. Stradiotto, N. R., Yamanaka, H., and Zanoni, M. V. B. (2003) Electrochemical sensors: a powerful tool in analytical chemistry. *Journal of the Brazilian Chemical Society*, **14**(2), 159-173.
15. Hitchman, M. L., and Berlouis, L. E. A. (1995) Electrochemical methods. In: *Process Analytical Chemistry*, McLennan, F., and Kowalski, B. D., Springer, Netherlands, pp. 217-258.
16. Janata, J. (2010) *Principles of Chemical Sensors*, 2nd edition, Springer Science & Business Media, USA.
17. Wang, J., Musameh, M., and Lin, Y. (2003) Solubilization of carbon nanotubes by Nafion toward the preparation of amperometric biosensors. *Journal of the American Chemical Society*, **125**(9), 2408-2409.
18. Numan, A., Shahid, M. M., Omar, F. S., Rafique, S., Bashir, S., Ramesh, K., and Ramesh, S. (2017) Binary nanocomposite based on Co_3O_4 nanocubes and multiwalled carbon nanotubes as an ultrasensitive platform for amperometric determination of dopamine. *Microchimica Acta*, **184**(8), 2739-2748.
19. Rahman, M. A., Kumar, P., Park, D.-S., and Shim, Y.-B. (2008) Electrochemical sensors based on organic conjugated polymers. *Sensors*, **8**(1), 118-141.
20. Bobacka, J., Ivaska, A., and Lewenstam, A. (2008) Potentiometric ion sensors. *Chemical Reviews*, **108**(2), 329-351.
21. Yang, C., Denno, M. E., Pyakurel, P., and Venton, B. J. (2015) Recent trends in carbon nanomaterial-based electrochemical sensors for biomolecules: A review. *Analytica Chimica Acta*, **887**, 17-37.
22. Omar, F. S., Numan, A., Duraisamy, N., Bashir, S., Ramesh, K., and Ramesh, S. (2017) A promising binary nanocomposite of zinc cobaltite intercalated with polyaniline for supercapacitor and hydrazine sensor. *Journal of Alloys and Compounds*, **716**, 96-105.
23. Janaky, C., and Visy, C. (2013) Conducting polymer-based hybrid assemblies for electrochemical sensing: a materials science perspective. *Analytical and Bioanalytical Chemistry*, **405**(11), 3489-3511.
24. Hou, Y., Cheng, Y., Hobson, T., and Liu, J. (2010) Design and synthesis of hierarchical MnO2 nanospheres/carbon nanotubes/conducting polymer ternary composite for high performance electrochemical electrodes. *Nano Letters*, **10**(7), 2727-2733.

25. Hatchett, D. W., Millick, N. M., Kinyanjui, J. M., Pookpanratana, S., Bar, M., Hofmann, T., Luinetti, A., and Heske, C. (2011) The electrochemical reduction of PdCl42− and PdCl62− in polyaniline: Influence of Pd deposit morphology on methanol oxidation in alkaline solution. *Electrochimica Acta*, **56**(17), 6060-6070.
26. Holzhauser, P., Bouzek, K., and Bastl, Z. (2005) Electrocatalytic properties of polypyrrole films prepared with platinate (II) counter-ions. *Synthetic Metals*, **155**(3), 501-508.
27. Shahid, M. M., Ismail, A. H., AL-Mokaram, A. M. A. A., Vikneswaran, R., Ahmad, S., Hamza, A., and Numan, A. (2017) A single-step synthesis of nitrogen-doped graphene sheets decorated with cobalt hydroxide nanoflakes for the determination of dopamine. *Progress in Natural Science: Materials International*, **27**(5), 582-587.
28. Lahiff, E., Lynam, C., Gilmartin, N., O'Kennedy, R., and Diamond, D. (2010) The increasing importance of carbon nanotubes and nanostructured conducting polymers in biosensors. *Analytical and Bioanalytical Chemistry*, **398**(4), 1575-1589.
29. Omar, F. S., Duraisamy, N., Ramesh, K., and Ramesh, S. (2016) Conducting polymer and its composite materials based electrochemical sensor for Nicotinamide Adenine Dinucleotide (NADH). *Biosensors and Bioelectronics*, **79**, 763-775.
30. Ding, L., Li, Q., Zhou, D., Cui, H., An, H., and Zhai, J. (2012) Modification of glassy carbon electrode with polyaniline/multi-walled carbon nanotubes composite: Application to electro-reduction of bromate. *Journal of Electroanalytical Chemistry*, **668**, 44-50.
31. Mai, N., Liu, X., Zeng, X., Xing, L., Wei, W., and Luo, S. (2010) Electrocatalytic oxidation of the reduced nicotinamide adenine dinucleotide at carbon ionic liquid electrode modified with polythionine/multi-walled carbon nanotubes composite. *Microchimica Acta*, **168**(3-4), 215-220.
32. Ozoner, S. K., Yalvac, M., and Erhan, E. (2010) Flow injection determination of catechol based on polypyrrole-carbon nanotube-tyrosinase biocomposite detector. *Current Applied Physics*, **10**(1), 323-328.
33. Hui, N., Chai, F., Lin, P., Song, Z., Sun, X., Li, Y., Niu, S., and Luo, X. (2016) Electrodeposited conducting polyaniline nanowire arrays aligned on carbon nanotubes network for high performance supercapacitors and sensors. *Electrochimica Acta*, **199**, 234-241.
34. Zhong, H., Yuan, R., Chai, Y., Zhang, Y., Wang, C., and Jia, F. (2012) Non-enzymatic hydrogen peroxide amperometric sensor based on a glassy carbon electrode modified with an MWCNT/polyaniline composite film and platinum nanoparticles. *Microchimica Acta*, **176**(3-4), 389-395.
35. Numan, A., Shahid, M. M., Omar, F. S., Ramesh, K., and Ramesh, S. (2017) Facile fabrication of cobalt oxide nanograin-decorated redu-

ced graphene oxide composite as ultrasensitive platform for dopamine detection. *Sensors and Actuators B: Chemical*, **238**, 1043-1051.

36. Shan, C., Yang, H., Han, D., Zhang, Q., Ivaska, A., and Niu, L. (2009) Water-soluble graphene covalently functionalized by biocompatible poly-L-lysine. *Langmuir*, **25**(20), 12030-12033.
37. Raj, M., Gupta, P., Goyal, R. N., and Shim, Y.-B. (2017) Graphene/conducting polymer nano-composite loaded screen printed carbon sensor for simultaneous determination of dopamine and 5-hydroxytryptamine. *Sensors and Actuators B: Chemical*, **239**, 993-1002.
38. Park, M. O., Noh, H. B., Park, D. S., Yoon, J. H., and Shim, Y. B. (2017) Long-life heavy metal ions sensor based on graphene oxide-anchored conducting polymer. *Electroanalysis*, **29**(2), 514-520.
39. Li, M., Wang, W., Chen, Z., Song, Z., and Luo, X. (2018) Electrochemical determination of paracetamol based on Au@ graphene core-shell nanoparticles doped conducting polymer PEDOT nanocomposite. *Sensors and Actuators B: Chemical*, **260**, 778-785.
40. Bai, S., Zhao, Y., Sun, J., Tian, Y., Luo, R., Li, D., and Chen, A. (2015) Ultrasensitive room temperature NH 3 sensor based on a graphene–polyaniline hybrid loaded on PET thin film. *Chemical Communications*, **51**(35), 7524-7527.
41. Hangarter, C. M., Chartuprayoon, N., Hernandez, S. C., Choa, Y., and Myung, N. V. (2013) Hybridized conducting polymer chemiresistive nano-sensors. *Nano Today*, **8**(1), 39-55.
42. Malinauskas, A., Malinauskiene, J., and Ramanavicius, A. (2005) Conducting polymer-based nanostructurized materials: Electrochemical aspects. *Nanotechnology*, **16**(10), R51-62.
43. Jiang, S., Chen, J., Tang, J., Jin, E., Kong, L., Zhang, W., and Wang, C. (2009) Au nanoparticles-functionalized two-dimensional patterned conducting PANI nanobowl monolayer for gas sensor. *Sensors and Actuators B: Chemical*, **140**(2), 520-524.
44. Liu, C., Hayashi, K., and Toko, K. (2012) Au nanoparticles decorated polyaniline nanofiber sensor for detecting volatile sulfur compounds in expired breath. *Sensors and Actuators B: Chemical*, **161**(1), 504-509.
45. Harish, S., Mathiyarasu, J., Phani, K., and Yegnaraman, V. (2008) PEDOT/Palladium composite material: synthesis, characterization and application to simultaneous determination of dopamine and uric acid. *Journal of Applied Electrochemistry*, **38**(11), 1583-1588.
46. Huang, K.-J., Zhang, J.-Z., Liu, Y.-J., and Wang, L.-L. (2014) Novel electrochemical sensing platform based on molybdenum disulfide nanosheets-polyaniline composites and Au nanoparticles. *Sensors and Actuators B: Chemical*, **194**, 303-310.
47. Saha, S., Sarkar, P., and Turner, A. P. (2014) Interference-free electrochemical detection of nanomolar dopamine using doped polypyrrole and silver nanoparticles. *Electroanalysis*, **26**(10), 2197-

2206.

48. Manesh, K., Santhosh, P., Gopalan, A., and Lee, K. (2008) Electrocatalytic oxidation of NADH at gold nanoparticles loaded poly (3, 4-ethylenedioxythiophene)–poly (styrene sulfonic acid) film modified electrode and integration of alcohol dehydrogenase for alcohol sensing. *Talanta*, **75**(5), 1307-1314.
49. Duraisamy, N., Muhammad, N. M., Ali, A., Jo, J., and Choi, K.-H. (2012) Characterization of poly (3, 4-ethylenedioxythiophene): poly (styrenesulfonate) thin film deposited through electrohydrodynamic atomization technique. *Materials Letters*, **83**, 80-83.
50. Harish, S., Mathiyarasu, J., Phani, K., and Yegnaraman, V. (2009) Synthesis of conducting polymer supported Pd nanoparticles in aqueous medium and catalytic activity towards 4-nitrophenol reduction. *Catalysis Letters*, **128**(1-2), 197.
51. Jin, W., Han, L., Han, X., Zhang, B., and Xu, P. (2016) Interfacial synthesis of lollipop-like Au–polyaniline nanocomposites for catalytic applications. *RSC Advances*, **6**(85), 81983-81988.
52. Chi, M., Nie, G., Jiang, Y., Yang, Z., Zhang, Z., Wang, C., and Lu, X. (2016) Self-assembly fabrication of coaxial Te@poly(3,4-ethylenedioxythiophene) nanocables and their conversion to Pd@poly (3,4-ethylenedioxythiophene) nanocables with a high peroxidase-like activity. *ACS Applied Materials & Interfaces*, **8**(1), 1041-1049.
53. Yang, Z., Sheng, Q., Zhang, S., Zheng, X., and Zheng, J. (2017) One-pot synthesis of Fe_3O_4/polypyrrole/graphene oxide nanocomposites for electrochemical sensing of hydrazine. *Microchimica Acta*, **184**(7), 2219-2226.
54. Tsele, T. P., Adekunle, A. S., Fayemi, O. E., and Ebenso, E. E. (2017) Electrochemical detection of epinephrine using polyaniline nanocomposite films doped with TiO_2 and RuO_2 nanoparticles on multiwalled carbon nanotube. *Electrochimica Acta*, **243**, 331-348.
55. Nezhadali, A., Pirayesh, S., and Shadmehri, R. (2013) Computer-assisted sensor design and analysis of 2-aminobenzimidazole in biological model samples based on electropolymerized-molecularly imprinted polypyrrole modified pencil graphite electrode. *Sensors and Actuators B: Chemical*, **185**, 17-23.
56. Nezhadali, A., and Shadmehri, R. (2013) Computer-aided sensor design and analysis of thiocarbohydrazide in biological matrices using electropolymerized-molecularly imprinted polypyrrole modified pencil graphite electrode. *Sensors and Actuators B: Chemical*, **177**, 871-878.
57. Zare, E. N., Motahari, A., and Sillanpaa, M. (2018) Nanoadsorbents based on conducting polymer nanocomposites with main focus on polyaniline and its derivatives for removal of heavy metal ions/dyes: A review. *Environmental Research*, **162**, 173-195.
58. Dey, A. (2018) Semiconductor metal oxide gas sensors: A review.

Materials Science and Engineering: B, **229**, 206-217.

59. Ahmed, J., Rahman, M. M., Siddiquey, I. A., Asiri, A. M., and Hasnat, M. A. (2017) Efficient bisphenol-A detection based on the ternary metal oxide (TMO) composite by electrochemical approaches. *Electrochimica Acta*, **246**, 597-605.
60. Al-Mokaram, A., Amir, M. A., Yahya, R., Abdi, M. M., and Mahmud, H. N. M. E. (2017) The development of non-enzymatic glucose biosensors based on electrochemically prepared polypyrrole-chitosan-titanium dioxide nanocomposite films. *Nanomaterials*, **7**(6), 129.
61. Emre, F. B., Kesik, M., Kanik, F. E., Akpinar, H. Z., Aslan-Gurel, E., Rossi, R. M., and Toppare, L. (2015) A benzimidazole-based conducting polymer and a PMMA-clay nanocomposite containing biosensor platform for glucose sensing. *Synthetic Metals*, **207**, 102-109.
62. Zanardi, C., Terzi, F., and Seeber, R. (2013) Polythiophenes and polythiophene-based composites in amperometric sensing. *Analytical and Bioanalytical Chemistry*, **405**(2-3), 509-531.
63. Aradilla, D., Estrany, F., Azambuja, D. S., Casas, M. T., Puiggali, J., Ferreira, C. A., and Aleman, C. (2010) Conducting poly (3,4-ethylenedioxythiophene)-montmorillonite exfoliated nanocomposites. *European Polymer Journal*, **46**(5), 977-983.
64. Shao, L., Wang, X., Yang, B., Wang, Q., Tian, Q., Ji, Z., and Zhang, J. (2017) A highly sensitive ascorbic acid sensor based on hierarchical polyaniline coated halloysite nanotubes prepared by electrophoretic dDeposition. *Electrochimica Acta*, **255**, 286-297.
65. Hsu, S.-C., Cheng, H.-T., Wu, P.-X., Weng, C.-J., Santiago, K. S., and Yeh, J.-M. (2017) Electrochemical sensor constructed using a carbon paste electrode modified with mesoporous silica encapsulating PANI chains decorated with GNPs for detection of ascorbic acid. *Electrochimica Acta*, **238**, 246-256.
66. de Barros, A., Ferreira, M., Constantino, C. J., and Ferreira, M. (2014) Nanocomposites based on LbL films of polyaniline and sodium montmorillonite clay. *Synthetic Metals*, **197**, 119-125.
67. de Barros, A., Ferreira, M., Constantino, C. J. L., Bortoleto, J. R. R., and Ferreira, M. (2015) Synergy between polyaniline and OMt clay mineral in Langmuir–Blodgett films for the simultaneous detection of traces of metal ions. *ACS Applied Materials & Interfaces*, **7**(12), 6828-6834.
68. Li, J., Miao, D., Yang, R., Qu, L., and Harrington, P. d. B. (2014) Synthesis of poly (sodium 4-styrenesulfonate) functionalized graphene/cetyltrimethylammonium bromide (CTAB) nanocomposite and its application in electrochemical oxidation of 2,4-dichlorophenol. *Electrochimica Acta*, **125**, 1-8.
69. Pontes, L., De Souza, J., Galembeck, A., and de Melo, C. (2013) Gas sensor based on montmorillonite/polypyrrole composites prepared by in - situ polymerization in aqueous medium. *Sensors and*

Actuators B: Chemical, **177**, 1115-1121.

Chapter 3

Polyaniline-Graphene Composite Nanoparticle Pigments for Anti-Corrosion Coatings

Ali U. Chaudhry and Vikas Mittal*,**

Department of Chemical Engineering, The Petroleum Institute (part of Khalifa University of Science and Technology), Abu Dhabi, UAE
**Corresponding author*: vik.mittal@gmail.com
***Current address*: Bletchington, Wellington County, Australia

3.1 Introduction

Metallic corrosion is a result of electrochemical phenomenon that involves the transfer of charge from one species to another, thus, leading to impairment of metal. Figure 3.1 depicts the corrosion phenomenon in the presence of sodium chloride aqueous media where chloride ions accelerate the corrosion process. In this case, the corrosion process proceeds by the dissolution of iron oxide film with the aid of chloride ions as well as enhancement of the transportation of electrons by sodium and chloride ions. Among various methods developed to slow down the corrosion process, corrosion inhibitors and anti-corrosion pigments receive widespread use owing to cost effectiveness and easy application. In these cases, the corrosion inhibition of metals results due to interactions between

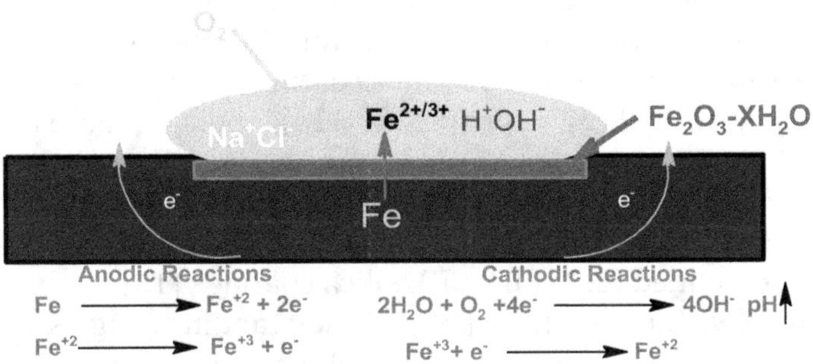

Figure 3.1 Schematic of corrosion process of iron.

Conducting Polymer Composites, edited by Vikas Mittal

free electrons of inhibitor and empty orbital of metals which leads to the formation of metal protective surface complexes [1]. These protective films slow down the deterioration process of metals by suppressing either cathodic, anodic or both types of corrosion processes.

The environmental constraints on using coatings incorporating chromium (VI) based corrosion inhibitors have driven the development of non-toxic organic and inorganic inhibitors and anti-corrosion pigments to be incorporated in polymer coatings. Organic inhibitors have polar groups such as sulfur, nitrogen, selenium, etc., which help them to chemisorb on metal surfaces resulting in formation of protective surface complexes. Examples of such inhibitors are amine- and imidazoline-derivatives, alkyl pyridines and thioureas [2]. On the other hand, inorganic corrosion inhibitors include sulfites, ferric salts, nitrates and calcium ions, with corrosion protection mechanisms based on the specific type of inorganic inhibitor [3-5]. For example, sulfites [3] consume the dissolved oxygen in solutions to reduce oxygen availability in order to retard corrosion; ferric salts and nitrates [4] foster passivity of metals in active-passive regions, and the calcium ions [5] reinforce the protective films formed on metal surfaces. In addition, inorganic pigments such as litharge, metallic lead, red lead, basic lead carbonate, hexavalent chromium compounds and zinc oxide have been extensively used in polymer coatings for the enhancement of anti-corrosion [6], gas barrier [7], mechanical [8], electrical and optical [9], rheological and adhesion properties as well as environmental degradation resistance [10]. Recent research studies have also reported about the potential of carbonaceous and other nanoparticles as anti-corrosion pigments in polymer coatings [11]. The nano-pigments exhibit advantages over conventional pigments due to improved barrier and anti-corrosion properties in polymer coatings [12] such as polylactic acid (PLA)/silica [13], epoxy-clay [14], polyaniline/TiO_2 [15], poly(methyl methacrylate) (PMMA)/graphene [16], poly(vinyl carbazole) (PVK)/multi-walled carbon nanotubes [17] anti-corrosion coatings. For instance, Figure 3.2 shows the Tafel plots of PLA/silica and chain extended PLA (PLA-CE)/silica coatings [13].

Specifically, the development of free-standing single-layer graphene, which is the building block of all graphitic forms, has resulted in its extensive use as reinforcement for polymer materials [18]. Graphene sheets are one-atom-thick two-dimensional layers of sp^2-bonded carbon atoms having a wide variety of remarkable features

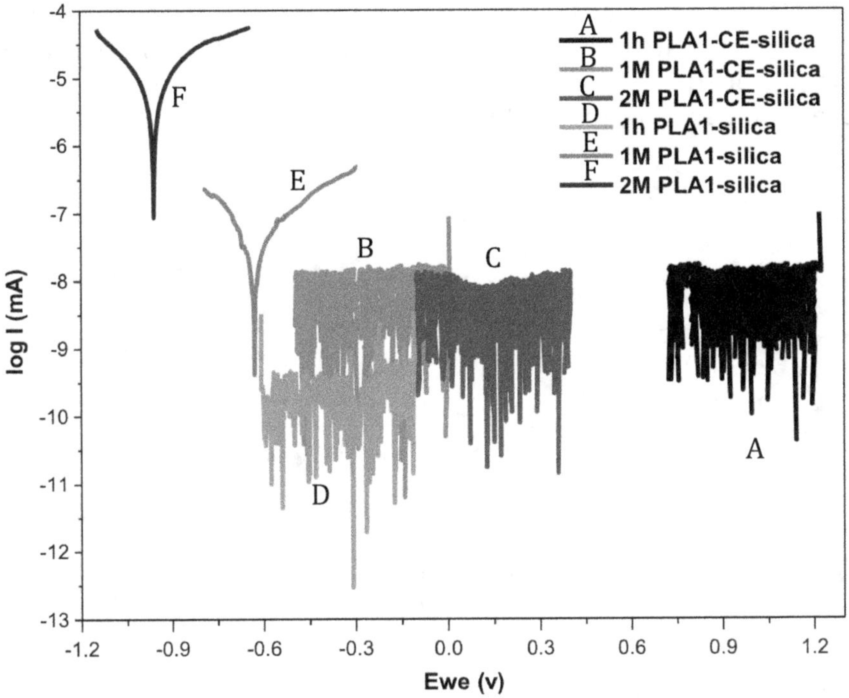

Figure 3.2 Tafel plots of PLA1-silica and PLA1-CE-silica coatings after 1 h, 1 month and 2 month exposure in 3.5 wt% sodium chloride solution. Reproduced from Reference 13.

and can enhance properties of polymers such as electrical and thermal conductivity, gas impermeability and mechanical performance [19]. For instance, Figure 3.3 presents the schematic of the protection provided by graphene in the composite coating [20]. Further, the role of conductive polymers, especially polyaniline (PANI), for corrosion protection of ferrous and non-ferrous metals has been vigorously studied [21]. For this purpose, PANI has been reported to provide corrosion protection of metals either as a neat film or as resin blended coatings [22,23]. Resin blending method is reported to be more convenient and acceptable method of coating on large structures as it leads to better mechanical properties of coatings [24]. Huh *et al.* reported PANI and poly-o-ethoxyaniline blended coatings on pure iron, which resulted in better corrosion protection in 3 wt% NaCI [25]. Armelin *et al.* generated a comparison between corrosion inhibitor $Zn_3(PO_4)_2$ at 10 % w/w and two forms of PANI (emeraldine base and emeraldine salt) mixed with epoxy at 0.3 %

w/w. Better behavior of emeraldine base at low concentration was attributed towards its ability to act as a barrier and to store charge due to oxidation and reduction in a reversible way as compared to emeraldine salt due to its fully oxidized state [26].

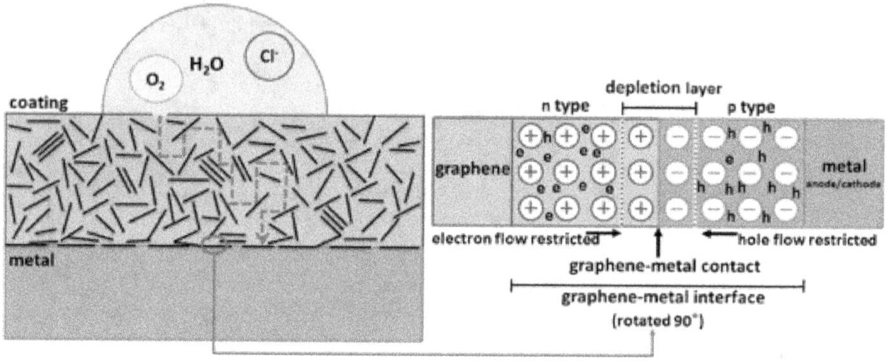

Figure 3.3 Pictorial representation of protection provided by graphene platelets in polyetherimide based nanocomposite coating. Reproduced from Reference 20.

In addition to resin blending, modification of inorganic pigments with PANI to achieve synergistic anti-corrosion effect has also been reported in literature [27]. Kalendova *et al.* proposed a method to combine the use of inorganic pigments and PANI so as to address the problems associated with resin blended coatings, i.e. inefficient PANI distribution, lack of excellent polymer-polymer contact, poor substrate adhesion and change in volume of PANI due to redox reaction. For this purpose, four pigments specularite (Fe_2O_3), goethite ($FeO(OH)$), talc ($Mg_3(OH)_2$-(Si_4O_{10}) and graphite were surface modified with PANI and subsequently blended with epoxy binder. Better corrosion resistance was observed in all PANI coated pigments and PANI modified graphite exhibited excellent corrosion inhibition due to improvement in conductivity which promoted redox reactions between iron and PANI or oxygen and PANI [6]. Sathiyanarayanan *et al.* modified TiO_2 and Fe_2O_3 with PANI and observed enhanced corrosion protection of steel as compared to pigments without polymer modification. The results were attributed towards the formation of passive film along with iron-phosphate salt film on the iron surface [28,29]. Brodinova *et al.* used surface modified ferrites with contents of Zn, Mg and Ca cations. The results showed that modified ferrites were better in terms of corrosion protection than the nonmodified versions. The corrosion protection mechanism was proposed to be a result of formation of cation soaps which lowered the tendency of the binder to saponify on exposure to the hydroxyl ions formed during corrosion of the metal. The presence of PANI filled the pinholes present in coating and also formed a better interconnection between inorganic pigments and resin [30]. Similarly, Wu *et al.* reported hybrid coating of PANI-layered zinc nickel ferrites and organically modified silicate. The film was deposited via spin coating on aluminum alloy. The anti-corrosion performance of the hybrid film was improved due to the denser configuration of organically modified silicate due to the incorporation of nickel zinc ferrites/PANI [31].

Modification of graphene based materials with PANI to generate anti-corrosion pigments is largely missing in literature and generation of such materials can result in anti-corrosion coatings with enhanced performance. In a recent study, Mittal *et al.* [32] studied the adsorption of chlorinated poly(ethylene-co-vinyl acetate)-*g*-maleic anhydride copolymer and *in-situ* generated polyaniline on thermally reduced graphene oxide (TRGO) platelets. The amount of copolymer adsorption reached a plateau of 0.22 g per g of TRGO, when the initial copolymer to TRGO weight ratio of 1 was used. In the case of PANI modification, much higher extent of adsorption of 0.92 g/g of TRGO (without reaching plateau) was observed due to *in-situ* polymer synthesis and the absence of any steric hindrance to the chains. This was also observed through EFTEM and EDX, where presence of N (along with other atoms) indicated layer of PANI on the surface of the platelets. PANI modification also led to a significant reduction in peak melting point from 175 °C to 140 °C owing to the hindrance in polymer crystallization. Furthermore, the basal plane signal of pristine TRGO was observed at 27° 2Θ, which was shifted to lower angle of 22.5° 2Θ after surface modification with the copolymer. In comparison, the basal plane peak corresponding to TRGO was not observed in the PANI-modified TRGO, which indicated that the TRGO platelets were delaminated completely and the periodicity was lost after polymer modification. It, thus, confirmed that the *in-situ* polymerization method of surface modification with PANI resulted in an overall higher amount of adsorption on the surface of graphene platelets.

In the current study, the use of similar nano-hybrids based on PANI-thermally reduced graphene (r-GO) has been explored as anti-corrosion pigments incorporated in poly(vinyl butyral) (PVB) resin.

The PANI/r-GO hybrids were generated with varying r-GO/aniline ratios during the hybrid generation, so as to relate the effect of graphene content to the coating performance. In addition, the r-GO/aniline ratios used in the current study were different from the ones reported in the earlier study [32]. Furthermore, the coatings of PVB-pigment formulations were applied on carbon steel panels and the corrosion performance of the nano-hybrid pigments was investigated in 4 % w/v sodium chloride (NaCl) solution.

3.2 Experimental

3.2.1 Materials

PVB (M_W 70,000-100,000) resin with broad molecular weight distribution, containing ~20 % vinyl alcohol and ~2 % vinyl acetate, was obtained as granules from Polysciences, Inc. (USA). Hydrochloric acid (38 %) and NaCl were supplied by Merck, Germany, whereas aniline monomer was procured from Mallinckrodt, USA. Hexamethylene tetramine, ammonium persulfate (APS) and acetone were received from Sigma-Aldrich, Germany. Graphite powder (325 mesh) was supplied by Alfa Aesur GmbH and Co., Germany. Concentrated sulfuric acid (H_2SO_4, 95-98%), sodium nitrate ($NaNO_3$) and potassium permanganate ($KMnO_4$) were purchased respectively from S. D. Fine Chemicals India, Eurostar Scientific UK, and Fisher Scientific UAE. Carbon steel coupons (elemental composition: C 0.07 wt%, Mn 1.36 wt%, Ti 0.008 wt%, S 0.003 wt%, P 0.004 wt% and rest Fe) were purchased locally and were machined to 30 x 16 x 5 mm dimensions.

3.2.2 Preparation of Graphite Oxide (GO) and r-GO

r-GO was prepared through thermal exfoliation of precursor graphite oxide using modified Hummer's method [33]. A short description for the preparation of graphite oxide and r-GO is as follows: graphite powder was added with concentrated H_2SO_4, followed by addition of $NaNO_3$ to the mixture. The mixture was stirred in an ice-bath (5 °C) and $KMnO_4$ was added to the mixture after 30 min. Gradually, the temperature was allowed to rise to 35 °C and the mixture was stirred for 2 h under these conditions. Deionized water was subsequently added while allowing the temperature to increase to 100 °C. After 15 min, the mixture was quenched and diluted by pouring it into deionized water. 30 % H_2O_2 was slowly added to the dilute solution and the solution was filtered to remove the non-graphite oxide waste. The residues were re-dispersed in deionized water and were added with dilute HCl to remove the SO_4^{-2} ions. The cleaning and filtration was continued until no SO_4^{-2} and Cl^- ions could be observed in the filtrate. Afterwards, the washed GO was dried under vacuum at 60 °C for 24 h. r-GO was generated via thermal exfoliation of dried graphite oxide by placing 1 g GO in a long quartz tube with 25 mm internal diameter and sealed at one end. The sample was flushed with nitrogen, followed by insertion of the tube in a tube furnace preheated to 1050 °C. The tube was held in the furnace for 30 s.

3.2.3 Preparation of Polyaniline and r-GO Composite Nanoparticles

Polyaniline modified r-GO was prepared using *in-situ* polymerization. To generate PANI/r-GO composites with different weight ratios, 0, 10, 20, 30, 40 and 50 mg of r-GO nanoparticles were added into a mixture of 1 ml of aniline and 90 ml of 1N HCl in a reaction vessel (Figure 3.4). The mixture was stirred in ice water bath (5 ^{0}C) for 1 h to obtain uniform dispersion of r-GO. To the mixture, 100 ml pre-cooled 1N HCl solution containing 2.5 g APS was added drop wise. The reaction was allowed to proceed in the ice bath for 4 h. The resulting product was washed with distilled water for several times and at last washed with methanol to remove low molecular weight oligomers along with other impurities. Subsequently, the sample was dried in a vacuum oven for 12 h. Based on the amount of r-GO, these hybrids were named as PANI/10r-GO, PANI/20r-GO, PANI/30r-GO, PANI/40r-GO and PANI/50r-GO.

3.2.4 Characterization of PANI/r-GO Composite Nanoparticle Pigments

Calorimetric properties of PANI and PANI/r-GO composite nanoparticle pigments were recorded on a Perkin-Elmer Pyris-1 differential scanning calorimeter (DSC) under nitrogen atmosphere. The scans were obtained from 50 to 250 °C at a heating rate of 5 °C/min. Thermogravimetric measurements were analyzed with Netzsch thermogravimetric analysis (TGA) using ASTM (American Society for Testing and Materials) standard E 1131. Nitrogen was used as a

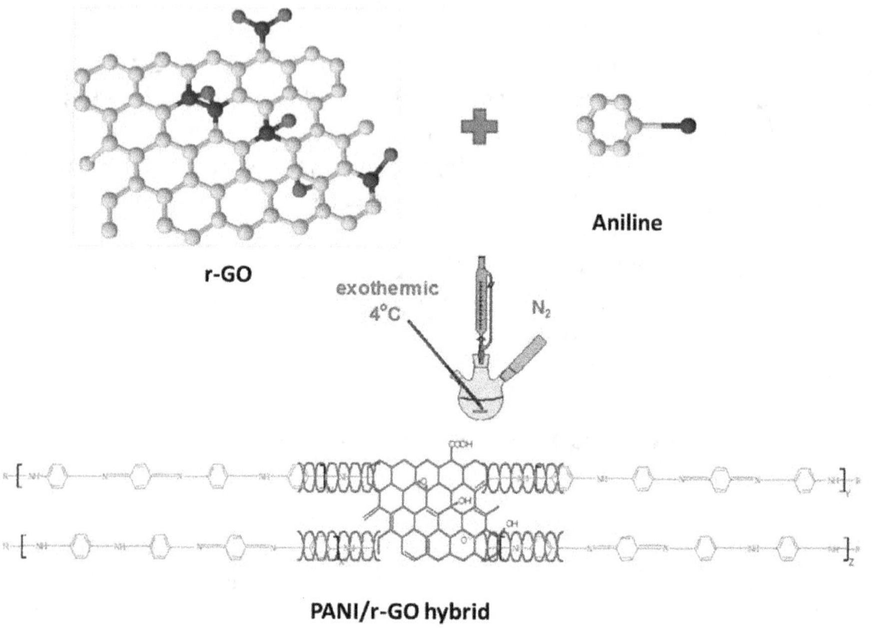

Figure 3.4 Schematic diagram of generation of PANI/r-GO composite particles.

carrier gas and the scans were obtained from 50 to 600 °C at a heating rate of 20 °C/min. Transmission electron microscopy (TEM) of hybrid samples was performed using EM 912 Omega (Zeiss, Oberkochen BRD) electron microscope at 120 kV and 200 kV accelerating voltage. Thin flakes of PANI/r-GO platelets were supported on 100 mesh grids sputter coated with a 3 nm thick carbon layer. Wide angle X-ray diffraction analysis of the samples was performed on Panalytical powder diffractometer (X'Pert PRO) using CuKα radiation (λ = 1.5406 A°) in reflection mode. Zero-background holder was used to minimize the noise. The samples were step-scanned from 10-40° 2Θ at room temperature using a step size of 0.02° 2Θ and a step time of 10 s [34].

3.2.5 Generation of PANI/r-GO/PVB Coatings

PVB was dissolved in methanol (20 ml) under stirring for 6 h, followed by sonication in a bath sonicator for 2 h. PANI/r-GO nanoparticles of different compositions were finely powdered with small quantity of methanol to obtain a uniform slurry. The slurry was added and mixed in the PVB solution and sonicated for 24 h to achieve uniform dispersion of nanoparticles in the PVB solution [35]. The amount of PANI/r-GO particles was fixed to 10 wt% of the amount of dry PVB. For coating of the generated PANI/r-GO/PVB formulations, carbon steel coupons with rounded corners and edges were polished by emery paper, washed with acetone, dried and weighed with an accuracy of ±1 mg. The substrates were brush coated with the PANI/r-GO/PVB formulations and dried at room temperature for 30 min. Subsequently, the coating was baked in air circulating oven at 60 °C for 6 h and coatings with thickness in the range of 15-20 μm were obtained. Care was taken that no bubbles were formed on the surface of the coatings.

3.2.6 Corrosion Performance Analysis

The anti-corrosion properties of the coatings were studied by immersion test (NaCl 4 % w/v) according to ASTM G31. The coated coupons were placed in a specially designed set up for 800 h at 25 °C. The function of this set-up was to maintain temperature, air flow and water level so as to attain similar conditions throughout the experiment [27]. After completing the immersion period, the samples were taken out of salt solution, cleaned and weighed again. Cleaning of the samples was achieved by removing the coating using acetone as a solvent and subsequently removing the corrosion products by a solution of 3.5 g of hexamethylenetetramine in 500 ml of hydrochloric acid diluted to 1000 ml by distilled water. The samples were dipped for 10 min in the amine solution at room temperature. The weight loss method was adopted to evaluate the corrosion rate, using the equation:

$$Corrosion\ rate, CR = \frac{87.6\,W}{DAT}$$

where CR is corrosion rate in mm/year, W is weight loss in mg, D is the substrate density, A is the exposed surface area and T is exposure time in salt solution in hours. The area of the exposed surface of the coated substrate was 9.60 cm^2, whereas the density of substrate was taken as 7.85 g/cm^3. The corrosion protection efficiency (CE) of anti-corrosion pigment was calculated using the following equation:

$$CE = \left[\frac{CR - CR'}{CR} \right]$$

where CR and CR' are the corrosion rates without and with the pigment respectively.

3.3 Results and Discussion

In the current study, PANI/r-GO hybrids with varying amount of graphene were incorporated in PVB matrix so as to achieve anticorrosion coatings where both PANI and graphene synergistically contributed to the coating's performance. Use of *in-situ* polymerization was chosen to generate the hybrid nanoparticles as the growing PANI chains would be able to extensively adsorb on graphene surface without steric hindrance. In addition, organic modification of graphene platelets with PANI was also expected to lead to nanoscale dispersion of graphene platelets in PVB matrix.

3.3.1 Characterization of PANI/r-GO Hybrid Nanoparticles

X-ray diffraction patterns of PANI, r-GO and PANI/r-GO hybrids are shown in Figure 3.5. The characteristic broad diffraction peak for r-GO was observed at 2θ°~27 and (002) plane, which can be attributed towards short-range order in piled stacks having interlayer spacing 0.34 nm. During thermal reduction, i.e. removal of intercalated water molecules and the oxide groups of hydroxyl and carboxyl groups, graphite oxide restores its crystalline state by reinstating conjugated graphene network (sp^2 carbon) having structural defects such as residual oxygen-containing functional groups, etc. [36,37]. The crystalline nature of pure PANI was observed due the presence of sharp crystalline diffraction peaks at 2θ~25.2° (interlayer spacing 0.35 nm) along with other peaks at 2θ = 15 ° and 21.9° [38]. In the hybrids, where PANI intercalation took place between r-GO interlayers, slight broadness could be observed in the PANI diffraction peaks. On the other hand, r-GO peak was absent in the spectra of the PANI/r-GO hybrids [39]. It indicated that extent of PANI intercalation in the graphene interlayers was significant, which resulted in the delamination of the graphene platelets. Even for the hybrid with highest amount of graphene, the diffractogram did not show any diffraction peak corresponding to pristine graphene. However, it has

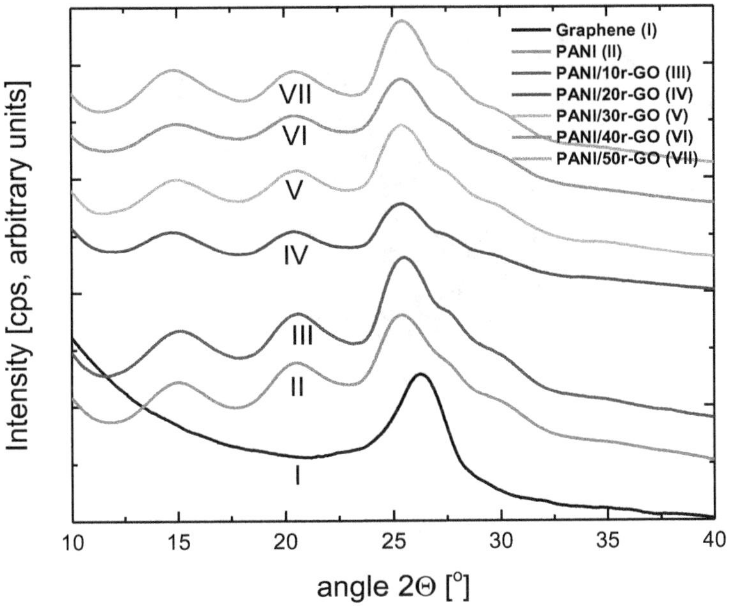

Figure 3.5 X-ray diffractograms of graphene, PANI and PANI/r-GO hybrids.

to be noted that the diffraction peaks of PANI and graphene overlap to some extent. Thus, microscopy analysis was also carried out to confirm the intercalation of PANI in the graphene interlayers. Figure 3.6 shows the TEM micrographs of pristine r-GO and PANI/30r-GO. The morphology of the graphene platelets altered completely with PANI intercalation in the interlayers. As is evident in Figure 3.6(b), PANI was observed to uniformly cover the surface of the platelets, which was also the reason for absence of any graphene diffraction signal in the X-ray diffractograms. Even decreasing or increasing the amount of graphene in the hybrid exhibited similar morphology. PANI chains are expected to physically adsorb on the surface of graphene nanoparticles due to interactions of N atoms with functional groups on graphene surface as well as due to van der Waals forces between PANI chains and graphene surface. This is, thus, also expected to affect the polymer crystallinity as the PANI chains would adsorb on the surface of platelets simultaneously during their synthesis, which will hinder their effective crystallization.

DSC melting curves of pure PANI and PANI/r-GO hybrids are shown in Figure 3.7. As is evident, the addition of r-GO in polyaniline resulted in significant reduction in the peak melting temperature of

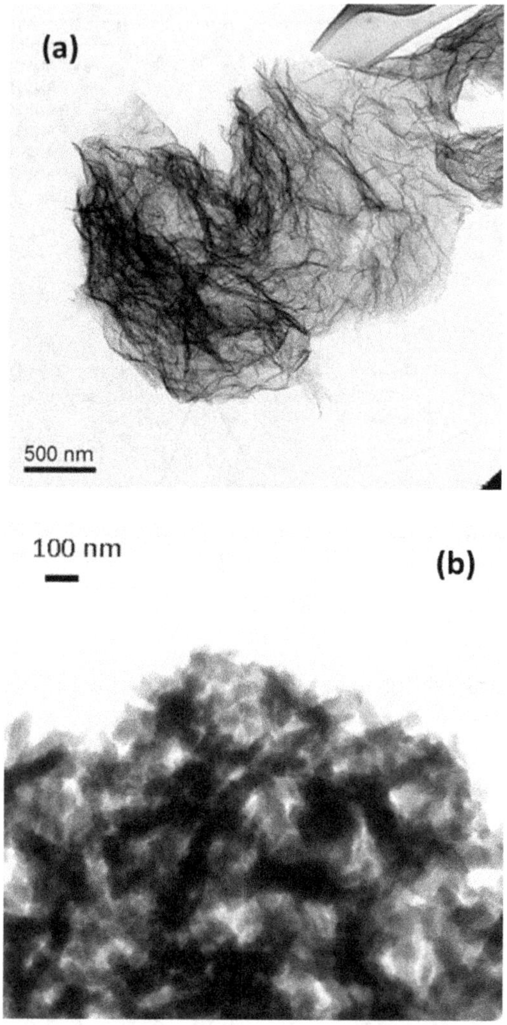

Figure 3.6 TEM micrographs of (a) pristine graphene and (c) PANI/30r-GO.

the polymer (Table 3.1). Moreover, the decrease in the melting point was proportional to the amount of graphene in the hybrid. As compared to 178 °C for pure polymer, the peak meting point of the polymer in PANI/50r-GO hybrid was reduced to 111 °C. It indicated that the crystallization of the polymer was affected with increasing amount of r-GO and the graphene platelets hindered the chain mobility and packing into the crystal structure [40]. It further indicated that the polymer chains were strongly immobilized on the surface of

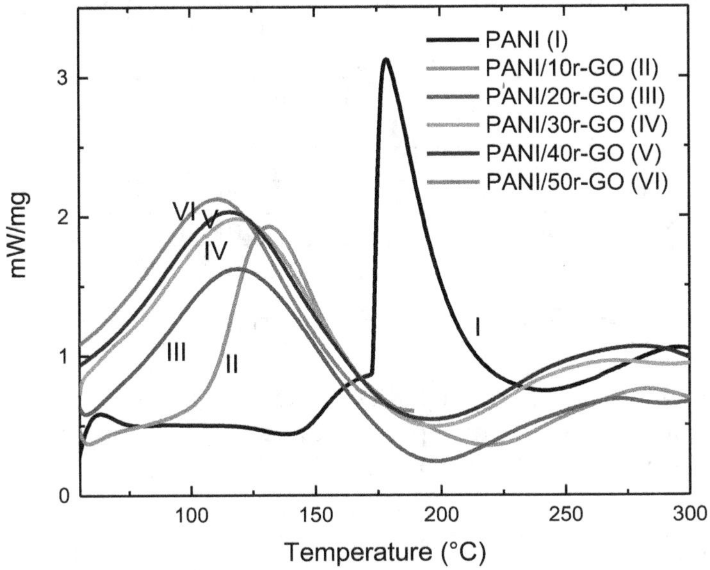

Figure 3.7 DSC melting curves for PANI/r-GO hybrids with different content of r-GO.

the graphene platelets and was not just present as intercalated material inside the interlayers. If the polymer was not immobilized to the surface, the excess polymer away from the interface would have been able to crystallize normally and would not have resulted in significant reduction in the melting point (e.g. in PANI/10r-GO, where large excess of PANI would be available). With increasing amount of graphene in the hybrid, more surface was available for the growing PANI chains to adsorb, which resulted in increased extent of adsorption and subsequently gradual decrease in the peak melting point and overall crystallinity. Figure 3.8 also shows the TGA thermograms of the PANI/r-GO hybrids compared with pure PANI. Initial mass loss around 100 °C was attributed to the evaporation of water molecules absorbed or adsorbed on surface. The extent of moisture in the samples was minimal as the samples were dried overnight under vacuum before TGA analysis. For pure PANI, the weight loss from 100-350 °C could be attributed to the deprotonation of the PANI, whereas, before 600 °C, major weight loss transition was related to the degradation and decomposition of PANI [41]. The hybrids exhibited similar degradation pattern as pure PANI. However, the extent of weight loss was decreased as the amount of

Table 3.1 Calorimetric properties of PANI and PANI/r-GO composite particles

Polymer/Composite	Peak melting temp. (°C)
PANI	178
PANI/10r-GO	132
PANI/20r-GO	118
PANI/30r-GO	118
PANI/40r-GO	115
PANI/50r-GO	111

graphene was enhanced in the hybrids, indicating increased thermal stability, though the thermal performance of the composites was very similar beyond PANI/10r-GO hybrid.

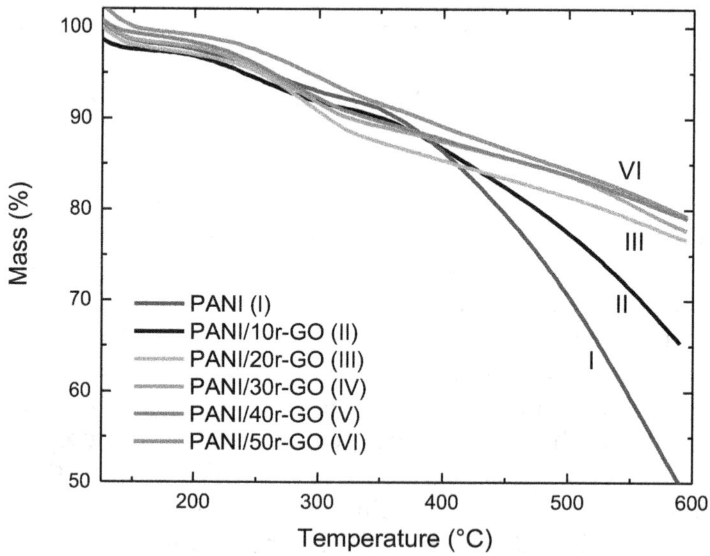

Figure 3.8 TGA thermograms of PANI and PANI/r-GO hybrids.

3.3.2 Anti-corrosion Performance and Protection Mechanism

Table 3.2 shows the corrosion rate of the PVB coatings reinforced with PNAI/r-GO pigments, when immersed in aqueous solution of 4 % w/v NaCl for 800 h. Under the test conditions used, pure PANI pigment in PVB did not provide any superior corrosion protection than pure PVB coating itself, though both improved the protection of bare metal substrate. The corrosion protection efficiency (CE) was observed to increase with increasing amount of r-GO in the coatings. These results can be attributed to the additional protection provided by r-GO, along with the intrinsic protection provided by PANI.

Table 3.2 Weight loss and corrosion rate after 800 h immersion in 4% NaCl

Material/Coating	Weight loss (mg)	Corrosion rate (mm/y)	CE (%)
Blank substrate	26.7	0.039	-
Pure PVB	16.4	0.024	-
PANI	16.2	0.024	-
PANI/10r-GO	15.6	0.023	4.8
PANI/20r-GO	12.4	0.018	24
PANI/30r-GO	10.7	0.016	34.7
PANI/40r-GO	7.9	0.012	51.6
PANI/50r-GO	6.9	0.01	58.2

For instance, coating with PANI/40r-GO was nearly 52 % more effective than the PVB coating without pigment. Further increase in graphene content in PANI/50r-GO exhibited marginal increase in protection efficiency to 58%, which indicated that the CE may have reached a plateau. To further confirm these findings, optical microscopy was employed to gain insights into the change in surface topography of the substrates after the corrosion test. Figure 3.9 shows the surface of the coated carbon steel substrates before corrosion testing. Similar to the CE results, the PANI/40r-GO and PANI/50r-GO incorporated coatings were very similar. Dispersion of pure PANI in PVB was difficult due to the aggregation of *in-situ* generated polymer in the form of solid lumps, thus, resulting in larger size distribution of PANI particles in PVB coatings. On the other hand, PANI/r-GO hybrids were more susceptible to uniform dispersion in the PVB matrix, and resulted in much smaller particle sizes in the formed coatings. Moreover, higher concentration of graphene in the hybrids did not pose any hindrance to the dispersion of the PANI/r-GO hybrids. The surface of the substrates was uniformly coated with the generated formulations, however, due to the brush coating method used, some surface roughness could be observed in the coatings. Figure 3.10 demonstrates the surface of the coatings after

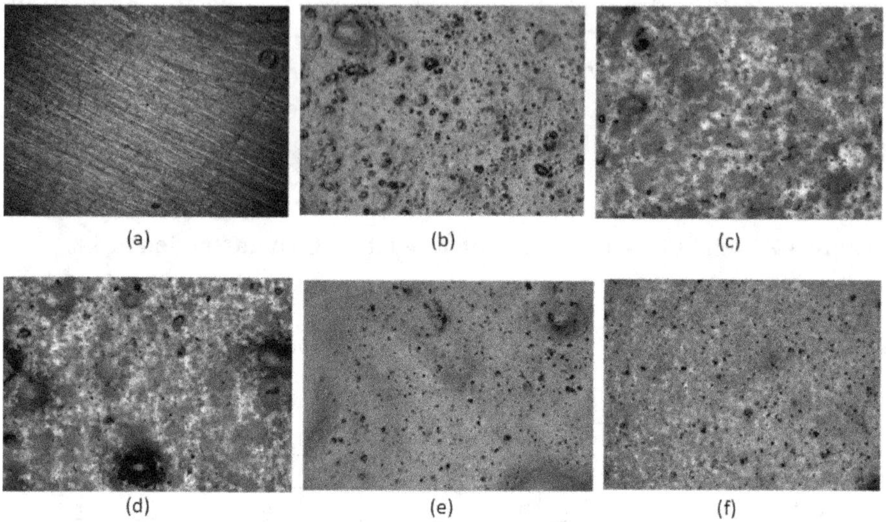

Figure 3.9 Surface of the coated substrates before corrosion testing; (a) blank (pure PVB), (b) PANI, (c) PANI/10r-GO, (d) PANI/20r-GO, (e) PANI/30r-GO and (f) PANI/40r-GO. The width of the images equals 200 μm.

corrosion test as well as removal of the coating. Substrates coated with pure PVB (Figure 3.10(a)) and pure PANI in PVB (Figure 3.10(b)) formulations exhibited extensive surface corrosion, confirming the findings depicted in Table 3.2. It indicated that the corrosion media could penetrate faster through the coatings, thus, initiating the corrosion process. The substrate coated with PANI/10r-GO (Figure 3.10(c)) in PVB exhibited improved anti-corrosion performance as the extent of the surface corrosion was markedly reduced in comparison with Figure 3.10(a) and (b). Similar to the findings in Table 3.2, further increase in the graphene content enhanced the performance further. PANI/20r-GO, PANI/30r-GO and PANI/40r-GO coated substrates optically exhibited high level of stability to the immersion conditions. The similar testing of the substrates was also carried out at a relatively higher temperature of 40 °C and the results are demonstrated in Figure 3.11 after removal of the surface coating. As expected, pure PANI was observed to extensively corrode on the surface (Figure 3.11(a)). Similar to room temperature testing conditions, coating incorporating PANI/10r-GO exhibited decreased extent of corrosion, which was further reduced in PANI/20r-GO system. The formulation having PANI/40r-GO did not

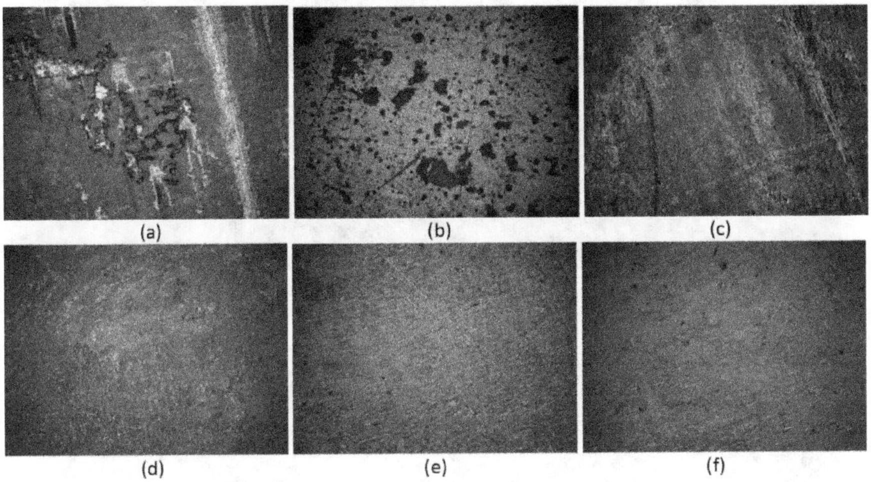

Figure 3.10 Surface of the substrates after corrosion testing and coating removal; (a) blank (pure PVB), (b) PANI, (c) PANI/10r-GO, (d) PANI/20r-GO, (e) PANI/30r-GO and (f) PANI/40r-GO. The width of the images equals 200 μm.

exhibit any significant initiation of the corrosion process. It indicated that the generated PANI/r-GO based pigments were also effective at higher temperature. Interestingly, the conductivity of the coatings decreased as the amount of reduced graphene was increased. It has been reported earlier that the lower conductivity of PANI/graphene composites results probably due to decrease in the degree of doping in PANI and change in the morphology of the composites [39,42]. Similarly, Sun *et al.* [43] have also reported graphene/pernigraniline composite (GPCs) coatings in PVB with reduced conductivity for the corrosion protection of copper. The GPCs were generated by *in-situ* polymerization-reduction/doping process and the authors concluded that pernigraniline modified graphene in PVB was able to provide effective corrosion protection.

The proposed phenomenon of corrosion protection by pure PVB coating and PANI/r-GO incorporated coatings is compared pictorially in Figure 3.12. The corrosion protection by the coating is due to the generation of disconnection between the bare metal and corrosive environment, thus, impeding the transportation of corrosive material. However, as shown in Figure 3.12(a), in the case of pure resin coating, the presence of free volume and coating defects result in the faster penetration of the corrosive solution, thereby, resulting

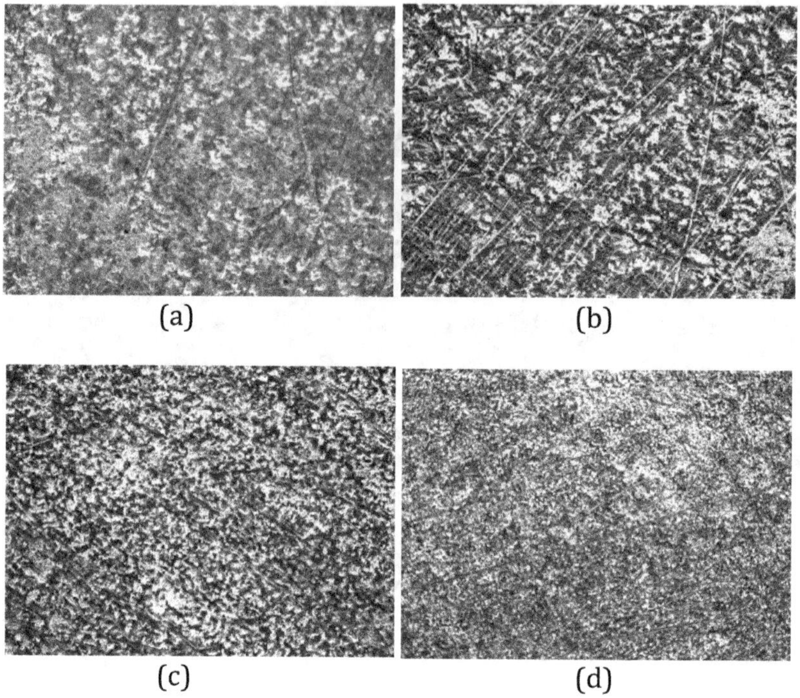

Figure 3.11 Surface of the substrates after corrosion testing at 40 °C and coating removal; (a) PANI, (b) PANI/10r-GO, (c) PANI/20r-GO and (d) PANI/40r-GO. The width of the images equals 200 μm.

in corrosion at the surface. Fundamentally, it is similar to the corrosion occurring in case of bare substrate (Figure 3.1) where continuous formation and degradation of iron oxide film occurs in the presence of water and oxygen along with additional accelerated effect caused by sodium chloride. Figure 3.12(b) shows the mechanism of corrosion protection in the presence of PANI/r-GO pigments embedded in the PVB coatings. The protection behavior of PANI at the interface of metal/electrolyte has been described earlier in the literature [24]. In this phenomenon, the formation of metal oxide layer occurs with the aid of more noble emeraldine salt (ES) of PANI. This form of PANI further leads to lower energy state and reduces to different forms such as non-conductive leucoemeralfine base (LEB) and emeraldine base (EB). This cycle is continuous and LE form turns to ES via EB form by oxygen. This cycle can be continuous only if the barrier properties of applied coating are able to avoid the removal of H^+ (maintained acidic pH) by the surroundings. This slows

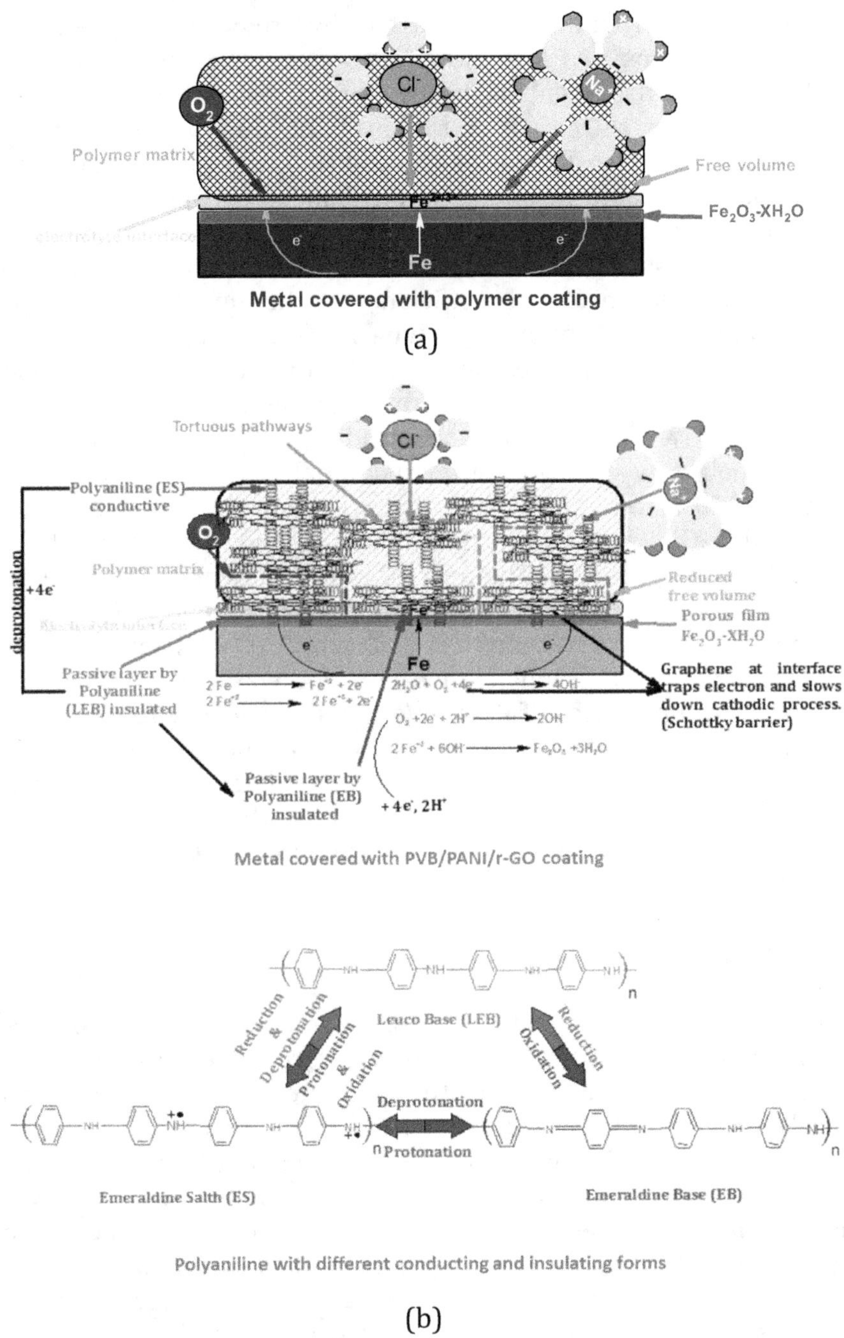

Figure 3.12 Comparison of corrosion protection mechanisms in the (a) absence and (b) of PANI/r-GO pigment nanoparticles incorporated in PVB.

the formation of rust at the iron surface by providing passive layers of different forms of PANI. However, as observed in the current study, pure PANI present in the PVB matrix was unable to significantly slow down the corrosion rate probably due to lower extent of PANI-metal interface formation. In addition, the process of corrosion protection could be specifically enhanced by the addition of r-GO, as it resulted in tortuous path for the corrosive media, thus, delaying the corrosion process. This was also confirmed from the improving corrosion protection efficiency as the amount of graphene was enhanced in the coatings. Such barrier effect of graphene based reinforcement is called a 'passive' role. Further, the 'active' behavior of such material was realized in the coatings by the formation of 'Schottky barrier' at interface, thus, leading to the depletion of electrons, which slowed down the corrosion half-cell reactions. Composites of PANI acted like hetro- junction, where PANI behaves as p-type while r-GO being n-type, thus, hindering the anodic and cathodic reactions respectively. The decreased conductivity of composites also impedes the transport of electron from graphene to PANI [35]. Furthermore, due to the physical adsorption of PANI on graphene surface, the effect of graphene incorporation was synergistically enhanced by the presence of PANI due to its better interface with the metal. It has also to be noted that with these coatings systems, the corrosion reaction was not completely eliminated, however, it was significantly slowed down, thus, underlining the need of the functional anti-corrosion coatings in order to achieve superior material performance [44,45].

3.4 Conclusions

In the current study, polyaniline/graphene composite nanoparticle pigments with different compositions were generated by *in-situ* polymerization of aniline in the presence of graphene nanoplatelets. In the PANI/r-GO hybrids, PANI intercalation took place between r-GO interlayers, which resulted in slight broadness in the PANI diffraction peaks. Diffraction peak corresponding to r-GO was also absent in the spectra of the PANI/r-GO hybrids, irrespective of the hybrid composition. The morphology of the graphene platelets altered significantly with PANI intercalation in the interlayers. The platelets were observed to be uniformly covered on the surface with PANI. The peak meting point of the polymer was observed to reduce as a function of graphene content in the hybrids (e.g. 111 °C for

PANI/50r-GO hybrid as compared to 178 °C for pure polymer), which indicated that the crystallization of the polymer was affected and the graphene platelets hindered the chain mobility and packing into the crystal structure. Significant reduction in the melting point also indicated that the polymer was strongly adsorbed on the surface of the graphene platelets. The hybrids exhibited similar degradation pattern as pure PANI, however, the extent of weight loss was decreased as the amount of graphene was enhanced in the hybrids, indicating increased thermal stability. The PVB coatings incorporating PANI/r-GO hybrids exhibited superior anti-corrosion performance, which was enhanced as the amount of graphene in the hybrid particles was increased. For instance, coatings with PANI/40r-GO and PANI/50r-GO were nearly 52% and 58% more effective than the PVB coating without pigment. The optical analysis of the substrates immersed in NaCl solution for 800 h also confirmed the superior performance of the PANI/r-GO reinforced coatings, which resulted due to the tortuous diffusion path for corrosion media in the presence of graphene platelets. The synergistic effect of PANI/r-GO hybrids was also expected to have resulted from the formation of 'Schottky barrier' at interface thus leading to the depletion of electrons, which slowed down the corrosion half-cell reactions. The resulting PANI/r-GO nanoparticles represent functional pigments for organic/polymeric coatings to achieve effective corrosion protection of metals. In addition, the development of such coatings also presents opportunities in replacing the conventional chromate basted corrosion inhibitors, thus, presenting environmentally friendly approach.

Summary of the Results

Development of anti-corrosion coatings with incorporation of polyaniline adsorbed graphene nanoparticle pigments (PANI/r-GO) in poly(vinyl butyral) (PVB) was successfully demonstrated in the current study. PANI/r-GO hybrids were generated through *in-situ* polymerization of aniline and hybrid composition was varied to evaluate the effect on the corrosion performance of the coatings. The diffraction studies of the hybrids exhibited slight broadness in the PANI diffraction peaks and did not reveal any diffraction peak corresponding to r-GO. Furthermore, crystallization of PANI was significantly affected in the presence of graphene platelets as the graphene platelets hindered the chain mobility and packing into the crystal structure, which also indicated that the polymer was strongly adsorbed on the surface of the graphene platelets. Enhancement of thermal stability of the polymer with increasing graphene content further confirmed the physical interaction of polymer and graphene surface. Irrespective of the graphene content in the hybrids, the morphology of the graphene platelets altered due to uniform coverage on the surface with PANI. PVB coatings with PANI/40r-GO and PANI/50r-GO exhibited 52% and 58% higher corrosion protection efficiency respectively as compared to PVB coating without pigment, which was also confirmed optically. The coatings were also equally effective when immersed in the salt solution at 40 °C. The conductivity of the coatings decreased as the amount of reduced graphene was increased probably due to decrease in the degree of doping. Thus, due to synergistic combinations of factors like tortuous path for the corrosive media in the presence of graphene platelets as well as formation of PANI/graphene 'Schottky barrier' at interface leading to the depletion of electrons corrosion half-cell reactions could be slowed down. The PANI/r-GO nanoparticle pigments represent functional materials capable to enhance the performance of corrosion protection coatings at low weight fractions.

References

1. Wahdan, M. H., and Gomma, G. K. (1997) Effect of copper cation on electrochemical behavior of steel in presence of imidazole in acid medium. *Materials Chemistry and Physics*, **47**, 176-183.
2. Denny, A. J. (1996) *Principles and Prevention of Corrosion*, Prentice-Hall Inc., USA.
3. Siegell, S. D., and Gaden, E. L. (1962) Automatic control of dissolved oxygen levels in fermentations. *Biotechnology and Bioengineering*, **4**, 345-356.
4. Hayyan, M., Sameh, S. A., Hayyan, A., and AlNashef, I. M. (2012) Utilizing of sodium nitrite as inhibitor for protection of carbon steel in salt solution. *International Journal of Electrochemical Science*, **7**, 6941-6950.
5. Shao, H. B., Wang, J. M., Zhang, Z., Zhang, J. Q., and Cao, C. N. (2003) The cooperative effect of calcium ions and tartrate ions on the corrosion inhibition of pure aluminum in an alkaline solution. *Materials Chemistry and Physics*, **77**, 305-309.
6. Kalendova, A., Sapurina, I., Stejskal, J., and Vesely, D. (2008) Anticorrosion properties of polyaniline-coated pigments in organic coatings. *Corrosion Science*, **50**, 3549-3560.

7. Zhu, Y. D., Allen, G. C., Adams, J. M., Gittins, D. I., Hooper, J. J., and Skuse, D. R. (2013) Barrier properties of latex/kaolin coatings. *Polymer Chemistry*, **4**, 4386-4395.
8. Behler, K. D., Stravato, A., Mochalin, V., Korneva, G., Yushin, G., and Gogotsi, Y. (2009) Nanodiamond-polymer composite fibers and coatings. *ACS Nano*, **3**, 363-369.
9. Sun, J., Gerberich, W. W., and Francis, L. F. (2003) Electrical and optical properties of ceramic–polymer nanocomposite coatings. *Journal of Polymer Science, Part B: Polymer Physics*, **41**, 1744-1761.
10. Mittal, V., Chaudhry, A. U., and Khan, M. I. (2011) Comparison of anti-corrosion performance of polyaniline modified ferrites. *Journal of Dispersion Science and Technology*, **33**, 1452-1457.
11. Chaudhry, A. U., Bhola, R., Mittal, V., and Mishra, B. (2014) $Ni_{0.5}Zn_{0.5}Fe_2O_4$ as a potential corrosion inhibitor for API 5L X80 steel in acidic environment. *International Journal of Electrochemical Science*, **9**, 4478-4492.
12. Saji, V. S. (2010) A review on recent patents in corrosion inhibitors. *Recent Patents in Corrosion Science*, **2**, 6-12.
13. Akhtar, T., Luckachan, G. E., and Mittal, V. (2018) Eco-friendly polylactic acid-silica coatings for corrosion protection of carbon steel. In: *Polymer Nanomaterials for Specialty Applications*, Mittal, V. (ed.), Central West Publishing, Australia, pp. 289-317.
14. Bagherzadeh, M. R., and Mahdavi, F. (2007) Preparation of epoxy–clay nanocomposite and investigation on its anti-corrosive behavior in epoxy coating. *Progress in Organic Coatings*, **60**, 117-20.
15. Mahulikar, P. P., Jadhav, R. S., and Hundiwale, D. G. (2011) Performance of polyaniline/TiO_2 nanocomposites in epoxy for corrosion resistant coatings. *Iran Polymer Journal*, **20**, 367-376.
16. Chang, K.-C., Ji, W.-F., Lai, M.-C., Hsiao, Y.-R., Hsu, C.-H., Chuang, T.-L., Wei, Y., Yeh, J.-M., and Liu W.-R. (2014) Synergistic effects of hydrophobicity and gas barrier properties on the anticorrosion property of PMMA nanocomposite coatings embedded with graphene nanosheets. *Polymer Chemistry*, **5**, 1049-1056.
17. Santos, C. M., Milagros Cui, K., Ahmed, F., Tria, M. C. R., Vergara, R. A. M. V., de Leon, A. C., Advincula, R. C., and Rodrigues, D. F. (2012) Bactericidal and anticorrosion properties in PVK/MWNT nanocomposite coatings on stainless steel. *Macromolecular Materials and Engineering*, **297**, 807-813.
18. Kuilla, T., Bhadra, S., Yao, D., Kim, N. H., Bose, S., and Lee, J. H. (2010) Recent advances in graphene based polymer composites. *Progress in Polymer Science*, **35**, 1350-1375.
19. Kim, J., Cote, L. J., Kim, F., Yuan, W., Shull, K. R., and Huang, J. (2010) Graphene oxide sheets at interfaces. *Journal of the American Chemical Society*, **132**, 8180-8186.
20. Akhtar, T., Luckachan, G. E., and Mittal, V. (2018) Inhibition of corr-

osion by polyetherimide-graphene nanocomposite coatings for long term protection of carbon steel. In: *Polymer Nanomaterials for Specialty Applications*, Mittal, V. (ed.), Central West Publishing, Australia, pp. 263-288.
21. Khan, M., Chaudhry, A. U., Hashim, S., Zahoor, M., and Iqbal, M. (2010) Recent developments in intrinsically conductive polymer coatings for corrosion protection. *Chemical Engineering Research Bulletin*, **14**, 73-86.
22. Spinks, G. M., Dominis, A. J., Wallace, G. G., and Tallman, D. E. (2002) Electroactive conducting polymers for corrosion control. *Journal of Solid State Electronics*, **6**, 85-100.
23. Tallman, D. E., Springs, G., Dominis, A., and Wallace, G. G. (2002) Electroactive conducting polymers for corrosion control, Part 1. General introduction and a review of non-ferrous metals. *Journal of Solid State Electronics*, **6**(2), 73-84.
24. Wessling, B. (1999) Scientific engineering of anti-corrosion coating systems based on organic metals (polyaniline). *Journal of Corrosion Science & Engineering*, **1**, paper 15.
25. Huh, J. H., Oh, E. J., and Cho, J. H. (2003) Investigation of corrosion protection of iron by polyaniline blend coatings. *Synthetic Metals*, **137**, 965-966.
26. Armelin, E., Aleman, C., and Iribarren, J. I. (2009) Anticorrosion performances of epoxy coatings modified with polyaniline: A comparison between the emeraldine base and salt forms. *Progress in Organic Coatings*, **65**(1), 88-93.
27. Khan, M., Chaudhry, A., Hashim, S., and Iqbal, M. (2010) Investigation of corrosion-protective performance of polyaniline covered inorganic pigments. *Nucleus*, **47**, 287-293.
28. Sathiyanarayanan, S., Azim, S. S., and Venkatachari, G. (2007) A new corrosion protection coating with polyaniline-TiO_2 composite for steel. *Electrochimica Acta*, **52**, 2068-2074.
29. Sathiyanarayanan, S., Azim, S. S., and Venkatachari, G. (2007) Preparation of polyaniline-Fe_2O_3 composite and its anticorrosion performance. *Synthetic Metals*, **157**, 751-757.
30. Brodinova, J., Stejskalb, J., and Kalendova, A. (2007) Investigation of ferrites properties with polyaniline layer in anticorrosive coatings. *Journal of Solid State Electronics*, **68**, 1091-1095.
31. Wu, K. H., Chao, C. M., Liu, C. H., and Chang, T. C. (2007) Characterization and corrosion resistance of organically modified silicate-NiZn ferrite/polyaniline hybrid coatings on aluminum alloys. *Corrosion Science*, **49**, 3001-3014.
32. Mittal, V., Chaudhry, A. U., and Matsko, N. B. (2016) Organic functionalization of thermally reduced graphene oxide nanoplatelets by adsorption: structural and morphological characterization. *Philosophical Magazine*, **96**(20), 2143-2160.

33. Hummers, W. S., and Offeman, R. E. (1958) Preparation of graphitic oxide. *Journal of the American Chemical Society*, **80**, 1339.
34. Mittal, V., Chaudhry, A. U., and Matsko, N. B. (2014) "True" biocomposites with biopolyesters and date seed powder: Mechanical, thermal, and degradation properties. *Journal of Applied Polymer Science*, **131**(19), doi: 10.1002/app.40816.
35. Radhakrishnan, S., Siju, C. R., Mahanta, D., Patil, S., and Madras, G. (2009) Conducting polyaniline-nano-TiO_2 composites for smart corrosion resistant coatings. *Electrochimica Acta*, **54**, 1249-1254.
36. Harish, C., Sreeharsha, V. S., Santhosh, C., Ramachandran, R., Saranya, M., Vanchinathan, T. M., Govardhan, K., and Grace A. N. (2013) Synthesis of polyaniline/graphene nanocomposites and its optical, electrical and electrochemical properties. *Advanced Science, Engineering and Medicine*, **5**(2), 140-148.
37. Fu, C., Zhao, G., Zhang, H., and Li, S. (2013) Evaluation and characterization of reduced graphene oxide nanosheets as anode materials for lithium-ion batteries. *International Journal of Electrochemical Science*, **8**, 6269-6280.
38. Chaudhari, H. K., and Kelkar, D. S. (1996) X-ray diffraction study of doped polyaniline. *Journal of Applied Polymer Science*, **62**, 15-18.
39. Zhang, K., Zhang, L. L., Zhao, X. S., and Wu, J. (2010) Graphene/polyaniline nanofiber composites as supercapacitor electrodes. *Chemistry of Materials*, **22**, 1392-1401.
40. Lagrenee, M., Mernari, B., Bouanis, M., Traisnel, M., and Bentiss, F. (2002) Study of the mechanism and inhibiting efficiency of 3,5-bis(4-methylthiophenyl)-4H-1,2,4-triazole on mild steel corrosion in acidic media. *Corrosion Science*, **44**(3), 573-588.
41. Liu, Y., Deng, R., Wang, Z., and Liu, H. (2012) Carboxyl-functionalized graphene oxide-polyaniline composite as a promising supercapacitor material. *Journal of Materials Chemistry*, **22**, 13619-13624.
42. Das, T. K., and Prusty, S. (2013) Graphene-based polymer composites and their applications. *Polymer-Plastics Technology and Engineering*, **52**, 319-331.
43. Sun, W., Wang, L., Wu, T., Pan, Y., and Liu, G. (2014) Synthesis of low-electrical-conductivity graphene/pernigraniline composites and their application in corrosion protection. *Carbon*, **79**, 605-614.
44. Yu, Y.-H., Lin, Y.-Y., Lin, C.-H., Chan, C.-C., and Huang, Y.-C. (2014) High-performance polystyrene/graphene-based nanocomposites with excellent anti-corrosion properties. *Polymer Chemistry*, **5**, 535-550.
45. Lu, W.-K., Elsenbaumer, R. L., and Wessling, B. (1995) Corrosion protection of mild steel by coatings containing polyaniline. *Synthetic Metals*, **71**, 2163-2166.

Chapter 4

Conductive Polymer Composites for Electromagnetic Interference Shielding Applications

A. Pangajam and K. Dinakaran*

Department of Chemistry, Thiruvalluvar University, Vellore 632115, India
**Corresponding author*: kdinakaran.tvu@gmail.com

4.1 Introduction

The electromagnetic radiations emitted by radars, mobile phones and various electronic appliances are found to interfere with other electronic devices, which may degrade the device performance and even adversely affect the human health [1-3]. The interference of electromagnetic frequencies (EMF) with electronic appliances, medical equipment and living beings is called electromagnetic interference (EMI). EM shielding is the practice of encapsulating electronic devices, medical equipment and cables with conductive or magnetic materials to guard against incoming or outgoing emissions of EMF. To enhance the performance and security of electronic systems, effective EM shielding is advised. Conventionally, physical isolation of components from each other within a device has been considered to avoid electromagnetic interference. The metallic mesh is commonly used as a shield to protect the components from affecting each other inside a particular device.

Following are the different materials and techniques are used for EM shielding:

- Electric wires may be encapsulated by a metallic mesh foil to shield EMI from the cased wires.
- Audio speakers often incorporated with inner metallic casing to prevent EMI produced by the drivers, so as to safeguard the performance of TVs and other electronics
- Special grade conductive paints may be used to block the EMF escaping from the networks to prevent wireless attacks and other hazards

Conducting Polymer Composites, edited by Vikas Mittal

4.2 Sources of EMI

EMI sources can be divided into two major categories described below.

4.2.1 Ambient EMI

- Arc welders
- Brush motors
- Computer circuits
- Solar magnetic storms
- Lightning
- Earth's magnetic field flux

4.2.2 Power Quality Problems

- Brownouts and blackouts
- Electrical fast transitions (EFT)
- Electrical noise on the mains power line
- Power line faults
- Voltage sags and spikes

Railroad and mass transit systems may experience EMI because of application-specific factors such as:

- Emissions from train control and propulsion systems
- High voltage contact switching
- Third rail shoes
- Train signaling and control systems

Medical equipment are susceptible to EMI too. Application-specific sources include:

- Electrical and electronic equipment in surgical units
- Life support devices such as ventilators and infusion pumps
- Patient telemetry and assistance equipment
- X-ray machines for diagnostics and therapeutics

In addition to IEMI (colloquially called "electronic warfare"), military assets and critical infrastructure face EMI threats such as:

- High-altitude nuclear electromagnetic pulse (HNEMP)
- High-power microwave weapons
- E-bombs
- EMP cannons

Electromagnetic interference shielding is an undesirable and uncontrolled off-shoot of explosive growth of electronics and widespread use of transient power sources. Among various materials employed for EMI shielding, conducting polymers (CPs) have drawn significant interest of researchers for more than 30 years because of their economic importance, superior stability, lightweight, superior workability, resistance to corrosion and satisfactory electrical conductivity. Conducting polymers have been employed in variety of applications in the industrial, medical and scientific fields. In addition, the nanoparticles embedded conducting polymer materials possess unique combination of electrical, thermal and dielectric properties, making them suitable for wide ranging applications for EMI shielding. The following are the conducting polymers commonly studied for EMI shielding:

- polyacetylene (PA)
- polyaniline (PANI)
- polypyrrole (PPy)
- polythiophene (PTh)
- polyphenylenevinylene (PPV)
- polyfluorene (PF)
- polycarbazole (PCZ)
- polyvinylidiene fluoride (PVDF)

4.3 EMI Shielding using Conducting Polymers based Materials

4.3.1 Polyacetylene

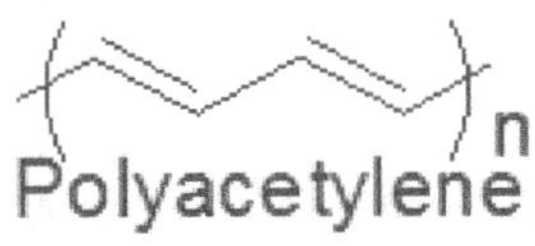

Trans-polyacetylene (trans-PA) is an organic polymer with the repeating unit $(C_2H_2)_n$. The invention of highly conducting PA led to a rapid rise in research directed towards the synthesis of new CPs. The high electrical conductivity of such polymers leads to their extensive use in microelectronics. The molecular nature of these polymers produces electronic movement around the individual macromolecules. The nature of procedures creating high conductivity in these polymers is different from inorganic semiconductors. The elevated conductivities depend upon doping and are related to the creation of self-localized excitons such as solutions, polar and bipolar. In CPs having degenerate ground state, for example trans-PA, charged solutions are the main charge carriers [1].

4.3.2 Polyaniline

Polyaniline

In the family of intrinsically conducting polymers (ICPs), polyaniline represents an effective choice for EMI shielding applications due to its corrosion protecting nature, lightweight, easy processing and tunable conductivity as compared to metals. Furthermore, PANI is especially promising for commercial applications because of environmental stability, facile synthesis, unique reversible proton doping, high electrical conductivity, ease of forming composites with metal nanoparticles and low cost as compared to other conducting polymers. The electromagnetic interference shielding effectiveness of conducting PANI-acrylonitrile butadiene styrene (ABS) composites increases with enhancing the loading levels of the conducting polymer doped with hybrid dopants. The lower amount doped PANI in the conducting composites can be effectively used for the dissipation of electrostatic charge. However, with higher loadings, a shielding effectiveness of 60 dB at 101 GHz has been reported, which makes the PANI conducting composites a potential EMI shielding material [2]. The polyaniline coated films (based on dodecylbenzenesulfonic acid) on a polyethylene terephthalate substrate have also been investigated for EMI shielding by two different methods: (i) flanged coaxial test and (ii) test bed [3]. In addition, shielding efficiencies for a number of PANI blends containing a variety of conductive fillers have been studied in both near field and far field shielding [4]. For far field shielding, efficiencies as high as 70 dB at 1GHz were reported. Polyaniline-coated bagasse fiber (BF) composites exhibited higher dielectric values corresponding to their dielectric constant, loss tangent and Cole-Cole plots [5]. The composite exhibited higher electrical conductivity, significantly higher than the unreinforced polyaniline. The total absorption-dominated EMI shielding effectiveness of 28.8 dB for the sample having a thickness of 0.4 mm indicated the usefulness of the composite for EMI shielding applications. Overall, the comparison of electrical and EMI SE of reinforced and unreinforced polyaniline revealed that the enhancement of EMI SE values in the composite was due to the improved electrical conductivity and the core-shell architecture [5]. The incorporation of silver was also observed to enhance the EMI SE values [6]. For instance, the electromagnetic interference of hydrochloric acid and silver (Ag) powder doped polyaniline was ~46 dB at room temperature, which revealed that HCl-Ag-PANI can be commercially applied to protect electronic equipment from the electromagnetic radiation interference. The authors further explained the theory/model for the calculation of the electromagnetic radiation interference shielding effectiveness of the multilayers of conducting polymers in terms of boundary conditions of electromagnetic plane waves. The frequency dependence of SE of monolayer polymer films was found to be constant at low frequency with an oscillating behavior at high frequency. The SE of multilayer polymer films, consisting of the high H and low L conducting layers, increased with increasing dc conductivity. The authors observed that the effect of coherent multiple reflections of EM radiation decreased when the thickness of the polymer layer increased, hence, the electrical conductivity and EMI shielding effectiveness increased [6].

Highly conducting polyaniline-multi-walled carbon nanotubes (MWCNT) nanocomposites were reported by Saini *et al.* [7]. The electrical conductivity of PANI-MWCNT composite was observed to be 19.7 S cm^{-1}, which was higher than MWCNT (19.1 S cm^{-1}) or PANI (2.0 S cm^{-1}). The absorption dominated total shielding effectiveness (SE) of 27.5 to 39.2 dB for these composites indicated their usefulness for shielding microwave in the Ku-band (12.4-18.0 GHz). The polyaniline coated MWCNTs composites were also reported to be suitable for use as hybrid conductive fillers in thermoplastics. In another study, polyaniline doped with β-naphthalene sulphonic acid (β-NSA) and reinforced with different carbon fillers such as carbon fibers (CF), MWCNTs and reduced graphene oxide (rGO) exhibited effective improvement in thermal stability [8]. The highest EMI SE obtained for polyaniline composites reinforced with CF, MWCNTs and rGO were 31, 37 and 39 dB respectively in the frequency range of 8.2-12.4 GHz. Polyaniline nanocomposite containing reduced grapheme oxide and iron oxide nanoparticles have also been studied by Mathew *et al.* [9]. The nanocomposite samples with 10% rGO showed the maximum electrical conductivity and the one with 10% RGO and 10% Fe_3O_4 exhibited the maximum EMI shielding effectiveness of 7.5 dB for a 1 mm thick sample.

The multiphase approach of polyaniline based nanocomposites loaded with MWCNTs/grapheme leads to an enhancement in the electric conductivity due to graphene acting as a bridge between PANI and MWCNTs [10]. It has been reported that EMI-SE increases with increasing the MWCNTs content from 1 to 10 wt% in the multiphase nanocomposites. The maximum value of total EMISE was observed to be -98 dB for the nanocomposite with 10 wt% of MWCNTs content. The high value of EMISE was dominated by the absorption phenomenon which was due to the collective effect of an increase in the space charge polarization and a decrease in the carrier mobility. The decrease in the carrier mobility had a positive effect on the shore hardness value due to the strong interaction between the reinforcing constituents in the multiphase nanocomposites [10]. Polyaniline nanofiber graphite composites with a thickness of 1 mm exhibited excellent electromagnetic interference shielding of above 80 dB in the frequency range of 8.2-18 GHz [11]. EMI shielding polyaniline nanofibers having thickness of 0.1 mm and their composites were developed by an *in-situ* polymerization route. The developed functional fabrics with 0.1 mm thickness exhibited EMI shielding effectiveness in the range of 11-15 dB in the 8.2-18 GHz frequency range. In another work, PANI dispersed aramid fiber was reported to have higher EMISE values, where highly conductive PANI formed conductive network within aramid fibers and shielded EM radiation [12]. The nanocomposite film having a thickness of several micrometers displayed a shielding effectiveness as high as 30 dB, a mechanical strength of 179 MPa as well as good thermal stability. Similarly, nanostructured polyaniline and polyaniline-clay nanocomposite (PANICN) were blended with ethylene vinyl acetate (EVA) matrix [13]. The presence of exfoliated nano-clay and interaction between the conductive filler-host matrix in conductive films containing PANICNs was manifested from the measurement of rheological properties. Films with conductive filler (~15% loading) exhibited a shielding effectiveness of ~40-80 dB at 8 GHz frequency.

Kumar *et al.* [14] developed self-supported lightweight thin sheets of polyaniline carbon fiber (PANI-CF) composite by compression molding using phenolic novolac resin as a binder, for use in electromagnetic interference shielding. Polyaniline carbon fiber-novolac (PCN) sheets having thickness of ~0.6 mm were subsequently prepared by hot pressing and thermal annealing, using different amount of novolac resin in order to optimize the thermal, mechanical, electrical and shielding properties. The sheets exhibited flexural strength of 33.28-50.48 MPa, electrical conductivity of 0.39-0.98 S/cm and electromagnetic shielding effectiveness of 35.0 dB at 10.6 GHz frequency.

Core-shell nanocomposites composed of $Mn_{0.5}Zn_{0.5}Fe_2O_4$ nanocrystals and polyaniline were synthesized using *in-situ* chemical oxidative polymerization of aniline by Dar *et al.* [15]. The inorganic nanocrystals were in the micellar solution of dodecyl benzene sulfonic acid, which acted both as a surfactant and dopant. The complex permittivity, permeability and microwave absorption performance of the composites was analyzed in the 8.2-12.2 GHz (X-band) frequency range. The nanocomposites were observed to exhibit high shielding effectiveness (31.2 dB), as compared to pure PANI polymer, which was attributed to dielectric loss, magnetic permeability and volume fraction of $Mn_{0.5}Zn_{0.5}Fe_2O_4$ nanocrystals. Tetragonal $BaTiO_3$ (TBT) has also been loaded in PANI to obtain higher EMI shielding materials [16]. The Ku-Band (12.4-18 GHz) network analysis of TBT-PANI composites exhibited absorption dominated EMI shielding response with a total shielding effectiveness value of -71.5 dB.

Polyaniline doped with paratoluenesulphonic acid as well as cobalt ferrite ($CoFe_2O_4$) was observed to exhibit significant microwave-absorbing properties [17]. The reflection loss (dB), complex relative permittivity ($\varepsilon_r'-j\varepsilon_r''$) and complex relative permeability ($\mu_r'-j\mu_r''$) were measured in the X-band (8.2-12.4 GHz) region. The composites showed a maximum reflection loss of -28.4 dB at 8.1 GHz and -9.6 dB at 11.2 GHz. Qiu *et al.* [18] reported that PANI doped with hydrochloric acid and camphorsulfonic acid (CSA) exhibited "holothurian-like" morphology, while PANI doped with phosphoric acid demonstrated nanofiber like structure. Among these materials, PANI-CSA exhibited the highest electrical conductivity (1.28 ± 0.17 S cm^{-1}) and effective electromagnetic interference shielding effectiveness of 20.7 dB for the sample having a thickness of 0.35 mm. Table 4.1 also summarizes these results.

Table 4.1 Summary of EMI shielding of PANI based composites

Conducting Polymer	EMI shielding	Ref.
PANI	60 dB at 101 GHz	2
Intrinsically/PANI	70 dB at 1GHz	4
BF/PANI	28.8 dB at a thickness of 0.4 mm	5
Hydrochloric acid doped PANI containing Ag	~46~46 dB at room temperature	6
PANI/MWCNT	-27.5 to -39.2 dB	7
PANI/MWCNT	37 dB	8
PANI/CF	31 dB	8
PANI/rGO	39 dB	8
PANI/10% weight of MWCNTs in 10%	-98 dB	10
ferrite-grafted polyaniline	30 MHz to 1 GHz	17
PANI and hierarchical layering structure of aramid	30 dB	12
polyaniline (PANI) and polyaniline-clay nanocomposite (PANICN)	80 dB at 8 GHz	13
polyaniline-carbon fiber	35.0 dB at 10.6 GHz	14
PANI/$Mn_{0.5}Zn_{0.5}Fe_2O_4$	31.2 dB	15
polyaniline (PANI)-tetragonal $BaTiO_3$ (TBT)	-71.5 dB	16
In situ polymerization of polyaniline nanofiber graphite	11-15 dB in the 8.2-18 GHz	11
$CoFe_2O_4$/PANI-PTSA	-28.4 dB at 8.1 GHz	17
Polyaniline/Fe_3O_4-rGO	7.5 dB for a 1 mm	9
PANI-CSA	20.7 dB thickness of only 0.35 mm	18

4.3.3 Polypyrrole

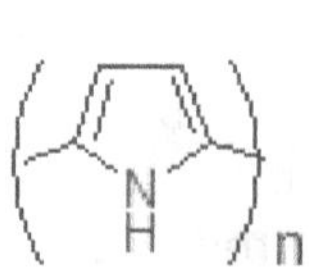

Polypyrrole

PPy is one of the most commonly used conducting polymers due to its excellent properties such as good environmental stability, excellent electrical conductivity and electromagnetic interference shielding effectiveness. Pyrrole monomer is commercially available, easily oxidized and water-soluble. Over the past three decades, polypyrrole has attracted interest for a number of applications including supercapacitors for energy storage and secondary batteries. For instance, poly(ethylene terephthalate)-polypyrrole composites exhibited a specific volume resistivity of as low as 0.2 Ω cm and EMISE of about 36 dB over a wide frequency range of up to 1.5 GHz [19]. Azadmanjiri *et al.* [20] also reported the development of magnetic nanocomposites by *in-situ* oxidative polymerization to encapsulate iron oxide nanoparticles with PPy, which were subsequently blended in an epoxy resin. In another study, electromagnetic interference shielding materials based on conductive PPy coated silver-palladium (AgPd) metal composite nanoparticles have also been reported [21]. The EMI shielding efficiency of PPy/fabric and AgPd/fabric complexes was in the range of 8~80 dB depending on the conductivity and additional Ag deposited by vacuum evaporation. The highest EMI shielding efficiency of PPy/fabric complexes with Ag was ~80 dB, indicating potential use for military applications.

Nano-stick shaped polypyrrole composites decorated with γ-Fe_2O_3 were reported to have a superparamagnetic nature and high electrical conductivity, which exhibited a shielding effectiveness up to 25.5 dB [22]. Epoxy nanocomposites reinforced with polypyrrole functionalized nano-magnetite (Fe_3O_4-PPy) were reported by Guo *et al.* [23]. Thr epoxy/(30.0 wt%)Fe_3O_4-PPy nanocomposite exhibited a minimum reflection loss (RL) value of -35.7 dB, which was much lower than epoxy/(7.5 wt%)PPy nanocomposites with a minimum RL value of -11.0 dB or epoxy/(30.0 wt%)Fe_3O_4 with a minimum RL value of -17.8 dB at the same thickness. The increased interface area, eddy current loss and anisotropic energy were reported to be essentially important to achieve higher reflection loss and broader absorption bandwidth for epoxy/(30.0 wt%)Fe_3O_4-PPy nanocomposites. In another study, pyrrole was polymerized both chemically and electrochemically on nylon 6 woven fabrics, to produce polypyrrole coated nylon 6 composite fabrics with a high electric conductivity [24]. The temperature dependence of the conductivity of the composites was verified over four heating and cooling cycles. The conductivity was affected by the interaction between the thermal stability of the dopant and the rearrangement of the PPy main chain. The electromagnetic interference shielding efficiency values of the composites were in the range 5-40 dB and depended on the conductivity and layer array sequence of the developed conductive fabric.

Core-shell micro/nanostructured and electromagnetic functionalized polypyrrole composites were reported by Li *et al.* [25]. The composites were generated by a self-assembly process in the presence of p-toluenesulfonate acid dopant. The generated core-shell micro/nanostructured morphology resulted in controllable electromagnetic properties of the composites. In another study, electromagnetic shielding efficiency of polypyrrole/polyester textile composites was also studied by Tunakova *et al.* [26]. The composite fabrics exhibited a EMISE value of 12 dB at 1.5 GHz. Multilayer sandwich heterostructural $Ti_3C_2T_x$ MXenes decorated with polypyrrole were reported by Tong *et al.* [27]. The composites consisting of 25 wt% $Ti_3C_2T_x$/PPy hybrids in a paraffin matrix exhibited a minimum reflection loss of -49.2 dB (~99.99% absorption) at the thickness of 3.2 mm and a maximum effective absorption bandwidth of 4.9 GHz (12.4-17.3 GHz) corresponding to an absorber thickness of 2.0 mm. Additionally, a broad effective absorption band width of 13.7 GHz (4.3-18.0 GHz) was reached by adjusting the thickness from 1.5 to 5.0 mm. Furthermore, the highest effective absorption bandwidth of 5.7 GHz was obtained when the mass fraction was 15 wt%.

Conductive polypyrrole/silver/poly(ethylene terephthalate) composite was reported by Erdogan *et al.* [28]. The authors studied the effect of $AgNO_3$/pyrrole mole ratio and concentrations of the reactants on the polypyrrole/silver content and surface resistivity of the composite. The electromagnetic shielding effectiveness as well as absorption loss, reflection loss with the values of absorbance and reflectance, showing their contributions to the electromagnetic shielding effectiveness value of the composite were studied within the frequency range of 15-3000 MHz. An absorption dominant shielding was obtained with the value of 13-15.5 dB for double coated composite. Bhardwaj *et al.* [29] reported polypyrrole composites incorporated amine functionalized multi-walled carbon nanotubes, graphene and hybrid carbon assemblage by *in-situ* emulsion polymerization of pyrrole with sodium lauryl sulphate as dopant. The fillers provided strong polarization due to the generation of a solid-state charge-transfer complex. The polarization resulted in appropriate impedance matching and dipole interaction, which inturn led to high microwave absorption. Electromagnetic radiation shielding measurement of the polypyrrole composites in Ku-band exhibited a shielding effectiveness from -28.8 dB to -79.9 dB, thus, confirming their potential for stealth technology.

Polypyrrole based composite encapsulated with barium strontium titanate (BST), reduced graphene oxide (RGO) and Fe_3O_4 nanoparticles was reported by Sambyal *et al.* [30]. The shielding effectiveness of 48 dB in the frequency range of 8.2-12.4 GHz (X-band) was observed. Moreover, the presence of dielectric and magnetic fillers increased the thermal and chemical stability of the composite material. The obtained shielding effectiveness value was above the recommended limit of 30-40 dB required for the commercial applications, therefore, these composite material exhibited strong potential as effective shield against EM pollution. Gahlout and Chaudhary [31] reported that the conductivity and shielding effectiveness of PPy was dependent on the concentration of oxidant and it was highest when the molar ratio of pyrrole : oxidant : dopant was 1:2:1. In a diffferent set of experiments, dopant concentration (0.6, 1, 1.4 and 1.8 mol per mole of pyrrole) was varied, while keeping the concentration of pyrrole : oxidant as 1:2, in order to determine the optimum concentration of dopant. The polymer prepared using pyrrole : oxidant : dopant ratio as 1:2:1 was observed to exhibit the highest conductivity as well as shielding effectiveness. Using these optimized conditions, polypyrrole was synthesized using different dopants (i.e. SLS, SDBS, LiSiPA and NDSA). Polypyrrole synthesized using optimized conditions and SLS as dopant had highest conductivity (42 S/cm) and electromagnetic shielding effectiveness of -126 dB in the X-band 8.2-12.4 GHz.

Textile based flexible shields were prepared by *in-situ* chemical polymerization pyrrole and their effectiveness by Maity and Chatterjee [31]. 3^3 Box-Behnken design was used to investigate the influence of manufacturing process parameters of neddle-punched non-woven fabric such as fabric gsm, punching density and depth of needle penetration on the polypyrrole add-on, surface resistivity and electro-magnetic shielding effectiveness. Surface resistivity as low as 18 Ω (with 16.28% polypyrrole add-on) was obtained for a typical composite fabric. The shielding effectiveness of the fabric was observed to be 20.07 dB at 4 GHz frequency.

Zhao *et al.* [33] reported multilayer structured cuprammonium fabric/polypyrrole/copper (CF/PPy/Cu) composite, for potential used as electromagnetic interference (EMI) shield due to its characteristics of both EM wave-absorption and EM wave-reflection. The results indicated that PPy and CF interacted by hydrogen bonding. Afterwards, Cu film was constructed on the as-made PPy coated CF (CF/PPy) by an electroless copper plating method. The shielding effectiveness (SE) values of copper plated CF (CF/Cu) and CF/PPy/Cu were recorded (30-1000 MHz). The resultant CF/PPy/Cu composite had a higher EMI SE (30.3-50.4 dB) with sheet resistance of 85.8 mΩ/sq.

In another study, highly conductive polypyrrole (PPy)/multi-walled carbon nanotube (MWCNT) composites were developed by *in-situ* polymerization of pyrrole using 5-sulfoisophthalic acid monolithium salt [lithiosulfoisophthalic acid (LiSiPA)] as dopant and ferric chloride as oxidant. The electrical conductivity of PPy increased with increasing amount of MWCNTs, and the maximum electrical conductivity was observed to be 52 S/cm for 5 wt% MWCNTs. Pure PPy, prepared under similar conditions, had a conductivity of 25 S/cm. Electromagnetic interference shielding effectiveness also showed a similar trend and average EMI shielding value of -108 dB (3 mm) was observed for the sample having 5 wt% MWCNTs in the frequency range of 8.2-12.4 GHz (X-band).

Overall, polypyrrole based composites have the advantage of superior electrical properties, lightweight and versatility in fabrication, thus, can be employed for electrostatic charge protection and EMI shielding.

4.3.4 Polythiophene

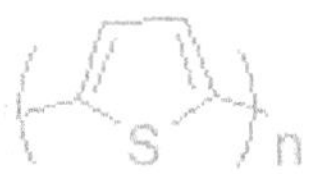

Polythiophene

PTh is an important CP for manufacturing environmentally and thermally stable materials. It is useful for electrical supercapacitors, non-linear optics, PLEDs, eletrochromics, photoresists, antistatic coatings, sensors, batteries, electromagnet shielding materials, solar cells, memory devices, transistors and imaging materials. Conjugated PTh materials in the non-doped and doped conducting state have an array of potential applications in the microelectronics industry. Thus, these can be used as effective discharge layers as well as conducting resists in electron beam lithography, along with metallization (electrolytic and electroless) of plated through-holes for printed circuit board technology, electrostatic discharge protection for packages and housings of electronic equipment, corrosion protection for metals, electromagnetic interference shielding, etc. [35].

Nanocomposite of polythiophene (PTh) with the photoadduct of sodium pentacyanonitrosylferrate(II) dihydrate and hexamine was synthesized by *in-situ* polymerization technique [36]. The nanocomposites had higher thermal stability compared to pure PTh. The nanocomposite exhibited high value of dielectric constant and conductivity useful for EMI shielding applications. The composites of poly(3,4-ethylenedioxythiophene)/multi-walled carbon nanotubes was prepared by Farukh *et al.* [37]. The composites exhibited EMISE of 58 dB in the Ku-band region. The microscopy analysis confirmed the complete encapsulation of multiwalled carbon nanotubes by poly(3,4-ethylenedioxthiophene), thus, resulting in strong interfacial, electronic and space charge polarizations. The authors observed that the EMI shielding of the composites were due to shielding effectiveness by absorption rather than reflection. In another study, Taka *et al.* [38] also reported EMI shielding performance of poly(3-octyl thiophene) (P3OT) blends with polystyrene (PS), polyvinyl chloride (PVC) and ethyl-vinyl acetate (EVA).

Polythiophene (PTh) nanofibers coated on $MnFe_2O_4/Fe_3O_4$ core-shell nanoparticles were generated by co-precipitation and *in-situ* polymerization [39]. $MnFe_2O_4/Fe_3O_4$ nanoparticles had a core-shell structure, with $MnFe_2O_4$ as the magnetic core and Fe_3O_4 as the shell. The microwave absorption properties of the nanocomposites were investigated in the frequency range 8.0-12.0 GHz. The structural, electronic, dielectric, magnetic and electromagnetic interference (EMI) shielding properties of $PTH/Ni_{0.5}Zn_{0.5}Fe_{2-x}Ce_xO_4$ composites were also analyzed by Dar *et al.* [40]. A total shielding effectiveness ($SE_T = SE_A + SE_R$) up to 34 dB (≈99.9% attenuation) was recorded for $PTH/Ni_{0.5}Zn_{0.5}Fe_{2-x}Ce_xO_4$ composites (x = 0.04) in a frequency range of 8.2-12.4 GHz (X-band), which surpassed the shielding criteria of SE_T> 30 dB for commercial purposes.

4.3.5 Polyvinylidiene Fluoride

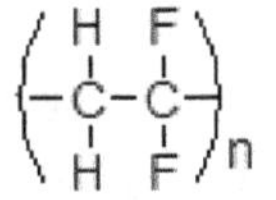

Polyvinylidienefluoride

Polyvinylidiene fluoride doped hexafluoropropylene (HFP), organic salt ($EMIMBF_4$), inorganic salt ($LiBF_4$), multi-walled carbon nanotubes (MWCNT) and magnetic (Fe_3O_4) nanoparticles based composites were used as electromagnetic shield materials [41]. The electrical conductivity value of as high as ~1.86 mS/cm was reported for MWCNT-PVdF-HFP- $EMIMBF_4$. This was due to the combined ionic conductivity of the ionic liquid $EMIMBF_4$ and electron conduction by the MWCNTs network. The total EMISE in the K_u band (12.4-18 GHz) of ion-conducting system alone was found to be ~19 dB and the polymer composites having both ion and electron conduction showed ~46 dB. In another study, nitrogen doped CNTs were synthesized using metallic catalysts such as Fe, Ni and Co by a chemical vapor deposition method [42]. The prepared N-CNTs based on Co and Fe showed better dispersion in PVDF as compared to Ni. The superior electrical properties of N-CNTs/polymer nanocomposites were attributed to a combination of high synthesis yield, low nitrogen content, high aspect ratio and crystallinity of N-CNTs, combined with a good state of dispersion. The order of electrical conductivity and electromagnetic interference shielding (from high to low) observed was Co > Fe > Ni. Nanocomposites of functionalized graphene (f-G) and polyvinylidene fluoride with high electrical conductivity and EMI shielding were reported by Eswaraiah *et al.* [43]. The electrical conductivity was found to be proportional to the concentration of f-G in PVDF matrix. A dramatic enhancement was observed from 10^{-16} Sm^{-1} for PVDF to 10^{-4} $S\,m^{-1}$ for 0.5 wt% f-G embedded PVDF composite, resulting from the formation of conductive network in the polymer and due to the high-aspect ratio of f-G. The f-G nanomaterial in PVDF foam composite exhibited an EMI shielding effectiveness of ≈20 dB at X-band (8-12 GHz) region and 18 dB in broadband (1-8 GHz) region for 5 wt% content.

PVDF composites with hybrid nanostructures of MWCNTs and FeCo anchored covalent rGO were reported by Arief *et al.* [44]. Conducting network of MWCNTs and rGO imparted higher electrical conductivity in the blend. Polymer matrix showed high total EMISE of -41.2 dB at 12 GHz, for a 10 wt% filler content. Graphene decorated with graphene quantum dots (G-D-GQDs) in PVDF matrix were reported by Lakshmi and Tambe [45]. A total EMI shielding effectiveness (SE_T) of 31 dB at 8 GHz was observed. The main mechanism of high EMI shielding effectiveness was reflection and absorption of EM radiation. The high absorption of EM radiation was due to tunneling of electrons from GQDs. Further, the decoration of G-D-

GQDs with conducting Ag nanoparticles (G-D-GQDsAg) enhanced the SE_T value of PVDF/G-D-GQDs Ag nanocomposite to 43 dB at 8 GHz, due to increase in electrical conductivity of PVDF/G-D-GQDs Ag nanocomposite and enhanced dispersion of G-D-GQDsAg in PVDF matrix. The incorporation of G-D-GQDs and G-D-GQDsAg in PVDF matrix also increased the thermal stability and crystallinity of PVDF. The increase in thermal stability and crystallinity were also observed to be higher for PVDF/G-D-GQDsAg nanocomposite as compare to PVDF/G-D-GQDs. Thus, PVDF/G-D-GQDsAg nanocomposites having high SE_T value were confirmed to have the potential to shield 99.9% of electromagnetic radiation in X-band range, making them suitable for EMI shielding. High electromagnetic interference shielding effectiveness of about 47 dB was also exhibited by highly conducting and ordered PVDF/grapheme nanocomposite films of thickness around 20 μm, in the frequency range 8-12 GHz [46].

The electrical conductivity and electromagnetic interference shielding properties of PVDF/CNT/Ni-chain and the PVDF/GNP/Ni-chain composites were observed to increase by enhancing the Ni-chain filler content [47]. The electrical conductivity of the PVDF/CNT/10 wt%Ni-chain composite was lower than the PVDF/CNT/6 wt%Ni-chain composite. The total shielding of the PVDF/CNT/6 wt%Ni-chain and the PVDF/GNP/8 wt%Ni-chain composite films increased from 23.6 to 57.3 dB and from 22.7 to 55.8 dB, as their thicknesses were increased from 0.3 mm to 0.6 mm. Joseph et al. [48] also reported the effect of silver incorporation into PVDF-barium titanate composites for EMI shielding applications. EMI shielding effectiveness of about 26 dB was observed for the PVDF-20 vol% nano-$BaTiO_3$-10 vol% Ag composite of thickness 1.2 mm. Kumaran *et al.* [49] reported Ag nanoparticles incorporated graphite/PVDF composites. The conductivity values were observed to significantly enhance up to 2.5 times (2.70 S/cm at 1 MHz) as compared to neat PVDF. The incorporation of Ag significantly improved the dielectric constant and lowered the dielectric loss of PVDF matrix. The electromagnetic interference shielding of the Ag-graphite-PVDF composites was found to be 29.1 dB at 12.4 GHz for the sample composed of 5 wt% Ag and 10 wt% graphite in PVDF. The authors also studied the EMI shielding effectiveness and conductivities of a flexible conducting polymer composite film of PVDF incorporated with Au-MWCNTs [50]. The values of dielectric constant were observed to be 12.11 and 13.89 at 1 MHz upon the incorporation of 1 and 3 wt% MWCNT in PVDF. The impedance and

EMI shielding effectiveness studies revealed that the MWCNT-PVDF possessed enhanced conductivity of 1.12×10^{-4} S/cm at 1 MHz, minimum return loss of 4.4 dB and effective EMI shielding of 26.7 dB at 12 GHz for 3 wt% Au NPs in 3 wt% MWCNT/PVDF thin film.

The electrical conductivity and the electromagnetic interference shielding of the PVDF/carbon composites was reported to increase by enhancing the conductive carbon filler amounts [51]. Also, the EMI shielding properties of PVDF/CNT/graphene composites exhibited higher values as compared to PVDF/CNT and PVDF/graphene composites. The EMI shielding values of 22.41, 18.70, and 27.58 dB were observed for PVDF/5 wt%-CNT, PVDF/10 wt%-graphene and PVDF/CNT/graphene composite films, respectively. The composite PVDF/CNT/graphene films also exhibited a thickness dependent EMISE values. For instance, the value was increased from 21.90 to 36.46 dB, when the thickness was increased from 0.06 mm to 0.25 mm. In particular, the PVDF/carbon composite films, with a thickness of 0.1 mm, achieved the highest specific shielding values of 1310 dB cm^2/g for the PVDF/5 wt %-CNT composite and 1557 dB cm^2/g for the PVDF/CNT/graphene composite.

The nanocomposites of PVDF/nickel prepared by blending in hot-molding process have been studied for dielectric and electromagnetic shielding effectiveness [52]. The complex permittivity, permeability and shielding effectiveness were measured in X-band 8.2-12.4 GHz. In another study, polymer-metal composites of nanocrystalline iron (n-Fe) dispersed PVDF were reported [53]. The PVDF/MWCNTs nanocomposites having 7 wt% MWCNTs content exhibited higher electrical conductivities and EMISE, having values of 0.06 S cm^{-1} and 30.89 dB in the X-band frequency region, which also increased across the frequency range as the sample thickness was increased from 0.6 to 3.0 mm [54].

Bera *et al.* [55] prepared PVDF/PFC (ferrosoferric oxide (Fe_3O_4) decorated polyaniline/single-walled carbon nano-horn (SWCNH)) nanocomposites by using a simple solution blending process, thus, resulting in the formation of porous composite having conducting network. The nanocomposites exhibited high conductivity of $\sim 2.5 \times 10^{-2}$ S cm^{-1} at 2 mm thickness. The composites had greater dielectric loss and permeability property, along with a high EMISE value of -29.7 dB at low loading of PFC (10 wt%) and SWCNH (1 wt%) in extended Ku-band region. The excellent EMI shielding effectiveness was attributed to the porous structure, formation of continuous conductive network (polyaniline-SWCNH) and ferromagnetic Fe_3O_4 nanoparticles. In another study, CNTs were melt-mixed with polyvinylidene fluoride matrix at various loadings followed by compression molding [56]. The composites exhibited lower percolation threshold and higher EMISE of 20.3 dB over the X-band for 3.5 wt% CNT and 1.1 mm thickness. The polyvinylidene fluoride nanocomposites with nickel oxide (NiO) and $BaTiO_3$ were also reported to have a high EMI shielding effectiveness of -11.5 dB by absorption in the Ku-band frequency range (8-12 Hz) [57].

Core-shell structure of PVDF/polystyrene (PS)/high density polyethylene (HDPE) ternary blends with, PS as the shell, HDPE as the core and this core-shell system dispersed in PVDF polymer was studied by Li *et al.* [58] In addition, multi-walled CNTs and Fe_3O_4 were incorporated selectively. F-F composites with MWCNTs was in PS shell and Fe_3O_4 was in PVDF matrix and E-F composites with MWCNTs was in PS shell and Fe_3O_4 was in HDPE core. It was observed that the electrical conductivity of the composites did not change with the incorporation of Fe_3O_4. Composites exhibited a EMI shielding effectiveness of 20 dB. The highest SE observed for the F-F composite with 1 vol% Fe_3O_4 and 1 vol% MWCNTs was 25 dB at 9.5 GHz. Table 4.2 summarizes the various studies reporting the development of PVDF composites for EMI shielding.

Table 4.2 PVDF composites for EMI shielding

Composite	EMI shielding	Ref.
PVDF/functionalized graphene	≈20 dB in X-band (8–12 GHz), 18 dB in broadband (1–8 GHz)	43
MWCNT/FeCo/PVDF	−41.2 dB at 12 GHz	44
Ag/incorporated graphite/PVDF	29.1 dB at 12.4 GHz	49
Au/MWCNT /PVDF	of 26.7 dB at 12 GHz	50
PVDF/carbon composites	1557 dB cm^2	51
PVDF/Ni composites	X-band 8.2-12.4 GHz	52
PVDF/MWCNT	30.89 dB	54
PVDF/PFC/SWCNH	-29.7 dB	55
CNT/PVDF	20.3 dB	56
G-D-GQDs/PVDF	43 dB at 8 GHz	45
PVDF/graphene	47 dB at 8-12 GHz	46
PVDF/CNT/Ni-chain	23.6 dB to 57.3 dB	47
PVDF/GNP/Ni-chain	22.7 dB to 55.8 dB	47
PVDF/NiO/$BaTiO_3$	-11.5 dB	57
PVDF/PS/HDPE/MWCNTs/Fe_3O_4	25 dB at 9.5 GHz	58

4.4 Conclusions

Conducting polymers are immensely useful materials for achieving electromagnetic shielding. The addition of nanoparticles to generate functional nanocomposites further improves their shielding capabilities. The development of a large variety of such nanocomposites in the recent years confirms the potential of such materials for diverse commercial applications.

References

1. Kumar, R., Singh, S., and Yadav, B. (2015) Conducting polymers: Synthesis, properties and applications. *International Advanced Research Journal in Science, Engineering and Technology*, **2**(11), 110-124.
2. Koul, S., Chandra, P., and Dhawan, S. (2000) Conducting polyaniline composite for ESD and EMI at 101 GHz. *Polymer,* **41**, 9305-9310.
3. Kim, B., Lee, H., Park, S., and Kim, H. (2011) Electromagnetic interference shielding characteristics and shielding effectiveness of polyaniline-coated films. *Thin Solid Films,* **519**, 3492-3496.
4. Colaneri, N., and Schacklette L. (1992) EMI shielding measurements of conductive polymer blends. *IEEE Transactions on Instrumentation and Measurement,* **41**(2), 291-297.
5. Zhang, Y., Qiu, M., Yu, Y., Wen, B., and Cheng, L. (2017) A novel polyaniline-coated bagasse fibre composite with core-shell heterostructure provides effective electromagnetic shielding performance. *Applied Materials and Interfaces,* **9**, 809-818.
6. Joo, J., and Lee, Y. (2000) High frequency electromagnetic interference shielding response of mixtures and multilayer films based on conducting polymers. *Journal of Applied Physics,* **88**, 513.
7. Saini, P., Choudhary, V., Singh, B. P., and Dhawan, S. K. (2009) Polyaniline-MWCNT nanocomposites for microwave absorption and EMI shielding. *Materials Chemistry and Physics,* **113**, 919-926.
8. Mishra, M., Singh, A. P., Gupta, V., Chandra, A. and Dhawan, S. K. (2016) Tunable EMI shielding effectiveness using new exotic carbon: Polymer composites. *Journal of Alloys and Compounds,* **688**, 399-403.
9. Mathew, J., Sathishkumar, M., Kothurkar, N., Senthilkumar, R., and Narayanan, B. (2018) Polyaniline/Fe_3O_4-RGO nanocomposites for microwave absorption. *IOP Conference Series: Materials Science and Engineering,* **310**, 012138.
10. Gupta, T., Singh, B., Mathur, R., and Dhakate, S. (2014) Multi-walled carbon nanotube grapheme polyaniline multiphase nanocomposite with superior electromagnetic shielding effectiveness. *Nanoscale,* **6**,

842-851.

11. Joseph, N., Varghese, J., and Sebastian, M. (2017) *In situ* polymerized polyaniline nanofibre-based functional cotton and nylon fabrics as millimeter-wave absorbers. *Polymer Journal,* **49**, 391-399.
12. Lyu, J., Zhao, X., Hou, X., Zhang, Y., Li, T., and Yan, Y. (2017) Electromagnetic interference shielding based on a high strength polyaniline-aramid nanocomposite. *Composites Science and Technology,* **149**, 159-165.
13. Sudha, J., Sivakala, S., Prasanth, R., Reena, V., and Nair, P. (2009) Development of electromagnetic shielding materials from the conductive blends of polyaniline and polyaniline-clay nanocomposite-EVA: Preparation and properties. *Composites Science and Technology,* **69**, 358-364.
14. Kumar, R., Joon, S., Singh, A., Singh, B., and Dhawan, S. (2015) Self-supported lightweight polyaniline thin sheets for electromagnetic interference shielding with improved thermal and mechanical properties. *American Journal of Polymer Science,* **5**, 28-39.
15. Dar, M. A., Kotnala, R. K., Verma, V., Shah, J., Siddiqui, W. A., and Alam, M. (2012) High magneto-crystalline anisotropic core–shell structured $Mn_{0.5}Zn_{0.5}Fe_2O_4$/polyaniline nanocomposites prepared by in situ emulsion polymerization. *Journal of Physical Chemistry C,* **116**(9), 5277-5287.
16. Saini, P. Arora, M., Gupta, G., Gupta, B., Singh, V., and Choudhary, V. (2013) High permittivity polyaniline–barium titanate nanocomposites with excellent electromagnetic interference shielding response. *Nanoscale,* **5**, 4330-4336.
17. Wang, W., Gumfekar, P., Jiao, Q., and Zhao, B. (2013) Ferrite-grafted polyaniline nanofibres as electromagnetic shielding materials. *Journal of Material Chemistry C,* **1**, 2851-2859.
18. Qiu, M., Zhang, Y., and Wen, B. (2018) Facile synthesis of polyaniline nanostructures with effective electromagnetic interference shielding performance. *Journal of Materials Science: Materials in Electronics,* **29**, 10437-10444.
19. Kim, M., Kim, H., Byun, S., Jeong, S., Hong, Y., Joo, J., and Lee J. (2002) PET fabric/polypyrrole composite with high electrical conductivity for EMI shielding. *Synthetic Metals,* **126**, 233-239.
20. Azadmanjiri, J., Talemi, P., Simon, G., Suzuki, K., and Selomulya, C. (2010) Synthesis and electromagnetic interference shielding properties of iron oxide/polypyrrole nanocomposites. *Polymer Engineering & Science,* **51**, 247-253.
21. Lee, C., Lee, D., Jeong, C., Hong, Y., Shim, J., Joo, J., Kim, M., Lee, J., Jeong, S., Byun, S., Zang, D., and Yang, H. (2002) Electromagnetic interference shielding by using conductive polypyrrole and metal compound coated on fabrics. *Polymers for Advanced Technologies,* **13**, 577-583.

22. Varshney, S., Ohlan, A., Jain, V., Dutta, V., and Dhawan, S. (2014) In situ synthesis of polypyrrole-γ-Fe_2O_3-fly ash nanocomposites for protection against EMI pollution. *Industrial and Engineering Chemistry Research*, **53**, 14282-14290.
23. Guo, J., Song, H., Liu, H., Luo, C., Ren, Y., Ding, T., Khan, M., Young, D., Liu, X., Zhang, X., Kong, J., and Guo, Z. (2017) Polypyrrole-interface-functionalized nano-magnetite epoxy nanocomposites as electromagnetic wave absorbers with enhanced flame retardancy. *Journal of Material Chemistry C*, **5**, 5334-5344.
24. Kim, S., Jang, S., Byun, S., Lee, J., and Joo, J. (2003) Electrical properties and EMI shielding characteristics of polypyrrole–nylon 6 composite fabrics. *Journal of Applied Polymer Science*, **87**, 1969-1974.
25. Li, X., Wan, M., Wei, Y., Shen, J., and Chen, Z. (2006) Electromagnetic functionalized and core-shell micro/nanostructured polypyrrole composites. *Journal of Physical Chemistry B*, **110**, 14623-14626.
26. Tunakova, V., Gregr, J., Tunak, M., and Dohnal, G., (2018) Functional polyester fabric/polypyrrole polymer composites for electromagnetic shielding: Optimization of process parameters. *Journal of Industrial Textiles*, **47**, 686-711.
27. Tong, Y., He, M., Zhou, Y., Zhong, X., Fan, L., Huang, T., Liao, Q., and Wang, Y. (2018) Hybridizing polypyrrole chains with laminated and two-dimensional $Ti_3C_2T_x$ toward high-performance electromagnetic wave absorption. *Applied Surface Science*, **434**, 283-293.
28. Erdogan, M., Karakisla, M., and Sacak, M. (2018) Polypyrrole and silver particles coated poly(ethylene terephthalate) nonwoven composite for electromagnetic interference shielding. *Journal of Composite Materials*, **52**, 1353-1362.
29. Bhardwaj, P., Kaushik, S., Gairola, P., and Gairola, S. (2018) Exceptional electromagnetic radiation shielding performance and dielectric properties of surfactant assisted polypyrrole-carbon allotropes composites. *Radiation Physics and Chemistry*, **151**, 156-163.
30. Sambyal, P., Dhawan, S., Gairola, P., Chauhan, S., and Gairola, S. (2018) Synergistic effect of polypyrrole/BST/RGO/Fe_3O_4 composite for enhanced microwave absorption and EMI shielding in X-band. *Current Applied Physics*, **18**, 611-618.
31. Gahlout, P., and Choudhary, V. (2016) Tailoring of polypyrrole backbone by optimizing synthesis parameters for efficient EMI shielding properties in X-band (8.2–12.4 GHz). *Synthetic Metals*, **222**, 170-179.
32. Maity, S., and Chatterjee, A. (2017) Polypyrrole functionalized polyester needle punched nonwoven fabrics for electro-magnetic interference shielding. *Polymer Composites*, https://doi.org/10.1002/pc.24399.

33. Zhao, H., Hou, L., and Lu, Y. (2016) Electromagnetic shielding effectiveness and serviceability of the multilayer structured cuprammonium fabric/polypyrrole/copper (CF/PPy/Cu) composite. *Chemical Engineering Journal*, **297**, 170-179.
34. Gahlout, P., and Choudhary, V. (2017) 5-Sulfoisophthalic acid monolithium salt doped polypyrrole/multiwalled carbon nanotubes composites for EMI shielding application in X-band (8.2-12.4 GHz). *Journal of Applied Polymer Science*, **134**, 45370.
35. Angelopoulos, M. (2002) Conducting polymers in microelectronics. *Journal of Research and Development*, **45,** 57-75.
36. Najar, M. H., and Majid, K. (2015) Synthesis and characterization of nanocomposite of polythiophene with $Na_2[Fe(CN)_3(OH)(NO)C_6H_{12}N_4]$ H_2O: a potent material for EMI shielding applications. *Journal of Materials Science: materials in Electronics*, **26**(9), 6458-6470.
37. Farukh, M., Singh, A. P., and Dhawan, S. K. (2015) Enhanced electromagnetic shielding behavior of multi-walled carbon nanotube entrenched poly (3, 4-ethylenedioxythiophene) nanocomposites. *Composites Science and Technology*, **114**, 94-102.
38. Taka, T. (1991) EMI shielding measurements on poly(3-octyl thiophene) blends. *Synthetic Metals*, **41**, 1177-1180.
39. Hosseini, S., Moghimi, A., and Moloudi, M. (2014) Magnetic, conductive, and microwave absorption properties of polythiophene nanofibres layered on $MnFe_2O_4/Fe_3O_4$ core–shell structures. *Materials Science in Semiconductor Processing*, **24**, 272 -277.
40. Dar, M. A., Majid, K., Najar, M., Kotnala, R. Shah, J., Dhawan, S., and Farukh, M. (2017) Surfactant-assisted synthesis of polythiophene/$Ni_{0.5}Zn_{0.5}Fe_{2-x}Ce_xO_4$ ferrite composites: study of structural, dielectric and magnetic properties for EMI-shielding applications. *Physical Chemistry Chemical Physics*, **16,** 10629-10643.
41. Vyas, M., and Chandra, A. (2016) Ion–electron-conducting polymer composites: promising electromagnetic interference shielding material. *ACS Applied Materials & Interfaces*, **8**, 18450-18461.
42. Arjmand, M., Chizari, K., Krause, B., Postchke, P., and Sundararaj, U. (2016) Effect of synthesis catalyst on structure of nitrogen-doped carbon nanotubes and electrical conductivity and electromagnetic interference shielding of their polymeric nanocomposites. *Carbon*, **98**, 358-372.
43. Eswaraiah, V., Sankaranarayanan, V., and Ramaprabhu, S. (2011) Functionalized graphene–PVDF foam composites for EMI shielding. *Macromolecular Materials and Engineering*, **296**, 894-898.
44. Arief, I., Biswas, S., and Bose. S. (2017) FeCo-Anchored reduced graphene oxide framework-based soft composites containing carbon nanotubes as highly efficient microwave absorbers with excellent heat dissipation ability. *ACS Applied Materials & Interfaces*, **9**,

19202-19214.

45. Lakshmi, N., and Tambe, P. (2017) EMI shielding effectiveness of graphene decorated with graphene quantum dots and silver nanoparticles reinforced PVDF nanocomposites. *Composite Interfaces,* **24**, 861-882.
46. Sabira, K., Jayakrishnan, M., Saheeda, P., and Jayalekshmi, S. (2018) On the absorption dominated EMI shielding effects in free standing and flexible films of poly(vinylidene fluoride)/grapheme nanocomposite. *European Polymer Journal*, **99**, 437-444.
47. Zhao, B., Wang, S., Chongxiang, Z., Ruosong, L. S., Hamidinejad, M., Kazemi, Y., and Park, C. K., (2018) Synergism between carbon materials and Ni chains in flexible poly(vinylidene fluoride) composite films with high heat dissipation to improve electromagnetic shielding properties. *Carbon*, **127**, 496-478.
48. Joseph, N., Singh, S. K., Sirugudu, R. K., Murthy, V. R. K., Ananthakumar, S., and Sebastian, M. T. (2013) Effect of silver incorporation into PVDF-barium titanate composites for EMI shielding applications. *Materials Research Bulletin*, **48**(4), 1681-1687.
49. Kumaran, R., Alagar, M., Kumar S. D., Subramanian, V., and Dinakaran, K. (2015) Ag induced EMI shielding of Ag-graphite/PVDF flexible nanocomposites thin films. *Applied Physics Letters*, **107**, 113107.
50. Kumaran, R. Kumar, S. D., Balasubramanian, N., Alagar, M., Subramanian, V., and Dinakaran, K. (2016) Enhanced conductivity and electromagnetic interference shielding in an Au-MWCNT dispersed PVDF flexible thin films. *Journal of Physical Chemistry C*, **120,** 13771-13778.
51. Zhao, B., Zhao, C., Li, R., Hamidinejad, S., and Park. C. (2017) Flexible, ultrathin, and high-efficiency electromagnetic shielding properties of Poly(Vinylidene Fluoride)/carbon composite films. *ACS Applied Materials & Interfaces*, **9**, 20873-20884.
52. Gargama, H., Thakur, *A., and* Chaturvedi, S. (2015) Polyvinylidene fluoride/nickel composite materials for charge storing, electromagnetic interference absorption, and shielding applications. *Journal of Applied Physics*, **117**, 224903.
53. Gargama, H., Thakur, A., and Chaturvedi, S. (2016) Polyvinylidene fluoride/nanocrystalline iron composite materials for EMI shielding and absorption applications. *Journal of Alloys and Compounds,* **654**, 209-215.
54. Wang, H., Zheng, K., Zhang, X., Du, T., Xiao, C., Ding, X., Bao, C., Chen, L., and Tian, X. (2016) Segregated poly(vinylidene fluoride)/MWCNTs composites for high-performance electromagnetic interference shielding. *Composites*, **90**, 606 -613.
55. Bera, R., Das, A. K., Maitra, A., Paria, S., Kumar, S., Bhanu, K., and Khatua, B. (2017) Salt leached viable porous Fe_3O_4 decorated pol-

yaniline – SWCNH/PVDF composite spectacles as an admirable electromagnetic shielding efficiency in extended Ku-band region. *Composites Part B: Engineering*, **129**, 210-220.

56. Mirkhani, S., Arjmand, M., Sadeghi, S., Krause, B., Postchke, P., and Sundararaj, U. (2017) Impact of synthesis temperature on morphology, rheology and electromagnetic interference shielding of CVD-grown carbon nanotube/polyvinylidene fluoride nanocomposites. *Synthetic Metals,* **230**, 49-50.
57. Muzaffar, A., Ahamed, M., Deshmukh, K., Faisal, M., and Pasha, S. (2018) Enhanced electromagnetic absorption in NiO and $BaTiO_3$ based polyvinylidenefluoride nanocomposites. *Materials Letters,* **218**. 217-220.
58. Li, L., Li, S., Shao, Y., Dou, R., Yin, B., and Yang, M. (2018), PVDF/PS/HDPE/MWCNTs/Fe_3O_4 nanocomposites: Effective and lightweight electromagnetic interference shielding material through the synergetic effect of MWCNTs and Fe_3O_4 nanoparticles. *Current Applied Physics,* **18**, 388-396.

Chapter 5

Conducting Polymer Composites for Energy Storage Systems

Qingli Hao* and Sadaf Mutahir

School of Chemical Engineering, Nanjing University of Science and Technology, Nanjing, 210094, China
**Corresponding author*: qinglihao@njust.edu.cn

5.1 Introduction

In the 21st century, energy is one of the most important concerns all over the world. As the world economy as well as the population are emerging, the worldwide need for energy resources has been increasing dramatically, giving rise to the urgent and pressing concerns for the sustainability of fuel reserves and environmental issues, including harsh changes in climate, melting of glaciers and rise in the sea levels. Hence, the development of renewable, safe, clean and sustainable energy storage technology has become an important research area both in fundamental research and industry now-a-days. Immense efforts have been devoted by scientists for mitigating the strong reliance on fossil fuels and solving the environmental issues [1,2]. Solar energy, wind energy, hydropower and biofuels are the virtuous sources of renewable energy. However, they endure from the shortcomings that they are erratic due to their intermittent nature [3].

The capturing of energy produced at one time and for use at a later time is simply called energy storage. Energy storage systems are basically the set of methods and technologies used to store various forms of energy [4]. Electrical energy storage system (ESS) is one of the most prevalent and reliable ways to store electric energy from the intermittent renewable sources to ensure timely and reliable delivery of ample amount of electric power for portable electronic devices, electric power vehicles and other applications. Large EES can also supply energy in power systems for hours in inaccessible areas as well. Another advantage of EES is that it can supply large amounts of

Conducting Polymer Composites, edited by Vikas Mittal

power in a brief period of time in places such as security installations. Batteries (Bs) and supercapacitors (SCs) belong to ESSs, or electrochemical energy storage systems. Other energy storage systems such as pumped hydro storage, thermal energy storage, compressed air energy storage and flywheel energy storage constitute non-electrochemical storage systems. The electrodes are the main parts of the electrochemical storage systems. These are mainly made of carbonaceous materials, metal-oxides and conducting hybrid materials. These materials are still not the ideal ones because scientists face several challenges that have to be overcome for the large-scale production and industrialization of the EES [5]. In the last decade, extraordinary new materials have been explored for the development of electrodes and electrolyte, among them conducting polymers (CPs), especially conducting polymer composites, have gained much attention [6-8]. In this chapter, an up-to-date account of the latest progress in the development of conducting polymer composites for advanced applications as an electrode material for Bs and SCs has been briefly presented.

A variety of electric energy storage systems, such as lithium-ion batteries (LIBs), lithium sulfur (Li/S) batteries and SCs, have been anticipated to address the high appraised energy issues. Presently, neither Bs (LIBs) nor SCs have so far satisfied the demand of electric vehicles because of the prevalent challenges. For instance, for LIBs, currently used anode and cathode materials exhibit a low energy density and poor rate performance. Although the SCs have high-rate performance, but the energy density is too low to power future electric vehicles. Due to this, large research efforts have been devoted for the design and fabrication of novel electrode materials with both high energy densities and high power, for the next-generation LIBs and SCs. As mentioned earlier, during the growth of recent energy innovations, CPs, in particular the conducting polymer composite materials, have been extensively used to obtain materials for electrochemical energy applications owing to their unique and tunable properties. For instance, conducting polymer composites can be used as active materials for pseudo-capacitive SCs and LIBs. These can also be employed as carbon or other heteroatom precursors for preparing the doped carbon materials used in various energy storage and conversion systems.

The conducting polymers, which are contiguous analogues of metals, have comparable electronic, magnetic and optical properties due to the exceedingly π-conjugation in their backbone, mainly composed of organic monomers, e.g., polyaniline (PANI), polypyrrole (PPy), polythiophene (PTh), poly (3,4-ethylenedioxythiophene) (PEDOT), polyacetylene (PA), polyparaphenylene (PPP), among other organic monoers.

The association of polymeric and inorganic components can give rise to new materials with remarkable properties that can be exploited in various fields and particularly in energy storage applications. So far, carbonaceous materials, metals, metal oxides and ferromagnetic materials have been extensively utilized as filler materials in the literature studies. Synthesis and other aspects of conducting polymer composites have been discussed in several excellent reviews. In electric energy storage systems, the conducting polymer based composites are expected to be the hybrids of two or more materials combined on a macroscopic level, which exhibit divergent physical and chemical properties from individual components and possible synergetic effects among CPs and other components, in order to obtain the high-performance electrode materials.

5.2 Lithium Batteries

5.2.1 Lithium-ion Batteries

LIBs, first designed by Sony Corporation in 1990, have garnered enormous attention from the scientists and engineers in the progression of solving the energy storage problems. LIBs have advantages over conventional batteries, including operating voltage of up to 4 V, specific energy density range of ~120 to 170 Wh kg^{-1} and capacity ranging from ~ 700 to 2400 mAh g^{-1} for a single cell (battery), which permit them to be applied in a varied range of applications from consumer electronics to hybrid vehicles [9,10]. Generally, an anode (negative), a cathode (positive), an electrolyte and a separator are the main components of a typical LIB. Typically, the materials used as positive electrode are mainly Li-containing metal oxides either with a layered structure (such as Li cobalt oxide) or the tunnel-shaped materials (such as Li manganese oxide), whereas insertion-type materials (such as carbonaceous materials, $Li_4Ti_5O_{12}$, TiO_2, etc.), conversion-type materials (mainly the oxides of transition metals like iron oxides, nickel oxides, cobalt oxides, etc.) and alloying-type materials (such as Sn, Si, etc.) are utilized as negative electrode materials. The electrolyte is also the integral part of a LIB, thus, the selection of electrolyte is as important as the electrodes, as it must be a good ionic conductor as well as an insulator for electronic conduction. Mostly, the electrolytes are based on the solution of inorganic Li salts such as $LiClO_4$, $LiPF_6$, $LiAsF_6$, $LiCF_3SO_3$ and $LiBF_4$, dissolved in a mixture of two or more organic solvents. The main function of the separator is to inhibit the short circuiting between the positive and negative electrodes and to provide copious channels for the transfer of Li ions during charging/discharging [11].

Cathodes for Li-ion Batteries

Nevertheless, the current LIB devices have been inadequate due to inferior power density, cyclic life and protection. The major limitation faced by LIBs is to develop the high capacity cathode material along with high energy density because of the limited accessibility and availability of Li metals and transition metal compounds for the cathode materials. Recently, carbonaceous materials and organic compounds, especially the π-conjugated polymers, have shown superior redox tunability and electrochemical solidity which outperform the features of inorganic semiconductors [6]. Most prevalent CPs utilized as cathode materials are polyaniline (PANI) [12], polypyrrole (PPy) [13], polythiophene (PTh) and their composites [14]. Conducting polymers generally show instability and low ionic conductivity for Li ions, when used alone as active material for cathode in LIBs [15]. To overcome these limitations, conducting polymer composites can be used as cathode material because these show much superior properties due to the synergic interaction between CPs and other components [16]. Vanadium oxide (V_2O_5) is a prevalent intercalation compound used in LIBs as cathode due to its extraordinary morphology having layered structure, ease of synthesis, low cost, high theoretical capacity (440 mAh g^{-1}), high abundance and low toxicity [17,18]. Yet, in practice, there are numerous challenges associated with it, such as poor ionic and electronic conductivity, low cyclic stability and slow reaction kinetics [19]. Park *et al.* [20] prepared a V_2O_5/PANI composite films by a two-step electrochemical procedure and evaluated their potential in LIBs by using as a cathode. The material presented a large specific capacity (ca. 270 mAh g^{-1}) and enhanced cyclability with an extremely low extent of capacity fading (ca. 3.4%) during repeated charge/discharge cycles. The cyclic voltametric results showed that in the resulting hybrid, PANI acted not only as a conducting material, but also as an electroactive material (binder) which was predominantly accountable for the enhancement of electrochemical response [20]. Later on, Chen *et al.* [21] improved the synthesis of V_2O_5/PANI nanocomposites by reporting an efficient, one pot and hydrothermal method via *in-situ* intercalation-polymerization-exfoliation and also evaluated as cathode material. The authors reported that the acidic environment and optimized reaction temperature are the important factors to obtain the resulting hybrid. The samples obtained at 140 °C exhibited highest cycling ability and improved reversible capacity, as compared to pristine V_2O_5. The capacity of the cell fabricated with this sample decreased up to 209 mAh g^{-1} after 30 cycles at a rate of 29.5 mAh g^{-1}, which was 87% of its preliminary capacity. The comparative analysis clearly suggested that the morphology of the composite (in this case, fungus like) was one of the promising factors to make it suitable cathode material for high-energy-density rechargeable lithium micro-batteries. In this study, PANI layer intercalated into the V_2O_5 layers and was responsible for controlling the morphology of V_2O_5 sheets.

Zhao *et al.* [22] reported the polymer nanocomposites of V_2O_5/PPy via *in-situ* polymerization of pyrrole (Py) on the surface of V_2O_5 by the hydrothermal reduction, followed by the thermal treatment in air atmosphere. The electrochemical results showed that the composite (V_2O_5/PPy) had improved lithium storage ability and cyclic performance as compared to neat V_2O_5. The PPy layer acted as a conductive plastic protecting shell on the surface of V_2O_5 and also effectively prevented its collapse due to expansion in the volume during the charge and discharge course [22]. Chao *et al.* [23] synthesized the nano-belt array (NBA) of V_2O_5 directly on the 3D graphite foam (UGF), followed by the thin layer coating of PEDOT (V_2O_5/PEDOT (core/shell)/UGF). The resulting hybrid showed high performance, excellent rate capability (297 mAh g^{-1}) and ultra-stable cycling up to 1000 cycles as compared to bare V_2O_5. PEDOT shell was the key of the enhanced electrochemical performance as it not only facilitated the charge transfer, but was also responsible for maintaining the special morphology of NBA.

The other widely used commercial cathode materials for LIBs are LiV_3O_8, $LiCoO_2$, $LiFePO_4$ and $LiMn_2O_4$, however, these slow electron diffusion and high decaying capacity, which are the major disadvantages associated with these [24]. Gao *et al.* [25] coated the uniform layer of PANI on LiV_3O_8 nanorods via chemical oxidative polymerization using sodium dodecyl benzene sulfate as anionic surfactant. The composite having 12 wt% PANI retained a discharge capacity of 204 mAh g^{-1} after 100 cycles and had better rate capability

(175 mAh g^{-1} at 2 C and 145 mAh g^{-1} at 4 C) than the pristine LiV_3O_8 electrode in the potential range of 1.5-4.0 V. PANI coating significantly increased the electrical conductivity between the LiV_3O_8 nanorods. PANI not only buffered the dissolution of LiV_3O_8 into the electrolyte, but also reduced the contact resistance among nanorods, which significantly improved the performance of the composite [25].

Tavorite flourophosphates ($LiVPO_4F$) are also used as cathode materials in LIBs. Yan *et al.* [26] reported the unique core-shell $LiVPO_4F$/PANI nanocomposite showing remarkable increase in the electrochemical performance of $LiVPO_4F$, generated by sol-gel synthesis followed by the method of self-assembly. The composite electrode displayed an initial discharge capacity of 149.3 mAh g^{-1} in the potential range of 3.0-4.5 V vs. Li^+/Li, which was prominently higher than bare $LiVPO_4F$ (127.5 mAh g^{-1}). The uniform coating of PANI layer enhanced the electronic conductivity and diffusion of Li-ions from $LiVPO_4F$. The excellent electrochemical performance of the composite made this hybrid as 4 V-class electrode, a propitious auxiliary for the next generation LIBs. Many other polymer nanocomposites with the above mentioned inorganic components have also been revealed as cathode material for the LIBs such as PANI/Co_3O_4 [27], PPy/$LiFeO_2$ [28], $LiFePO_4$/PEDOT [29], PPy/MnO_2 [13], PTh/MnO_2 nanosheet [14] and $LiFePO_4$/PTh [30].

The CPs and their composites with carbonaceous materials are widely employed as cathode material in LIBs. The importance of this class mainly lies in their light weight, ease of synthesis, low cost and easily tunable properties. Nevertheless, this class of materials is still facing numerous drawbacks to achieve the goal of "flexible plastic battery", among them the most notable are low energy density, poor Li-ion conductivity, low working potential space and degradation in the presence of non-aqueous electrolytes [31]. Recently, Liu *et al.* [32] synthesized a core-shell composite of PANI with multi-walled carbon nanotube (MWCNTs) via chemical oxidative polymerization, in order to enhance the electrochemical performance of pristine PANI. The resulting hybrid attained the highest specific capacity, up to 181.8 mAh g^{-1}, and the capacity retention rate was 76.75% at a constant current density of 0.2 C for 100 cycles. Here, the multi-walled CNTs not only enhanced conductivity, but also tuned the structure of PANI [32].

Anodes for Li-ion Batteries

Anode materials for LIBs are mainly accountable for both energy and power density, as well as the cyclic life of the battery. Commercially, lithiated graphite has presented the excellent electrochemical performance as an anode material for LIBs. However, it considerably shows lower capacity as related to the highest theoretical capacity of Li (4,000 mA h g^{-1}). In LIBs, during the first cycle, at the anode-electrolyte interface, the decomposition of non-aqueous electrolyte occurs, which creates a protective layer called as solid-electrolyte interphase (SEI), thus, protecting the further decomposition of the electrolyte [33]. Even though it is useful in one sense, however, the SEI is also liable for prompting the irreversibility in an intercalation/deintercalation route that leads to a slow loss of energy density during cyclic charging and discharging process. Moreover, the diffusion rate of Li ions into the carbon based materials is low, which lowers down the power density of Bs [34].

Use of CPs has been a protracted approach to upgrade the electronic or ionic conductivity of LIBs via the formation of very thin film/layer (preferably at nanoscale) on the surface of the anode materials [6,35,36]. The CPs film/layer provides the conducting pathway for the ions or electrons, which advances the ionic or electronic conductivity, avoids the interfacial resistance between the nanoparticles and also separates the electrolyte from the anode material. Thus, it subsequently strengthens the formation of SEI film on the surface of the anode, which attributes the improved reversibility and cyclic life of the anodes. The reduced packing density of the electrodes is one of the main limitations of this system, which slows down the volumetric energy density [34]. Among the CPs, PANI [37], PEDOT [38] and PTh [39] have been mainly explored as the conducting coating to improve the electrochemical performance of anodes, due to their low cost, easy synthesis, longstanding stability and safety features during the charging and discharging course.

Spinel materials, such as $Li_4Ti_5O_{12}$, are one of the plausible replacements for the graphite anode materials in LIBs. Recently, Xu *et al.* [39] reported a core-shell hybrid of $Li_4Ti_5O_{12}$ with PTh, which was wrapped on its surface by an *in-situ* oxidative polymerization method, and utilized it as anode material for LIBs. The composite exhibited the specific discharge capacity of 172 mAh g^{-1} at the 0.1 C discharge rate, which was far better than bare $Li_4Ti_5O_{12}$ or even PANI/$Li_4Ti_5O_{12}$ [37]. The new anode material presented excellent stability and cycle life, even at a very high discharge rate. The PTh coating on the surface of $Li_4Ti_5O_{12}$ was mainly attributed to the excellent electrochemical performance of the material, which facilitated the electron conduction during the charge/discharge process and inhibited the dissolution of active material.

Liu *et al.* [40] reported a ternary composite, having sandwich like morphology, as an anode material for LIBs. SnO_2 nanoparticles were grown on the layered graphene oxide (2D) and subsequently covered with a layer of PANI, which resulted in the formation of GO/SnO_2/PANI composite. The composite was found to be more flexible and showed very high energy density (397 mAh g^{-1}) at high discharge rate of 10 A g^{-1}. The nanostructured composite not only allowed the volume expansion during the lithiation of SnO_2, but also prevented the particle agglomeration.

Titanium oxide (TiO_2), another oxide material, exhibits high potential to substitute graphite as an anode material in LIBs. The high Li insertion/extraction potential, longer cyclic stability, high reversible capacity and very low volume expansion during the charging/discharging process make it suitable candidate for anode material. TiO_2 suffers from some limitations such as very low theoretical capacity (330 mAh g^{-1}) as compared to graphite as well as intrinsically low electrical conductivity. Zheng *et al.* [41] prepared TiO_2/PANI nanocomposites by solid coating method, with different concentrations of PANI. The composites exhibited better reversible capacity as compared to neat TiO_2. The electrochemical performance showed that PANI might be responsible to keep the TiO_2 particles electrically connected as well as provided the conducting pathway. The authors confirmed PANI as an encouraging matrix material for LIBs.

Silicon (Si) has a high theoretical specific capacity (4200 mAh g^{-1}) as an anode material for LIBs, but it undergoes mechanical and electrical degradation due to the stress generated by large volume expansion (>300%) during the intercalation/deintercalation processes, which is responsible for the poor cyclic stability [42]. Therefore, the design of nanostructured Si-based composite materials has become one of the most effective solutions to overcome these problems. Wu *et al.* [43] reported a new methodology for the incorporation of Si nanoparticles (SiNP) in PANI hydrogel framework. The composite exhibited a well-connected 3D network structure in which Si nanoparticles were fully coated by the PANI hydrogel. In this work, the hierarchical PANI hydrogel framework exhibited multiple advantages: providing a continuous electrically conductive network of PANI, containing abundant functional sites in the positively charged polymer chains for binding with Si surface by hydrogen bonding or electrostatic interaction and offering extensive porous space for buffering the volume expansion of Si particles. Therefore, the resulting composite anode exhibited stable capacity of ~ 12,00 mAh g^{-1} after 1,000 cycles and good rate performance. Its cycle life could reach over 90% capacity retention after 5,000 cycles at a current density of 6.0 A g^{-1}.

Kummer *et al.* [44] also synthesized Si/PANI composites by the *in-situ* polymerization of aniline with SiNPs. The author attributed the improvement in the electrochemical performance to the presence of CP (PANI), which not only increased the electrical conductivity, but also enhanced the flexibility of the structure by buffering the stress due to volume expansion and contraction. Xiao *et al.* [45] reported a novel ternary composite of Si with CNTs and PANI as anode material in LIBs. The SiNPs in gaseous phase were introduced on CNTs and PANI using the same method described above. The composite showed a high reversible capacity of 2430 mAh g^{-1} with good retention over 500 cycles compared to neat Si. In Si-CNT/PANI composite, PANI further improved the electrical conductivity in the presence of CNTs as well as protected the Si Structure. The hybrid showed improved columbic efficiency (CE) and rate capabilities. Lawes *et al.* [46] reported the SiNPs based anode for LIBs by inkjet-printing with commercially available binders. Among these, PEDOT:PSS binder showed highest electrochemical performance due to its conductive network, which promoted the quick electron transfer and flexibility, thus, accommodating the greater volume expansion of SiNPs during charging and discharging process.

5.2.2 Lithium-Sulfur Batteries

Sulfur (S) is one of the most extensively studied materials due to its availability and low cost, along with both higher capacity and energy density of Li/S batteries as compared to LIBs. In a typical Li/S battery, elemental sulfur acts as the active cathode material and Li-metal as the anode. Sulfur undergoes reduction by Li to form a series of polysulfides (Li_2S_8-Li_2S_2) and subsequently Li_2S, which relates to 100% depth of discharge [47]. Li/S battery is an attractive system due to its highest theoretical specific capacity (1675 mA h g^{-1}) and high theoretical specific energy of (2600 W h kg^{-1}) among all solid cathode LIB systems [48]. However, poor electrical conductivity of S results in relatively low capacity, thus, restricting its usage for commercial applications [49]. Another important parameter which limits its use are the polysulfides, the discharged products which leak out from the electrode and cover the cathode, thus, resulting in squat capacity and reduced re-chargeability. These also damage the anode by diffusing and migrating towards anode via electrolyte and react with Li, thus, causing the depletion of the anode material which ultimately decreases the discharge capacity and open circuit potential. Many approaches have been adopted by researchers to overcome these challenges, such as the use of carbonaceous materials to form the network, however, the discharge of polysulfides is inescapable from the electrode. CPs have also been explored and promoted as suitable substitute matrices to carbon for S entrapment.

Li *et al.* [50] systematically investigated the effect of different CPs (PANI, PPy and PEDOT) coated with high degree of uniformity over hollow sulfur nanoparticles synthesized by *in-situ* polymerization as shown in Figure 5.1. All composites exhibited specific morphology, with comparison to S nanospheres without coating. The rate capability of the cells based on the three CPs coated hollow sulfur nanospheres was investigated at different current rates from C/10 to 4C

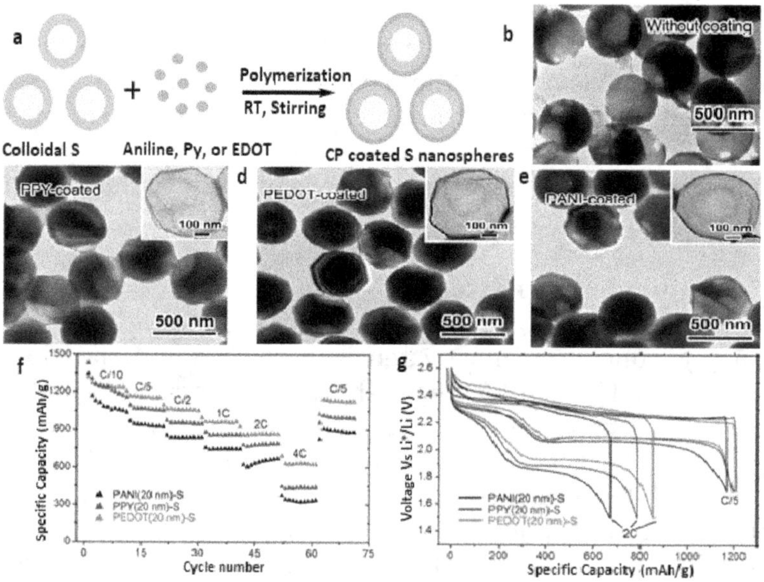

Figure 5.1 (a) Schematic illustration of the fabrication process of conductive polymer-coated hollow sulfur nanospheres. TEM images of nanospheres (b) without-coating, (c) PPy-, (d) PEDOT- and (e) PANI-coated hollow sulfur nanospheres. (f) Rate capability of the cells discharged at various current rates from C/10 to 4C. (g) Typical discharge-charge voltage profiles of the cells at C/5 and 2C (1C = 1673 mA g^{-1}). Reproduced from Reference 50 with permission of American Chemical Society.

(1C = 1673 mA g^{-1}). The competency of the three polymers in improving the long-term cycling stability and high-rate performance of the sulfur cathode decreased in the order of PEDOT > PPY > PANI.

Yang *et al.* [51] made an effort to trap polysulfides by encapsulation of carbon/sulfur particles with PEDOT:PSS. The synthesized particles were thermally stable and moderately rigid in the electrochemical environment, as demonstrated in Figure 5.2. The decay in absolute discharge capacity against cycle number was fast in bare samples, while it slowed down in the polymer composites. This strategy provided an effective way to entrap polysulfides and reduced the dissolution of polysulfides from the cathode. The discharge capacity was significantly enhanced which remained at ~ 600 mA h g^{-1} [51].

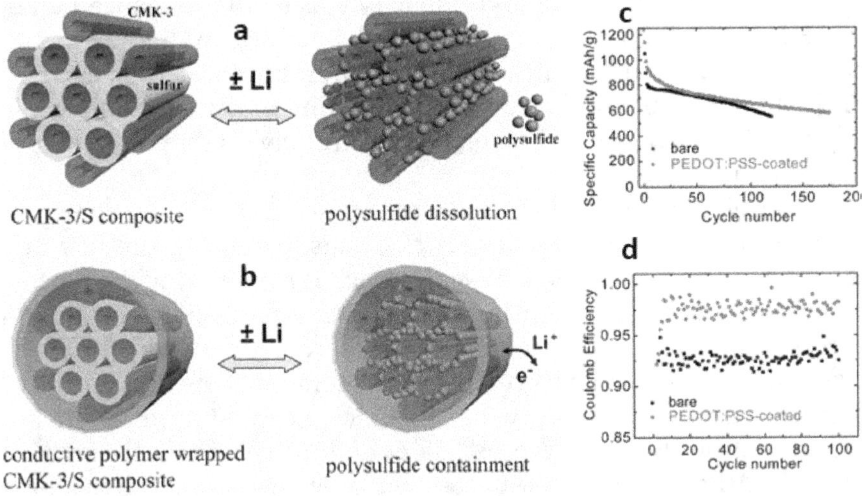

Figure 5.2 Illustration of the encapsulation of carbon/sulfur particles with PEDOT:PSS for improving polysulfides encapsulation: (a) without PEDOT:PSS coating (b) with a PEDOT:PSS coating; (c) absolute discharge capacity and (d) normalized discharge capacity against cycle number. Reproduced from Reference 51 with permission from American Chemical Society.

Chen *et al.* [52] reported a membrane assisted precipitation technique to generate the ultra-fine sulfur particles and subsequently coated these with PEDOT, resulting in the formation of Si/PEDOT core/shell nanocomposites. The fine ultra-small sulfur particles increased the electric conduction of the material and also increased the level of its utilization. PEDOT provided the capability to encapsulated core/shell composites to trap polysulfides, along with the conductive matrix for electron transport which also improved the cycling stability. The composite exhibited an excellent electrochemical performance, e.g. initial discharge capacity of 1,117 mA h g^{-1} and stable capacity of 83% over 50 cycles.

Many researchers have utilized PPy and its derivatives to enhance the efficiency of cathode materials for Li/S batteries [53]. Liang *et al.* [54] reported the core-shell composite of PPy with MWCNTs as a cathode material for Li/S batteries. The MWCNTs provided the high electronic conductive network in the composite. In the meantime, PPy offered high surface area to ensure a uniform dispersion of sulfur on the conducting matrix. S/PPy-MWCNTs having 25 wt% PPy showed the best discharge capacity of 725.8 mAh g^{-1} after 100 cycles. In another study, the discharge capacity was further enhanced up to 937.8 mAh g^{-1} at 20 cycles, and reached about 860 mAh g^{-1} after 100 cycles at 0.2C for sulfur cathode with 3D mesoporous carbon and PPy [55]. The 3D carbon provided the conducting matrix to sulfur while PPy acted as the wrapping layer, which strongly inhibited the migration of polysulfide.

As far as Li-anode in Li/S battery is concerned, it faces some unembellished and complex drawbacks, thus, exploring the CPs and their composites could be a new approach for improving the performance of Li/S batteries. The utilization of CPs in this field is still at its early stage and needs greater input of researchers and industry. However, there are a few reports on the enhancement of the lithium anode performance in Li/S battery. For instance, keeping in mind the advancement of PEDOT in the Bs, Ma *et al.* [56] coated a thin layer of PEDOT-PEG as a functional protective layer on Li anode by using a simple and facile method. The thin protecting layer of PEDOT/PEG created strong interaction with Li electrode, which was beneficial to form a stable and less resistive SEI between the anode and electrolyte, which also suppressed the dendrite growth and shuttle effect on the surface of the anode, thus, resulting in the excellent cycling stability and high CE. Additionally, it might restrict the physical contact of the lithium polysulfides with the Li anode which is responsible for the corrosion of the electrode. The results revealed that the protecting strategy of Li anode can be anticipated with excellent electrochemical performance of the battery.

CPs composites offer exciting potentials to design the state-of-the-art electrode materials with high Li-ion conductivities, although the optimization of every single component of composite is technically challenging. Both, theoretical and experimental studies on the transport parameters such as ion mobility and transport correlation as well as on morphology are needed. Computational studies can also play a key role for the fundamental understanding of the interactions between the components.

Bottom-up synthesis of free standing electrodes is an emerging methodology for the fabrication of nanomaterials. In this method, current collector, electrochemically active materials and binder are the part of an identical multicomponent system, which permits the synergistic benefits coming from each component of the composite. For an ideal innovative electrode design, including the fabrication, electrode morphology, rheological studies, mechanical performance and thermal analysis of the composite are important. The usage of low-cost binders with eco-friendly solvents is under progress for the development of high capacity materials similar to Si.

Recent studies have revealed that the gravimetric energy density is one of the main advantage of Li/S batteries over LIBs especially for lightweight applications. The Li/S cells demonstrate energies of about 400 Wh/kg at cell level, which has attracted investment from renowned companies such as Robert Bosch Gmbh, Polyplus Battery Company, Oxis Energy Limited, BASF SE, Samsung SDI Co. Ltd., etc.

5.3 Supercapacitors

SCs are the leading class of modern energy storage devices. SCs offer high power output (10 kW/kg) at the cost of low energy density (5-10 Wh kg^{-1}). Consequently, these are utilized in applications where "power bursts" is requisite, i.e. it is required to deliver large amounts of electric energy in a very short period of time, such as in military applications, satellite launch, powering remote areas and hybrid electric vehicles, where SCs can also be coupled with batteries or fuel cells to deliver the needed power.

SCs provide a practical compromise between Bs (high storage capacity and low power output) and conventional dielectric capacitors (high power output and low energy storage), with fast discharge/charge process. SCs have many benefits over Bs, mainly due to the physical nature of the surface charge storage mechanism and absence of slow electrochemical reactions [57]. Depending on the way the charge is stored, SCs can be classified as (1) electrochemical double-layer capacitors (EDLCs) and (2) pseudo-capacitors [58]. The working principle of both types of supercapacitors is shown in Figure

5.3. The EDLCs store charge electrostatically through a reversible physical adsorption/desorption of ions from the electrolyte on the electrode/electrolyte interface [59]. This is essentially a non-faradic process. In case of EDLCs, as shown in Figure 5.3A, the charge is stored quickly at the interface of the active material (electrode) and electrolyte, and the electrode should have a high surface area per unit mass and high electric conductivity to facilitate the formation of double layers. At present, the commercial SCs are mainly based on activated carbon or other porous carbon materials, which show the major shortcoming of low energy density (typically 5-10Wh kg^{-1}) in the presence of aqueous electrolytes, although the cycling lifetime is quite acceptable.

In case of pseudo-capacitors (Figure 5.3B), capacitance originates from the fast and reversible faradic reaction of the electrochemically active electrode materials with the electrolyte [60]. Therefore, pseudo-capacitors have much higher specific capacitance and higher energy density than EDLCs based on carbon materials, with the active electrode materials usually transition metal oxides, CPs, etc., and reversible reactions occurring at the interface or near the surface of electrodes. However, poor cycling stability is the main drawback of the pseudo-capacitive materials because of the mechanical instability caused by multi-valence changing for metal oxides or swelling and shrinking of polymeric chains during the charge and discharge processes.

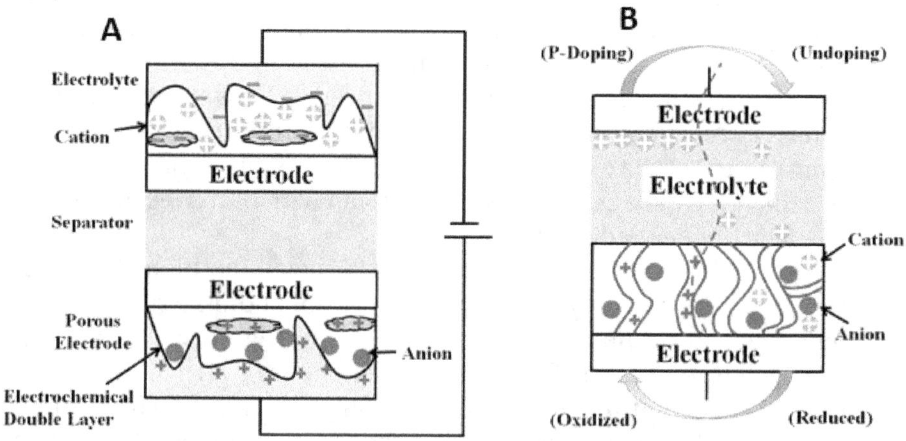

Figure 5.3 A schematic diagram illustrating the operational principles of (A) electric double layer capacitor (EDLC) and (B) pseudo-capacitor.

The performance of SCs is mainly determined by the electrochemical activity and kinetic features of the electrodes. To improve the power density of SCs, it is crucial to enhance the kinetics of ion and electron transport in the electrodes and at the electrode/electrolyte interface. Therefore, electrodes with proper pore structure and good electrical conductivity are highly desirable for fast kinetics contributing to high power density. On the other hand, by optimizing the structure of electrode materials, introducing pseudo-capacitive materials and designing energy storage devices, the energy density of SCs can be increased according to energy storage mechanisms.

The CPs, being an attractive active species, have been investigated for SCs, as they have a high charge density, low cost, extensive structural diversity and mechanical flexibility. Compared to EDLCs having carbon as the electrode material, CP-based SCs can store a greater amount of charges both in the electrical double layer and through a rapid faradic charge transfer, as the electrochemical process occurs both on the surface and in the bulk near surface of the electrode material. However, CPs used as electrode materials often suffer remarkable degradation because of the mechanical instability caused by swelling and shrinking during the charge and discharge processes. CPs also tend to irreversibly agglomerate during electrode preparation and may peel off from the current collector during charge/discharge processes. These drawbacks lead to the poor cycle life of the CP based electrodes, thereby, greatly restricting their practical applications. Hence, it is of great importance to combine CPs with other materials, which is beneficial for the development of electrode materials with high specific capacitance and excellent cycling stability. More specifically, the efficient combination of inorganic materials with CPs plays a key role in improving the performance because the CPs can interact synergistically with the inorganic materials to provide conducting backbone and electrical conductivity [7].

The most often studied CPs in SCs applications are PANI, PPy, PEDOT, PTh and their corresponding derivatives [61]. It has been demonstrated that the electrochemical performance of CPs could be remarkably enhanced when mixed with various materials such as carbon, metal or metal compounds [62-64].

5.3.1 Carbon/Conducting Polymer Binary Composites

All forms of carbon (0D to 3D) have received considerable attention as a reinforcement for CPs. Carbon dots or graphene quantum dots

(GQDs) as 0D carbon materials are often investigated for fabricating various novel devices due to their chemical/physical properties such as nanometer-size, rich edge defects and good surface grafting, stable photoluminescence, good electrical conductivity, high mobility, etc. Mondal *et al.* [65] reported that the GQDs doped PANI composites showed high specific capacitance of ~1044 F g^{-1} at a current density of 1 A g^{-1} with a retention of ~80% after 3000 cycles, as well as moderate cyclic stability. The composites of 1D carbon substrates with CPs, such as CNTs, carbon nanofibers (CNFs) and graphene nanoribbons (GNRs), have been extensively investigated as electrode materials for SCs [66,67]. The CP composites with 2D carbon materials, including graphene (GE), graphene oxide (GO), reduced graphene oxide (rGO) and graphite nanoflakes (GFs) have attracted considerable interest as active electrode materials for SCs, as these materials provide high thermal stability, large surface area, uneven nanostructure and adequate interlayer spacing for easy access of the electrolyte ions, in addition to the above said merits of carbon materials [68-70].

Various binary composites of PANI have been reported with different active carbon materials, such as PANI/GE [71-73], PANI/CNTs [74-76], PANI/CNFs [77], PANI/GO, PANI/rGO [78,79], PANI/carbon spheres [80,81] and PANI/carbon particles [82]. Cong *et al.* [73] prepared a binary composite of PANI with GE. In the first step, a free-standing GE paper (GEP) was prepared, and nanorods of PANI were subsequently electro-polymerized on GEP. The excellent specific capacitance (~763 Fg^{-1}) and good cycle stability (with 82% of the retention after 1000 cycles) showed that the prepared paper like composite had good flexibility and high electrical conductivity due to the synergistic effect of GE and PANI.

Nanostructures of PANI/carbon composites have a significant impact on the electrochemical properties and several strategies have been developed by varying the synthetic conditions. Doping in carbon or CPs is one of the successful strategy to increase the conductivity of the composite materials. Hao *et al.* [83] also reported the facile *in-situ* synthesis of sandwich-like boron-doped graphene (BG)/PANI hybrid for SCs (Figure 5.4). The hybrid exhibited a homogeneous morphology of wrinkled sheets with a relatively rough surface, indicating the successful coating of PANI on the surface of BG sheets. Moreover, the porous structure formed with the well-dispersed and wrinkled sheets provided more area for the contact of the electrolyte with active materials, facilitating the electrode transfer and charge storage. A thin PANI film of less than 2 nm was observed to be formed on the surface of BG sheets. Compared with the individual component and polyaniline/nondoped graphene, the sandwich-like PANI/BG exhibited remarkably enhanced electrochemical specific capacitance and long cycle stability in both acid and alkaline electrolytes (100% and 83% after 5000 cycles, respectively). The improved electrochemical performance of the hybrid was mainly attributed to the introduction of additional p-type carriers in carbon systems by boron-doping and the sound combination of pseudo-capacitive conducting polyaniline [83]. Other heteroatoms like N, S, P are also possible for doping into carbon materials to improve the electrochemical properties.

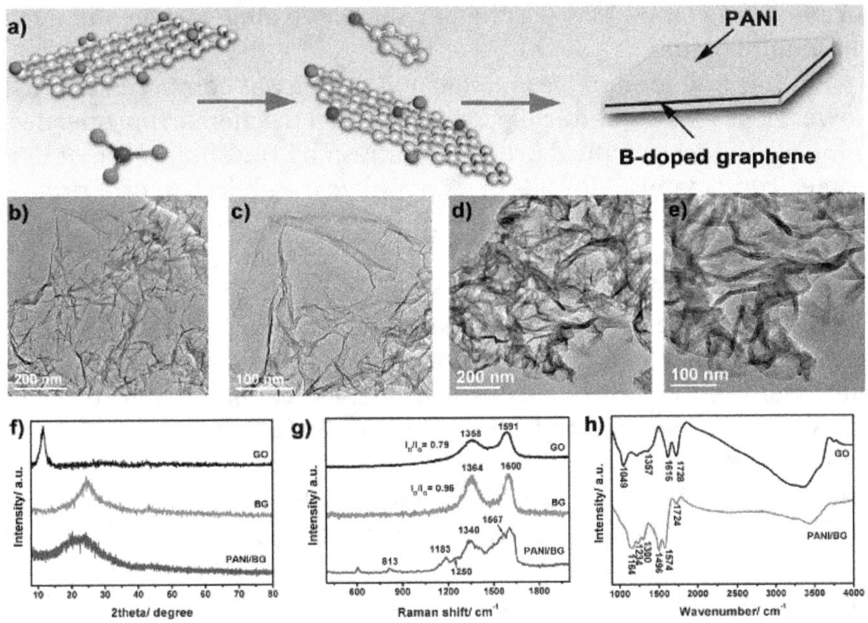

Figure 5.4 Scheme for the preparation route of PANI/BG (a); TEM images of BG (b, c) and PANI/BG (d, e); XRD patterns (f) and Raman spectra (g) of GO, BG, and PANI/BG; FT-IR spectra (h) of GO and PANI/BG. Reproduced from reference [83] with permission from Elsevier.

Moreover, the sulfonated graphene (SGE) composites show highly reactive nature in energy storage devices because the SGE nanosheets act as a macromolecule doping acid and the supporting substrate in the polymerization process [84,85]. Hao *et al.* [85] reported a novel strategy for the fabrication of GE/PANI composites with different morphologies by using SGE via liquid/liquid interfacial polymerization. As mentioned earlier, the SGE sheets could be used as both macromolecular acid dopant as well as substrate for the polymerization of PANI to form SGE/PANI nanocomposites. The obtained SGE/PANI nanocomposites exhibited good cycling stability with 96% capacitance retention after 100 cycles and good rate properties. The morphology of SGE/PANI composites could be tuned from nanorod to nanoparticle-type by additional dopant (hydrochloride acid). Moreover, the SGE/PANI composites obtained with HCl as an additional acid dopant with higher amount of PANI showed higher specific capacitances, but lower capacity retention, compared to the SGE/PANI prepared with single SGE dopant. The electrochemical performance of composites was higher than activated carbon and even ruthenium oxide.

PPy has low conductive and thermal stability as compared to PANI, however, it retains better electrical and electrochemical properties. All in all, the most critical problem restraining the application of PPy as electrode material for SCs is its poor cycle stability. In order to increase its cyclic stability, researchers have intensively focused on the PPy/carbon binary nanocomposites. The composites with carbon exhibit improved electrochemical and mechanical properties, as compared to pristine PPy. Among the different carbon substrates, CNTs are widely used as additives or templates to synthesize composite electrodes of PPy with improve electrochemical performance, especially electric conductivity and cycle stability due to the synergistic effect [86-92]. Zhou *et al.* [91] investigated the impact of short and long CNTs on the PPy/PSS-CNT composites. It was observed that the addition of both types of CNTs enhanced the capacitance of the PPy based electrodes to a large extent. The PPy/PSS composite with long CNTs showed significant superior capacitive behavior and cycling stability due to the core-shell porous surface morphology and interconnected conductive nano-network. Apart from PPy/CNTs, GE and their derivatives (GO and rGO) have also been used as ideal additives [93-96]. Researchers have also investigated other carbon materials to form binary composites with PPy, including PPy/active carbon [93], PPy/carbon cloth [94], PPy/GE sheets [95,96] and PPy/CNFs [97].

PEDOT is also utilized and investigated as an active electrode material in SCs due to good flexibility, low-cost and good electrical conductivity. However, the major limitation like other CPs is poor capacitance and stability, which implies an obvious capacitance drop after a certain number of charge/discharge cycles. From the industrial point of view, it also suffers from limited capacitance due to relatively stable chemical activity. To solve these issues, several approaches have been developed, including the composites with different carbon materials like CNTs, GN, GO, rGO, carbon cloth, etc., to increase the double layered capacitance and conductivity, along with mixing or doping of metal oxides to improve its pseudo-capacitance [98-105]. Chu *et al.* [100] synthesized PEDOT/GE/carbon cloth composites via filtration followed by electrochemical polymerization, and investigated as SCs electrode material. The nanocomposites revealed a specific capacitance of 714.93 F g^{-1}, where carbon cloth acted as 3D conducting matrix. The smooth surface of carbon cloth became wrinkled after GE deposition which further increased the surface area for the redox reactions of PEDOT [100].

PTh has a number of distinctive properties, however, using it alone cannot meet the practical application requirements for SCs electrodes, like other CPs. Thus, many research studies have focused on the composites of PTh with different forms of carbon and explored their potential for SCs. [106-111]. The composites of PTh with carbon materials show higher electrochemical performance and cyclic stability as compared to neat PTh via a synergistic effect. The enhanced properties are affected by a variety of factors, including effective mass loading of PTh on carbon materials, nature of carbon materials, morphology and synthetic routes.

All in all, compared with pristine CPs, the electrochemical performance of CP-carbon composites can be enhanced significantly, especially power density and cycling performance, however, the energy density of the composites is still lower than LIBs because of their charge accumulation ability [112]. Therefore, introduction of other pseudo-capacitive materials into CP-based composites has been taken into account for further development of electrodes materials for SCs.

5.3.2 Metal Oxide/Conducting Polymer Binary Composites

The carbon materials have low specific capacitance and their disagreements too show in their composites with CPs. Metal oxides, especially transition metal oxides, such as Co_3O_4 [113], NiO [114], MnO_2 [115], CuO [116], CoO [117], Fe_2O_3 [118], TiO_2 [119], WO_3 [120], VOx [120] and RuO_2[121], are idealistic pseudo-capacitive materials due to their high theoretical capacitance. Most transition metal oxides pertain to semiconductors which have wide band gap. The wide band gap gives rise to low electron and hole concentrations causing low electrical conductivity [122]. Thus, the composites of conducting polymer/metal oxides are beneficial in the field of SCs due to the good conductivity of CPs as well as the high capacitance of the polymer composites.

Manganese dioxide (MnO_2) is one of the most important pseudocapacitive materials due to its multi-valence reaction, low cost, high surface area, excellent charge storage capacity characteristic as well as eco-friendly nature. MnO_2 has extraordinary theoretical capacitance (~1370 F g^{-1}), but its low electrical conductivity is a major limitation towards its commercialization, which ultimately leads to its poor capacitance and low power output at high current density [123]. Furthermore, in the slightly acidic environment of SCs, it shows poor stability. Composites of CPs with MnO_2 can combine the qualities (good electric conductivity of CPs and relatively better cyclic stability of MnO_2) of both components through well-designed nanostructures. For instance, composites of MnO_2 with PANI have been reported by many researchers under different conditions, showing high capacitance and high cyclic stability [123-125]. Similarly, PPy/MnO_2 [126,127] and PTh/MnO_2 [14,128] composites also exhibit better performance than the individual polymer and MnO_2. Many researchers have also reported PEDOT/MnO_2 composites by adopting different strategies and by altering the source of MnO_2. PEDOT:PSS/MnO_2 composites [129] having different morphologies were also studied as electrode material in supercapacitors, where the composites had 3D morphology, which enhanced the ion diffusion and utilization of active materials.

The electrochemical performance of CP based composites with other metal oxides has also been investigated. For example, Ates *et al.* [116] prepared the binary composites of CuO with different CPs, i.e. PPy/CuO, PEDOT/CuO and PANI/CuO, via electrochemical method on glassy carbon electrode and studied their performance in H_2SO_4 solution. PANI/CuO nanocomposite presented the highest performance and was confirmed as an ideal candidate for SC electrodes as compared to PEDOT/CuO and PPy/CuO binary composites. The specific capacitance was 286.35 F/g at 20 mV/s for PANI/CuO, while the maximum values of PPy/CuO and PEDOT/CuO were 20.78 and 198.90 F/g at 5 mV/s, respectively.

Some traditional battery materials, such as NiO_2 or $NiCo_2O_4$, at nanoscale also exhibit the capacitive-like behavior. Spinel $NiCo_2O_4$, with two metal cations (Ni and Co), retains higher electrochemical activity and theoretical capacitance than corresponding mono metal oxides. However, the larger challenge for $NiCo_2O_4$ is the poor cycling stability and rate performance, consequentially due to its poor conductivity and degradation during long-term cycling. To solve this issue, many researchers have combined $NiCo_2O_4$ with CPs or carbon nanomaterials, which has been proved to be an effective approach. Xu *et al.* [130] reported the binary PANI/$NiCo_2O_4$ nanocomposites by *in-situ* chemical oxidation polymerization. The subsequent nanocomposite showed high specific capacitance of 439.4 F/g with a retention of ~66.11% after 1000 cycles at a current density of 5 mA/cm^{-2}, signifying that the binary PANI/$NiCo_2O_4$ nanocomposite retained good electrochemical performance as compared to its primary constitutes.

Apart from traditional carbon nanomaterials and metal oxide materials, other materials have also been used to prepare conducting polymer based composites for supercapacitor electrodes, such as PANI/$Ni(OH)_2$ [131] , PANI/MoS_2 [132] and PPy/MoS_2 [133].

5.3.3 Carbon/Metal Oxide/Conducting Polymer Ternary Composites

Binary composites have made much progress in the field of SCs, however, their properties still cannot meet the requirements such as high energy density, high capacitance and long cycling performance for EES. To overcome the limitations of binary composites, efforts have been made to develop ternary composites as electrode material for SCs. In this respect, CPs, conducting carbon and metal oxides (or other compounds) have been intensively investigated as ternary composites to accomplish the requirements of SCs. Each constituent in the composite plays an important role in uplifting and overcoming the drawbacks of other components by the synergistic effects. The electrode materials based on ternary composite are expected to significantly improve the capacitance of SCs for next generations. Graphene based ternary composites with PANI, PPy, PTh, and PEDOT with metal oxides have been widely studied till date [134].

Many researchers have reported the ternary composites of graphene and PANI with Fe_2O_3 [135,136], MoO_3, [137] and $NiMoO_4$. [138]. In each case, the specific capacitance increased as compared to pristine CPs or their binary composites. Hao and coworkers synthesized a ternary nanocomposite of graphene, Fe_2O_3 and PANI by a facile two-step methodology [135]. The developed GE/Fe_2O_3/PANI (GFP) composite overcame the limitations of the binary composites to a great extent. The GFP composite was obtained by the *in-situ* polymerization of aniline in the presence of graphene/Fe_2O_3 (GF), which was prepared by the reduction of GO by Fe^{2+}. A well-regulated thin film of conducting PANI could be finely deposited on the GF composite sheets. The structural data showed that GF sheets were effectively enveloped by PANI, as shown in Figure 5.5. The ternary composite

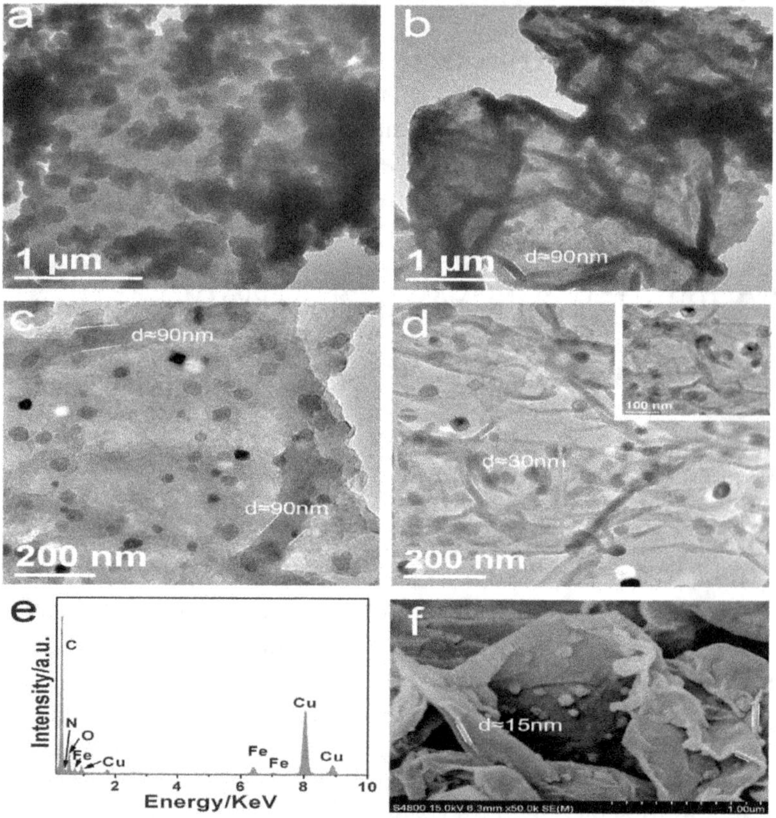

Figure 5.5 (a) Low-magnification TEM images of GFP_{10}, (b). and GFP_1 (c, d) High-magnification TEM images of GFP_1 and $GFP_{0.5}$ respectively .(d). The inset in (d) is the enlarged part of the region indicated by the rectangle. (e) EDX data from the regions of $GFP_{0.5}$ (f). SEM image of $GFP_{0.5}$ Reproduced from reference 135 with permission from Royal Society of Chemistry.

provided a direct short pathway for charge/electrode diffusion because of the finely dispersed nanoparticles of Fe_2O_3, attached firmly on the GE sheets and also due to the fabrication of PANI film. $GFP_{0.5}$, among the ternary composites of GE/Fe_2O_3 and PANI using different ratios, showed the excellent electrochemical performance, such as 107 Wh kg^{-1} of energy density at power density of 351 W kg^{-1} and even 17 Wh kg^{-1} at the high power density of 4407 W kg^{-1}. The resulting composite also attained the high capacitance of 638 F g^{-1} along with excellent stability with a fall of only 8% after 5000 cycles. In this ternary composite, the thin film of PANI not only increased the surface area, but also prevented the aggregation, dissolution and volume changes of Fe_2O_3 during charge/discharge process. Bolagam *et al.* [139] also reported a ternary composite having PANI, SGE and TiO_2 to improve the energy density and cyclic performance of CPs. The resulting composite had sphere-shaped morphology and exhibited enhanced electrochemical performance as compared to its constituents.

Ternary composites of PPy with MnO_2 have also been designed. With the introduction of EDL capacitive carbon and pseudocapacitive MnO_2, the ternary composite materials of MnO_2 with PPy/activated carbon have shown high capacitance and high cyclic stability [140-142]. Hou *et al.* [143] also designed a ternary nanocomposite consisting of MnO_2, CNTs and PEDOT-PSS. Maximum specific capacitance (427 F g^{-1}) was obtained from the ternary composite with excellent cyclic rate. The electrode showed good cycling stability, with 99% retention even after 1000 cycles. In ternary composite, CNTs not only enhanced the electrical conductivity and mechanical stability but also offered high surface for the deposition of hierarchical MnO_2 porous nanospheres. PEDOT-PSS acted both as an effective dispersant for CNTs/MnO_2 and a binder to improve the contact of adjacent CNTs/MnO_2 nanoparticles and adhesion of the composite to the substrate. MnO_2 nanospheres provided numerous active sites for contributing large pseudo-capacitance.

Hao and coworkers also reported the ternary electrode materials of both PPy [144] and PEDOT [145] based on GE and tin oxide (SnO_2) via one pot method at a moderate temperature. The excellent electrochemical performance, especially excellent cycling stability of the composites, was achieved due to the well-designed structure, good combination of components and synergistic effect between the three components.

In recent years, the composites of metals, especially metal nanoparticles, with CPs and active carbon materials is an emerging technique for elevating the electrochemical properties of the composites. Usually, metals such as Au and Ag are decorated on the surface of CPs, thus, improving electrical and thermal conductivity, impact strength, cycle life of the CPs, etc. [146-148].

Overall, the research on SCs indicates that the composites of CPs offer a great potential for the improvement in the cycling stability. Especially, the ternary composites of CPs have emerged as the superior candidates for the next generation SCs. Some important issues still need to be addressed, such as complex synthetic protocols, low mass loading of active materials, low specific capacitance and rate capability as well as poor cycling stability. Therefore, efforts are required for precisely optimizing the procedures for the achievement of environment friendly, cheap and multivariate composites for large-scale application as SCs.

5.4 Conclusions and Future Outlook

In this chapter, an attempt has been made to comprehensively summarize the recent studies reporting the development of conducting polymer composite-based electrode materials, for the electrochemical energy storage systems, with especial emphasis on their importance in LIBs, Li/S batteries and SCs. These electrochemical energy storage devices are pervasive in our daily routine.

In the last two decades, substantial efforts have been made by the researchers for the manufacturing and application of CPs based composites because of their high conductivity, ease of preparation, excellent flexibility, low cost and good environmental stability. Many CPs are used as active electrode materials for various energy storage systems, however, face many operational issues. Thus, to optimize the desired properties required for a dynamic material, CPs have been combined with other active materials like carbon-based materials, metal nanoparticles and metal oxides, etc. Many techniques have been employed for preparing the composites with appropriate morphology, dimensions and size, thus, resulting in significant advancements in the past two decades. These composites exhibit superior performance due to synergic interaction between their components.

CP based composites have enhanced electrochemical performance for Li-batteries and SCs, in which they generally act as a porous conductive support, protective network or/and connective matrix on the surface of active electrode materials. However, to receive the desired attributes in the composites, further theoretical and experimental research is needed to inaugurate the best possible structure property relationships and to design materials with a unique set of features. Fabrication of the binary or ternary composites with desired properties is the key in the development of high energy density and high power output materials for LIBs and SCs with long term cyclic stability for realizing the vision of reduced reliance on fossil fuels and safe energy for future. This emerging class also lightens up the igniter on the horizon to replace the precious materials with non-precious ones. Thus, as mentioned earlier, in order to promote the CPs based composite research to the next level, both synthetic and computational chemistry can play a vibrant role. Precision in the synthetic approaches will help in the development of the composites with controlled structure in good yield. The computational studies will also help in further understanding of the fundamental mechanisms of the composites, which would ultimately enable us to improve the performance of energy storage systems.

References

1. Whittingham, M. S. (2008) Materials challenges facing electrical energy storage. *MRS Bulletin*, **33**(4), 411-419.
2. Liu, C., Li, F., Ma, L. P., and Cheng, H. M. (2010) Advanced materials for energy storage. *Advanced materials*, **22**(8), E28-E62.
3. Kaygusuz, K. (2011) Energy services and energy poverty for sustainable rural development. *Renewable and Sustainable Energy Reviews*, **15**(2), 936-947.
4. Beaudin, M., Zareipour, H., Schellenberglabe, A., and Rosehart, W. (2010) Energy storage for mitigating the variability of renewable electricity sources: An updated review. *Energy for Sustainable Development*, **14**(4), 302-314.
5. Hannan, M., Hoque, M., Mohamed, A., and Ayob, A. (2017) Review of energy storage systems for electric vehicle applications: Issues and challenges. *Renewable and Sustainable Energy Reviews*, **69**, 771-789.
6. Das, T. K., and Prusty, S. (2012) Review on conducting polymers and their applications. *Polymer-Plastics Technology and Engineering*, **51**(14),1487-1500.
7. Yang, J., Liu, Y., Liu, S, Li, L, Zhang, C., and Liu, T. (2017) Conducting polymer composites: material synthesis and applications in electrochemical capacitive energy storage. *Materials Chemistry Frontiers*, **1**(2), 251-268.
8. Chan, K.-Y., Jia, B., Lin, H., Hameed, N., Lee, J.-H., and Lau, K.-T.(2018) A Critical review on multifunctional composites as structural capacitors for energy storage. *Composite Structures*, **188**, 126-142.
9. Long, L., Wang, S., Xiao, M., and Meng, Y. (2016) Polymer electrolytes for lithium polymer batteries. *Journal of Materials Chemistry A*, **4**(26), 10038-10069.

10. Goodenough, J. B., and Park, K.-S. (2013) The Li-ion rechargeable battery: a perspective. *Journal of American Chemical Society*, **135**(4),1167-1176.
11. Arya, A., and Sharma, A. (2017) Polymer electrolytes for lithium ion batteries: A critical study. *Ionics*, **23**(3),497-540.
12. Wang, H., Lin, J., and Shen, Z. X. (2016) Polyaniline (PANi) based electrode materials for energy storage and conversion. *Journal of Science: Advanced Materials and Devices*, **1**(3), 225-255.
13. Cuentas-Gallegos, A., and Gomez-Romero, P. (2005) In-situ synthesis of polypyrrole-MnO_2-x nanocomposite hybrids. *Journal of New Materials for Electrochemical Systems*, **8**(3), 181-188.
14. Xiao, W., Chen, J. S., Lu, Q., and Lou, X. W. (2010) Porous spheres assembled from polythiophene (PTh)-coated ultrathin MnO_2 nanosheets with enhanced lithium storage capabilities. *The Journal of Physical Chemistry C*, **114**(27), 12048-12051.
15. Otero, T., and Cantero, I. (1999) Conducting polymers as positive electrodes in rechargeable lithium-ion batteries. *Journal of Power Sources*, **81**, 838-841.
16. Sengodu, P., and Deshmukh, A. D. (2015) Conducting polymers and their inorganic composites for advanced Li-ion batteries: A review. *RSC Advances*, **5**(52), 42109-42130.
17. Li, G., Pang, S., Jiang, L., Guo, Z., and Zhang, Z. (2006) Environmentally friendly chemical route to vanadium oxide single-crystalline nanobelts as a cathode material for lithium-ion batteries. *Journal of Physical Chemistry B*, **110**(19), 9383-9386.
18. Wang, S., Li, S., Sun, Y., Feng, X., and Chen, C. (2011) Three-dimensional porous V_2O_5 cathode with ultra high rate capability. *Energy & Environmental Science*, **4**(8), 2854-2857.
19. Liu, D., and Cao, G. (2010) Engineering nanostructured electrodes and fabrication of film electrodes for efficient lithium ion intercalation. *Energy & Environmental Science*, **3**(9), 1218-1237.
20. Park, K.-I., Song, H.-M., Kim, Y., Mho, S.-i., Cho, W. I., and Yeo, I.-H. (2010) Electrochemical preparation and characterization of V_2O_5/polyaniline composite film cathodes for Li battery. *Electrochimica Acta*, **55**(27), 8023-8029.
21. Chen, Y., Yang, G., Zhang, Z., Yang, X., Hou, W., and Zhu, J.-J. (2010) Polyaniline-intercalated layered vanadium oxide nanocomposites - One-pot hydrothermal synthesis and application in lithium battery. *Nanoscale*, **2**(10), 2131-2138.
22. Zhao, H., Yuan, A., Liu, B., Xing, S., Wu, X., and Xu, J. (2012) High cyclic performance of V_2O_5@PPy composite as cathode of recharged lithium batteries. *Journal of Applied Electrochemistry*, **42**(3), 139-144.
23. Chao, D., Xia, X., Liu, J., Fan, Z., Ng, C. F., Lin, J., Zhang, H., Shen, Z. X., Fan, H. J. (2014) A V_2O_5/conductive-polymer core/shell nanobelt array on three-dimensional graphite foam: A high-rate, ultrastable,

and freestanding cathode for lithium-ion batteries. *Advanced Materials*, **26**(33), 5794-5800.

24. Chikkannanavar, S. B., Bernardi, D. M., and Liu, L. (2014) A review of blended cathode materials for use in Li-ion batteries. *Journal of Power Sources*, **248**, 91-100.
25. Gao, X.-W., Wang, J.-Z., Chou, S.-L., and Liu, H.-K. (2012) Synthesis and electrochemical performance of LiV_3O_8/polyaniline as cathode material for the lithium battery. *Journal of Power Sources*, **220**, 47-53.
26. Yan, H., Wu, X., and Li,Y. (2015) Preparation and characterization of conducting polyaniline-coated $LiVPO_4F$ nanocrystals with core-shell structure and its application in lithium-ion batteries. *Electrochimica Acta*, **182**, 437-444.
27. Qi, M., Xie, D., Zhong, Y., Chen, M., and Xia, X. (2017) Smart construction of polyaniline shell on cobalt oxides as integrated core-shell arrays for enhanced lithium ion batteries. *Electrochimica Acta*, **247**, 701-707.
28. Zhang, Z.-j., Wang, J.-Z., Chou, S.-L., Liu, H.-K., Ozawa, K., and Li, H.-j. (2013) Polypyrrole-coated α-$LiFeO_2$ nanocomposite with enhanced electrochemical properties for lithium-ion batteries. *Electrochimica Acta*, **108**, 820-826.
29. Trinh, N., Saulnier, M., Lepage, D., and Schougaard, S. (2013) Conductive polymer film supporting $LiFePO_4$ as composite cathode for lithium ion batteries. *Journal of Power Sources*, **221**, 284-289.
30. Bai, Y.-m., Qiu, P., Wen. Z. -l., and Han, S.-c. (2010) Improvement of electrochemical performances of $LiFePO_4$ cathode materials by coating of polythiophene. *Journal of Alloys and Compounds*, **508**(1), 1-4.
31. Novak, P., Muller, K., Santhanam, K., and Haas, O. (1997) Electrochemically active polymers for rechargeable batteries. *Chemical Reviews*, **97**(1), 207-282.
32. Liu, P., Han, J.-J., Jiang, L.-F., Li, Z.-Y., and Cheng, J.-N. (2017) Polyaniline/multi-walled carbon nanotubes composite with core-shell structures as a cathode material for rechargeable lithium-polymer cells. *Applied Surface Science*, **400**, 446-452.
33. An, S. J., Li, J., Daniel, C., Mohanty, D., Nagpure, S., and Wood III, D. L. (2016) The state of understanding of the lithium-ion-battery graphite solid electrolyte interphase (SEI) and its relationship to formation cycling. *Carbon*, **105**, 52-76.
34. Roy, P., and Srivastava, S. K. (2015) Nanostructured anode materials for lithium ion batteries. *Journal of Materials Chemistry A*, **3**(6), 2454-2484.
35. Gangopadhyay, R., and De, A. (2000) Conducting polymer nanocomposites: a brief overview. *Chemistry of Materials*, **12**(3), 608-622.
36. Ha, J. -S., Lee, J.-M., Lee, H. -R., Huh, P., and Jo, N. -J. (2015) Polymer

nanocomposite electrode consisting of polyaniline and modified multi-walled carbon nanotube for rechargeable battery. *Journal of Nanoscience and Nanotechnology*, **15**(11), 8977-8983.
37. He, Z. -Q., Xiong, L. -Z., Shang, C., Wu, X. -M., Liu, W. -P., and Huang, K. -L. (2010) In situ polymerization preparation and characterization of $Li_4Ti_5O_{12}$-polyaniline anode material. *Transactions of Nonferrous Metals Society of China*, **20**, s262-s266.
38. Wang, X., Shen, L., Li, H., Wang, J., Dou, H., and Zhang, X. (2014) PEDOT coated $Li_4Ti_5O_{12}$ nanorods: Soft chemistry approach synthesis and their lithium storage properties. *Electrochimica Acta*, **129**, 283-289.
39. Xu, D., Wang, P., and Yang, R. (2017) Conducting polythiophene-wrapped Li4Ti5O12 spinel anode material for ultralong cycle-life Li-ion batteries. *Ceramics International*, **43**(5), 4712-4715.
40. Liu, H., Liu, B. H., and Li, Z. P. (2016) A reduced graphene oxide/SnO_2/polyaniline nanocomposite for the anode material of Li-ion batteries. *Solid State Ionics*, **294**, 6-14.
41. Zheng, H., Ncube, N. M., Raju, K., Mphahlele, N., and Mathe, M. (2016) The effect of polyaniline on TiO_2 nanoparticles as anode materials for lithium ion batteries. *SpringerPlus*, **5**(1), 630.
42. Ma, D., Cao, Z., and Hu, A. (2014) Si-based anode materials for Li-ion batteries: A mini review. *Nano-Micro Letters*, **6**(4), 347-358.
43. Wu, H., Yu, G., Pan, L., Liu, N., McDowell, M. T., Bao, Z., and Cui, Y. (2013) Stable Li-ion battery anodes by in-situ polymerization of conducting hydrogel to conformally coat silicon nanoparticles. *Nature communications*, **4**, 1943.
44. Kummer, M., Badillo, J., Schmitz, A., Bremes, H. -G., Winter, M., Schulz, C., and Wiggers, H. (2014) Silicon/polyaniline nanocomposites as anode material for lithium ion batteries. *Journal of the Electrochemical Society*, **161**(1), A40-A45.
45. Xiao, L., Sehlleier, Y. H., Dobrowolny, S., Mahlendorf, F., Heinzel, A., Schulz, C., and Wiggers, H. (2017) Novel Si-CNT/polyaniline nanocomposites as Lithium-ion battery anodes for improved cycling performance. *Materials Today: Proceedings*, **4**, S263-S268.
46. Lawes, S., Sun, Q., Lushington, A., Xiao, B., Liu, Y., and Sun, X. (2017) Inkjet-printed silicon as high performance anodes for Li-ion batteries. *Nano Energy*, **36**, 313-321.
47. Jayaprakash, N., Shen, J., Moganty, S. S., Corona, A., and Archer, L. A. (2011) Porous hollow carbon@ sulfur composites for high-power lithium–sulfur batteries. *Angewandte Chemie*, **123**(26), 6026-6030.
48. Ji, X., and Nazar, L. F. (2010) Advances in Li–S batteries. *Journal of Materials Chemistry*, **20**(44), 9821-9826.
49. Abdelhamid, M. E., O'Mullane, A. P., and Snook, G. A. (2015) Storing energy in plastics: a review on conducting polymers & their role in electrochemical energy storage. *RSC Advances*, **5**(15), 11611-11626.

50. Li, W., Zhang, Q., Zheng, G., She, Z. W., Yao, H., and Cui, Y. (2013) Understanding the role of different conductive polymers in improving the nanostructured sulfur cathode performance. *Nano letters*, **13**(11), 5534-5540.
51. Yang, Y., Yu, G., Cha, J. J., Wu, H., Vosgueritchian, M., Yao, Y., Bao, Z., and Cui, Y. (2011) Improving the performance of lithium–sulfur batteries by conductive polymer coating. *ACS Nano*, **5**(11), 9187-9193.
52. Chen, H., Dong, W., Ge, J., Wang, C., Wu, X., Lu, W., and Chen, L. (2013) Ultrafine sulfur nanoparticles in conducting polymer shell as cathode materials for high performance lithium/sulfur batteries. *Scientific Reports*, **3**, 1910.
53. Fu, Y., and Manthiram, A. (2012) Enhanced cyclability of lithium-sulfur batteries by a polymer acid-doped polypyrrole mixed ionic-electronic conductor. *Chemistry of Materials*, **24**(15), 3081-3087.
54. Liang, X., Wen, Z., Liu, Y., Zhang, H., Jin, J., Wu, M., and Wu, X. (2012) A composite of sulfur and polypyrrole–multi walled carbon combinatorial nanotube as cathode for Li/S battery. *Journal of Power Sources*, **206**, 409-413.
55. Ma, G., Wen, Z., Jin, J., Lu, Y., Rui, K., Wu, X., Wu, M., and Zhang, J. (2014) Enhanced performance of lithium sulfur battery with polypyrrole warped mesoporous carbon/sulfur composite. *Journal of Power Sources*, **254**, 353-359.
56. Ma, G., Wen, Z., Wang, Q., Shen, C., Jin, J., and Wu, X. (2014) Enhanced cycle performance of a Li–S battery based on a protected lithium anode. *Journal of Materials Chemistry A*, **2**(45), 19355-19359.
57. Chen, G. Z. (2017) Supercapacitor and supercapattery as emerging electrochemical energy stores. *International Materials Reviews*, **62**(4), 173-202.
58. Gonzalez, A., Goikolea, E., Barrena, J. A., and Mysyk, R. (2016) Review on supercapacitors: technologies and materials. *Renewable and Sustainable Energy Reviews*, **58**, 1189-1206.
59. Simon, P., and Gogotsi, Y. (2008) Materials for electrochemical capacitors. *Nature Materials*, **7**(11), 845-854.
60. Meng, Q., Cai, K., Chen, Y., and Chen, L. (2017) Research progress on conducting polymer based supercapacitor electrode materials. *Nano Energy*, **36**, 268-285.
61. Shao, L., Jeon, J. -W., and Lutkenhaus, J. L. (2011) Polyaniline/vanadium pentoxide layer-by-layer electrodes for energy storage. *Chemistry of Materials*, **24**(1), 181-189.
62. Jiang, H., Ma, J., and Li, C. (2012) Polyaniline–MnO_2 coaxial nanofiber with hierarchical structure for high-performance supercapacitors. *Journal of Materials Chemistry*, **22**(33), 16939-16942.
63. Xue, M., Li, F., Zhu, J., Song, H., Zhang, M., and Cao, T. (2012) Structure-based enhanced capacitance: in situ growth of highly ordered polyaniline nanorods on reduced graphene oxide patterns. *Advan-*

ced Functional Materials, **22**(6), 1284-1290.

64. Lang, X., Zhang, L., Fujita, T., Ding, Y., and Chen, M. (2012) Three-dimensional bicontinuous nanoporous Au/polyaniline hybrid films for high-performance electrochemical supercapacitors. *Journal of Power Sources*, **197**, 325-329.
65. Mondal, S., Rana, U., and Malik, S. (2015) Graphene quantum dot-doped polyaniline nanofiber as high performance supercapacitor electrode materials. *Chemical Communications*, **51**(62), 12365-12368.
66. Hyder, M. N., Lee, S. W., Cebeci, F. C., Schmidt, D. J., Shao-Horn, Y., and Hammond, P. T. (2011) Layer-by-layer assembled polyaniline nanofiber/multiwall carbon nanotube thin film electrodes for high-power and high-energy storage applications. *ACS Nano*, **5**(11), 8552-8561.
67. Yu, G., Xie, X., Pan, L., Bao, Z., and Cui, Y. (2013) Hybrid nanostructured materials for high-performance electrochemical capacitors. *Nano Energy*, **2**(2), 213-234.
68. Yan, J., Wei, T., Shao, B., Fan, Z., Qian, W., Zhang, M., and Wei, F. (2010) Preparation of a graphene nanosheet/polyaniline composite with high specific capacitance. *Carbon*, **48**(2), 487-493.
69. Feng, X. M., Li, R. M., Ma, Y. W., Chen, R. F., Shi, N. E., Fan, Q. L., and Huang, W. (2011) One-step electrochemical synthesis of graphene/polyaniline composite film and its applications. *Advanced Functional Materials*, **21**(15), 2989-2996.
70. Zhang, Q., Li, Y., Feng, Y., and Feng, W. (2013) Electropolymerization of graphene oxide/polyaniline composite for high-performance supercapacitor. *Electrochimica Acta*, **90**, 95-100.
71. Zhang, K., Zhang, L. L., Zhao, X., and Wu, J. (2010) Graphene/polyaniline nanofiber composites as supercapacitor electrodes. *Chemistry of Materials*, **22**(4), 1392-1401.
72. Wang, S., Ma, L., Gan, M., Fu, S., Dai, W., Zhou, T., Sun, X., Wang, H., and Wang, H. (2015) Free-standing 3D graphene/polyaniline composite film electrodes for high-performance supercapacitors. *Journal of Power Sources*, **299**, 347-355.
73. Cong, H-. P., Ren, X. -C., Wang, P., and Yu, S.-H. (2013) Flexible graphene–polyaniline composite paper for high-performance supercapacitor. *Energy & Environmental Science*, **6**(4), 1185-1191.
74. Imani, A., and Farzi, G. (2015) Facile route for multi-walled carbon nanotube coating with polyaniline: tubular morphology nanocomposites for supercapacitor applications. *Journal of Materials Science: Materials in Electronics*, **26**(10), 7438-7444.
75. Niu, Z., Luan, P., Shao, Q., Dong, H., Li, J., Chen, J., Zhao, D., Cai, L., Zhou, W., and Chen, X. (2012) A "skeleton/skin" strategy for preparing ultrathin free-standing single-walled carbon nanotube/polyaniline films for high performance supercapacitor electrodes. *Energy & En-*

vironmental Science, **5**(9), 8726-8733.

76. Mi, H., Zhang, X., An, S., Ye, X., and Yang, S. (2007) Microwave-assisted synthesis and electrochemical capacitance of polyaniline/multi-wall carbon nanotubes composite. *Electrochemistry Communications*, **9**(12), 2859-2862.
77. Tran, C., Singhal, R., Lawrence, D., and Kalra, V. (2015) Polyaniline-coated freestanding porous carbon nanofibers as efficient hybrid electrodes for supercapacitors. *Journal of Power Sources*, **293**, 373-379.
78. Sun, H., She, P., Xu, K., Shang, Y., Yin, S., and Liu, Z. (2015) A self-standing nanocomposite foam of polyaniline@ reduced graphene oxide for flexible super-capacitors. *Synthetic Metals*, **209**, 68-73.
79. Wang, H., Hao, Q., Yang, X., Lu, L., and Wang, X. (2009) Graphene oxide doped polyaniline for supercapacitors. *Electrochemistry Communications*, **11**(6), 1158-1161.
80. Khosrozadeh, A., Xing, M., and Wang, Q. (2015) A high-capacitance solid-state supercapacitor based on free-standing film of polyaniline and carbon particles. *Applied Energy*, **153**, 87-93.
81. Shen, K., Ran, F., Zhang, X., Liu, C., Wang, N., Niu, X., Liu, Y., Zhang, D., Kong, L., and Kang, L. (2015) Supercapacitor electrodes based on nano-polyaniline deposited on hollow carbon spheres derived from cross-linked co-polymers. *Synthetic Metals*, **209**, 369-376.
82. Du, Y., Cai, K., and Shen, S. Z. (2013) Facile preparation and characterization of graphene nanosheet/polyaniline nanofiber thermoelectric composites. *Functional Materials Letters*, **6**(5), 1340002.
83. Hao, Q., Xia, X., Lei, W., Wang, W., and Qiu, J. (2015) Facile synthesis of sandwich-like polyaniline/boron-doped graphene nano hybrid for supercapacitors. *Carbon*, **81**, 552-563.
84. Hao, Q. L., Wang, H. L., Yang, X. J., Lu, L. D., and Wang, X. (2010) Liquid/liquid interfacial polymerization to fabricate sulfonated graphene/polyaniline nanocomposite for supercapacitors. *Applied Mechanics and Materials*, **29-32**, 1902-1906.
85. Hao, Q., Wang, H., Yang, X., Lu, L., and Wang, X. (2011) Morphology-controlled fabrication of sulfonated graphene/polyaniline nanocomposites by liquid/liquid interfacial polymerization and investigation of their electrochemical properties. *Nano Research*, **4**(4), 323-333.
86. Yang, L., Shi, Z., and Yang, W. (2015) Polypyrrole directly bonded to air-plasma activated carbon nanotube as electrode materials for high-performance supercapacitor. *Electrochimica Acta*, **153**, 76-82.
87. An, K. H., Jeon, K. K., Heo, J. K., Lim, S. C., Bae, D. J., and Lee, Y. H. (2002) High-capacitance supercapacitor using a nanocomposite electrode of single-walled carbon nanotube and polypyrrole. *Journal of the Electrochemical Society*, **149**(8), A1058-A1062.
88. Song, H., Cai, K., Wang, J., Shen, S. (2016,) Influence of polymeriza-

tion method on the thermoelectric properties of multi-walled carbon nanotubes/polypyrrole composites. *Synthetic Metals* **211**, 58-65.

89. Wang, J., Cai, K., Shen, S., and Yin, J. (2014) Preparation and thermoelectric properties of multi-walled carbon nanotubes/polypyrrole composites. *Synthetic Metals*, **195**, 132-136.
90. Chen, Y., Du, L., Yang, P., Sun, P., Yu, X., and Mai, W. (2015) Significantly enhanced robustness and electrochemical performance of flexible carbon nanotube-based supercapacitors by electrodepositing polypyrrole. *Journal of Power Sources*, **287**, 68-74.
91. Zhou, H., Han, G., Xiao, Y., Chang, Y., and Zhai, H.-J. (2015) A comparative study on long and short carbon nanotubes-incorporated polypyrrole/poly (sodium 4-styrenesulfonate) nanocomposites as high-performance supercapacitor electrodes. *Synthetic Metals*, **209**, 405-411.
92. Warren, R., Sammoura, F., The, K. S., Kozinda, A., Zang, X., and Lin, L. (2015) Electrochemically synthesized and vertically aligned carbon nanotube–polypyrrole nanolayers for high energy storage devices. *Sensors and Actuators A: Physical*, **231**, 65-73.
93. Keskinen, J., Tuurala, S., Sjodin, M., Kiri, K., Nyholm, L., Flyktman, T., Stromme, M., and Smolander, M. (2015) Asymmetric and symmetric supercapacitors based on polypyrrole and activated carbon electrodes. *Synthetic Metals* **203**, 192-199.
94. Gao, B., He, D., Yan, B., Suo, H., and Zhao, C. (2015) Flexible carbon cloth based polypyrrole for an electrochemical supercapacitor. *Journal of Materials Science: Materials in Electronics*, **26**(9), 6373-6379.
95. Raj, C. J., Kim, B. C., Cho, W.-J., Lee, W.-g., Jung, S.-D., Kim, Y. H., Park, S. Y., and Yu, K. H. (2015) Highly flexible and planar supercapacitors using graphite flakes/polypyrrole in polymer lapping film. *ACS Applied Materials & Interfaces*, **7**(24),13405-13414.
96. Tao, J., Ma, W., Liu, N., Ren, X., Shi, Y., Su, J., and Gao, Y. (2015) High-performance solid-state supercapacitors fabricated by pencil drawing and polypyrrole depositing on paper substrate. *Nano-Micro Letters*, **7**(3), 276-281.
97. Cai, J., Niu, H., Li, Z., Du, Y., Cizek, P., Xie, Z., Xiong, H., and Lin, T. (2015) High-performance supercapacitor electrode materials from cellulose-derived carbon nanofibers. *ACS Applied Materials & Interfaces*, **7**(27), 14946-14953.
98. Zhao, Z., Richardson, G. F., Meng, Q., Zhu, S., Kuan, H.-C., and Ma, J. (2015) PEDOT-based composites as electrode materials for supercapacitors. *Nanotechnology*, **27**(4), 042001.
99. Wen, J., Jiang, Y., Yang, Y., and Li, S. (2014) Conducting polymer and reduced graphene oxide Langmuir–Blodgett films: a hybrid nanostructure for high performance electrode applications. *Journal*

of Materials Science: Materials in Electronics, **25**(2), 1063-1071.

100. Chu, C.-Y., Tsai, J.-T., and Sun, C.-L. (2012) Synthesis of PEDOT-modified graphene composite materials as flexible electrodes for energy storage and conversion applications. *International Journal of Hydrogen Energy*, **37**(18), 13880-13886.
101. Lee, S., Cho, M. S., Lee, H., Nam, J.-D., and Lee, Y. (2012) A facile synthetic route for well defined multilayer films of graphene and PEDOT via an electrochemical method. *Journal of Materials Chemistry*, **22**(5), 1899-1903.
102. Sun, D., Jin, L., Chen, Y., Zhang, J. R., and Zhu, J. J. (2013) Microwave-assisted in situ synthesis of graphene/PEDOT hybrid and its application in supercapacitors. *ChemPlusChem*, **78**(3), 227-234.
103. Zhang, J., and Zhao, X. (2012) Conducting polymers directly coated on reduced graphene oxide sheets as high-performance supercapacitor electrodes. *The Journal of Physical Chemistry C*, **116**(9), 5420-5426.
104. Lehtimaki, S., Suominen, M., Damlin, P., Tuukkanen, S., Kvarnstrom, C., and Lupo, D. (2015) Preparation of supercapacitors on flexible substrates with electrodeposited PEDOT/graphene composites. *ACS Applied Materials & Interfaces*, **7**(40), 22137-22147.
105. He, X., Yang, W., Mao, X., Xu, L., Zhou, Y., Chen, Y., Zhao, Y., Yang, Y., and Xu, J. (2018) All-solid state symmetric supercapacitors based on compressible and flexible free-standing 3D carbon nanotubes (CNTs)/poly (3, 4-ethylenedioxythiophene) (PEDOT) sponge electrodes. *Journal of Power Sources*, **376**, 138-146.
106. Fu, C., Zhou, H., Liu, R., Huang, Z., Chen, J., and Kuang, Y. (2012) Supercapacitor based on electropolymerized polythiophene and multi-walled carbon nanotubes composites. *Materials Chemistry and Physics*, **132**(2-3), 596-600.
107. Zhang, H., Hu, Z., Li, M., Hu, L., and Jiao, S. (2014) A high-performance supercapacitor based on a polythiophene/multiwalled carbon nanotube composite by electropolymerization in an ionic liquid microemulsion. *Journal of Materials Chemistry A*, **2**(40), 17024-17030.
108. Wang, Y., Tao, S., An, Y., Wu, S., and Meng, C. (2013) Bio-inspired high performance electrochemical supercapacitors based on conducting polymer modified coral-like monolithic carbon. *Journal of Materials Chemistry A*, **1**(31), 8876-8887.
109. Hur, E., Varol, G. A., and Arslan, A. (2013) The study of polythiophene, poly (3-methylthiophene) and poly (3, 4-ethylenedioxythiophene) on pencil graphite electrode as an electrode active material for supercapacitor applications. *Synthetic Metals*, **184**, 16-22.
110. Kim, M., Lee, C., Seo, Y. D., Cho, S., Kim, J., Lee, G., Kim, Y. K., and Jang, J. (2015) Fabrication of various conducting polymers using graphene oxide as a chemical oxidant. *Chemistry of Materials*, **27**(18), 6238-6248.

111. Alabadi, A., Razzaque, S., Dong, Z., Wang, W., and Tan, B. (2016) Graphene oxide-polythiophene derivative hybrid nanosheet for enhancing performance of supercapacitor. *Journal of Power Sources*, **306**, 241-247.
112. Wang, H., Zhang, Y., Ang, H., Zhang, Y., Tan, H. T., Zhang, Y., Guo, Y., Franklin, J. B., Wu, X. L., and Srinivasan, M. (2016) A high-energy Lloparticle chain anode with a pyridine-derived porous nitrogen-doped carbon cathode. *Advanced Functional Materials*, **26**(18), 3082-3093.
113. Cao, L., Xu, F., Liang, Y. Y., and Li, H. L. (2004) Preparation of the novel nanocomposite $Co(OH)_2$/ultra-stable Y zeolite and its application as a supercapacitor with high energy density. *Advanced Materials*, **16**(20), 1853-1857.
114. Ji, W., Ji, J., Cui, X., Chen, J., Liu, D., Deng, H., and Fu, Q. (2015) Polypyrrole encapsulation on flower-like porous NiO for advanced high-performance supercapacitors. *Chemical Communications*, **51**(36), 7669-7672.
115. Ji, J., Zhang, X., Liu, J., Peng, L., Chen, C., Huang, Z., Li, L., Yu, X., and Shang, S. (2015) Assembly of polypyrrole nanotube@MnO_2 composites with an improved electrochemical capacitance. *Materials Science and Engineering: B*, **198**, 51-56.
116. Ates, M., Serin, M. A., Ekmen, I., and Ertas, Y. N. (2015) Supercapacitor behaviors of polyaniline/CuO, polypyrrole/CuO and PEDOT/CuO nanocomposites. *Polymer Bulletin*, **72**(10), 2573-2589.
117. Kandalkar, S., Dhawale, D., Kim, C.-K., and Lokhande, C. (2010) Chemical synthesis of cobalt oxide thin film electrode for supercapacitor application. *Synthetic Metals*, **160**(11-12), 1299-1302.
118. Lu, X.-F., Chen, X.-Y., Zhou, W., Tong, Y.-X., Li, G.-R. (2015) α-Fe_2O_3@PANI core–shell nanowire arrays as negative electrodes for asymmetric supercapacitors. *ACS Applied Materials & Interfaces*, **7**(27), 14843-14850.
119. Xia, X., Chao, D., Qi, X., Xiong, Q., Zhang, Y., Tu, J., Zhang, H., Fan, H. J. (2013) Controllable growth of conducting polymers shell for constructing high-quality organic/inorganic core/shell nanostructures and their optical-electrochemical properties. *Nano Letters*, **13**(9), 4562-4568.
120. Qiu, M., Sun, P., Shen, L., Wang, K., Song, S., Yu, X., Tan, S., Zhao, C., and Mai, W. (2016) WO 3 nanoflowers with excellent pseudo-capacitive performance and the capacitance contribution analysis. *Journal of Materials Chemistry A*, **4**(19), 7266-7273.
121. Sugimoto, W. (2014) Ruthenium oxides as supercapacitor electrodes. In: *Encyclopedia of Applied Electrochemistry*, Springer, Germany, pp. 1813-1821.
122. Jiang, J., Li, Y., Liu, J., Huang, X., Yuan, C., Lou, X. W. D. (2012) Recent advances in metal oxide-based electrode architecture design for

electrochemical energy storage. *Advanced Materials*, **24**(38), 5166-5180.

123. Sun, X., Gan, M., Ma, L., Wang, H., Zhou, T., Wang, S., Dai, W., Wang, H. (2015) Fabrication of PANI-coated honeycomb-like MnO2 nanospheres with enhanced electrochemical performance for energy storage. *Electrochimica Acta*, **180**, 977-982.
124. Zhang, J., Shu, D., Zhang, T., Chen, H., Zhao, H., Wang, Y., Sun, Z., Tang, S., Fang, X., and Cao, X. (2012) Capacitive properties of PANI/MnO2 synthesized via simultaneous-oxidation route. *Journal of Alloys and Compounds*, **532**, 1-9.
125. Chen, L., Song, Z., Liu, G., Qiu, J., Yu, C., Qin, J., Ma, L., Tian, F., and Liu, W. (2013) Synthesis and electrochemical performance of polyaniline-MnO_2 nanowire composites for supercapacitors. *Journal of Physics and Chemistry of Solids*, **74**(2), 360-365.
126. Bahloul, A., Nessark, B., Briot, E., Groult, H., Mauger, A., and Julien, C. M. (2013) Polypyrrole-covered MnO2 as electrode material for hydrid supercapacitor. *ECS Transactions*, **50**(43), 79-84.
127. Shayeh, J. S., Siadat, S. O. R., Sadeghnia, M., Niknam, K., Rezaei, M., and Aghamohammadi, N. (2016) Advanced studies of coupled conductive polymer/metal oxide nano wire composite as an efficient supercapacitor by common and fast Fourier electrochemical methods. *Journal of Molecular Liquids*, **220**, 489-494.
128. Lu, Q., and Zhou, Y. (2011) Synthesis of mesoporous polythiophene/MnO_2 nanocomposite and its enhanced pseudocapacitive properties. *Journal of Power Sources*, **196**(8), 4088-4094.
129. Su, Z., Yang, C., Xu, C., Wu, H., Zhang, Z., Liu, T., Zhang, C., Yang, Q., Li, B., and Kang, F. (2013) Co-electro-deposition of the MnO_2-PEDOT: PSS nanostructured composite for high areal mass, flexible asymmetric supercapacitor devices. *Journal of Materials Chemistry A*, **1**(40), 12432-12440.
130. Xu, H., Wu, J.-X., Chen, Y., Zhang, J.-L., and Zhang, B.-Q. (2015) Facile synthesis of polyaniline/$NiCo_2O_4$ nanocomposites with enhanced electrochemical properties for supercapacitors. *Ionics*, **21**(9), 2615-2622.
131. Zhang, J., Shi, L., Liu, H., Deng, Z., Huang, L., Mai, W., Tan, S., Cai, X. (2015) Utilizing polyaniline to dominate the crystal phase of $Ni(OH)_2$ and its effect on the electrochemical property of polyaniline/$Ni(OH)_2$ composite. *Journal of Alloys and Compounds*, **651**, 126-134.
132. Lei, J., Jiang, Z., Lu, X., Nie, G., and Wang, C. (2015) Synthesis of few-layer MoS_2 nanosheets-wrapped polyaniline hierarchical nanostructures for enhanced electrochemical capacitance performance. *Electrochimica Acta*, **176**, 149-155.
133. Ma, G., Peng, H., Mu, J., Huang, H., Zhou, X., and Lei, Z. (2013) In situ intercalative polymerization of pyrrole in graphene analogue of

MoS2 as advanced electrode material in supercapacitor. *Journal of Power Sources*, **229**, 72-78.

134. Azman, N. H. N., Nazir, M. S. M., Ngee, L. H., and Sulaiman, Y. (2018) Graphene-based ternary composites for supercapacitors. *International Journal of Energy Research*, **42**(6), 2104-2116.
135. Xia, X., Hao, Q., Lei, W., Wang, W., Sun, D., and Wang, X. (2012) Nanostructured ternary composites of graphene/Fe_2O_3/polyaniline for high-performance supercapacitors. *Journal of Materials Chemistry*, **22**(33), 16844-16850.
136. Mezgebe, M. M., Yan, Z., Wei, G., Gong, S., Zhang, F., Guang, S., and Xu, H. (2017) 3D graphene-Fe_3O_4-polyaniline, a novel ternary composite for supercapacitor electrodes with improved electrochemical properties. *Materials Today Energy*, **5**, 164-172.
137. Xia, X., Hao, Q., Lei, W., Wang, W., Wang, H., and Wang, X. (2012) Reduced-graphene oxide/molybdenum oxide/polyaniline ternary composite for high energy density supercapacitors: Synthesis and properties. *Journal of Materials Chemistry*, **22**(17), 8314-8320.
138. Chen, Y., Liu, B., Liu, Q., Wang, J., Liu, J., Zhang, H., Hu, S., and Jing, X. (2015) Flexible all-solid-state asymmetric supercapacitor assembled using coaxial $NiMoO_4$ nanowire arrays with chemically integrated conductive coating. *Electrochimica Acta*, **178**, 429-438.
139. Bolagam, R., Boddula, R., and Srinivasan, P. (2018) Design and synthesis of ternary composite of polyaniline-sulfonated graphene oxide-TiO2 nanorods: a highly stable electrode material for supercapacitor. *Journal of Solid State Electrochemistry*, **22**(1), 129-139.
140. Zhou, J., Zhao, H., Mu, X., Chen, J., Zhang, P., Wang, Y., He, Y., Zhang, Z., and Pan, X., Xie, E. (2015) Importance of polypyrrole in constructing 3D hierarchical carbon nanotube@MnO_2 perfect core–shell nanostructures for high-performance flexible supercapacitors. *Nanoscale*, **7**(35), 14697-14706.
141. Yun, T. G., Hwang, B. I., Kim, D., Hyun, S., and Han, S. M. (2015) Polypyrrole–MnO_2-coated textile-based flexible-stretchable supercapacitor with high electrochemical and mechanical reliability. *ACS Applied Materials & Interfaces*, **7**(17), 9228-9234.
142. Ng, C., Lim, H., Lim, Y., Chee, W., and Huang, N. (2015) Fabrication of flexible polypyrrole/graphene oxide/manganese oxide supercapacitor. *International Journal of Energy Research*, **39**(3), 344-355.
143. Hou, Y., Cheng, Y., Hobson, T., and Liu, J. (2010) Design and synthesis of hierarchical MnO_2 nanospheres/carbon nanotubes/conducting polymer ternary composite for high performance electrochemical electrodes. *Nano Letters*, **10**(7), 2727-2733.
144. Wang, W., Hao, Q., Lei, W., Xia, X., and Wang, X. (2012) Graphene/SnO_2/polypyrrole ternary nanocomposites as supercapacitor electrode materials. *RSC Advances*, **2**(27), 10268-10274.
145. Wang, W., Lei, W., Yao, T., Xia, X., Huang, W., Hao, Q., and Wang, X.

(2013) One-pot synthesis of graphene/SnO_2/PEDOT ternary electrode material for supercapacitors. *Electrochimica Acta*, **108**, 118-126.

146. Ehsani, A., Jaleh, B., and Nasrollahzadeh, M. (2014) Electrochemical properties and electrocatalytic activity of conducting polymer/copper nanoparticles supported on reduced graphene oxide composite. *Journal of Power Sources*, **257**, 300-307.
147. Kalambate, P. K., Dar, R. A., Karna, S. P., and Srivastava, A. K. (2015) High performance supercapacitor based on graphene-silver nanoparticles-polypyrrole nanocomposite coated on glassy carbon electrode. *Journal of Power Sources*, **276**, 262-270.
148. Shayeh, J. S., Ehsani, A., Ganjali, M., Norouzi, P., and Jaleh, B. (2015) Conductive polymer/reduced graphene oxide/Au nano particles as efficient composite materials in electrochemical supercapacitors. *Applied Surface Science*, **353**, 594-599.

Chapter 6

Rubber/Conducting Polymer Blends: A Review

Suneel Kumar Srivastava

Inorganic Materials and Polymer Nanocomposite Laboratory, Department of Chemistry, Indian Institute of Technology, Kharagpur, India
sunil111954@yahoo.co.uk

6.1 Introduction

There exists very limited number of polymers exhibiting appreciable conductivity in contrast to majority of polymers behaving like insulators. For example, intrinsically conducting polymers (ICPs), namely polyaniline (PANI), polypyrrole (PPy), polythiophene (PT), poly(*p*-phenylene) (PPP) and poly(*p*-phenylenevinylene) (PPV) have attracted an increasing interest since their discovery around 1977 [1]. They display distinct physical/electrical properties due to their unique chemical structures as compared to the non-conducting polymers [2]. These conductive polymers have been studied extensively in the last few decades due to their wide range of promising applications in the field of sensors, electromagnetic shielding, corrosion, protection actuators rechargeable batteries, conducting paints, conducting glues, anti-static formulations, electronic devices, protective coatings against corrosion, electrochromic devices, electrolytic capacitors, artificial muscles, conductive paints and adhesives for electronic applications, static charge dissipation, nanodevices, microactuators, etc. [1-10]. Among ICPs, PANI remains one of the most studied conducting polymers. It exists as leucoemeraldine (white/clear and colorless) (LB), emeralidine base (blue) (EB), emeralidine salt (green) (ES) and pernigraniline (blue/violet) depending on its oxidation and protonation state. Different salts of PANI are interchangeable depending on pH and redox switching [4]. Interestingly, only emerlidine salt is conducting, and its formation could be monitored by controlling the pH of the medium. Emerlidine base, though insulating, could also be transformed to conducting emerlidine salt in presence of protic acid as dopant. Therefore, considera-

Conducting Polymer Composites, edited by Vikas Mittal

ble amount of research effort has been devoted on PANI owing to its low cost of synthesis, long term environmental stability and high electrical conductivity which can be reversibly controlled by a change in the oxidation state and protonation of the imine nitrogen groups [5,6]. However, the great potential of highly conducting emeraldine salt of PANI from emeraldine base (EB) and a strong acid is masked by disadvantages such as poor infusibility. In addition, low solubility in inorganic solvents also restricts processability. Though, EB can be dissolved in *N*-methylpyrrolidone, concentrated sulfuric acid and other strong acids, these are almost unacceptable because of either high price or significant corrosion. Another major drawback includes intractable nature, especially in the doped form which remains the main disadvantage of conductive PANI for its many applications [7-9]. The low solubility of PANI can be improved using functionalized sulfonic acids, such as dodecylbenzenesulfonic acid (DBSA) and p-toluene sulfonic acid (PTSA). These protonating agents improve the compatibility of PANI with the host matrix as well as act as dopants and increase electrical conductivity of the material [10-12]. Doped PPy has very good environmental stability, and its degradation temperature varies from 150 to 300 °C or above [13,14]. Further, PANI exhibits inferior mechanical properties compared to conventional plastics. Thereafter, much research has been focused on the formation of blends and composites of the conductive polymers, such as PANI, with conventional polymers to achieve the desired mechanical properties and better processability. In this respect, several studies have been reported on the development of electrically conducting composites or blends that combine the elastomeric properties of a rubbery host polymer with the conductive properties of PANI [15]. Thus, the blend or composite materials of PANI have been prepared by introducing it into several elastomers. Such conductive polymer/elastomeric blend/composites exhibit interesting physical properties and many potential applications [15,16].

In view of this, the main objective of the present article is to review recent literature studies on the preparation and morphology as well as mechanical and thermal properties, followed by the applications of electrically conductive rubber blends prepared based on the combination of natural rubber (NR), styrene butadiene rubber (SBR), nitrile rubber (NBR), ethylene-vinyl acetate (EVA) and ethylene propylene diene monomer rubber (EPDM) with conducting polymers [17-92].

6.2 Preparative Methods for Elastomer/Conducting Polymer Blends

Conducting polymer blend/nanocomposites have been invariably prepared by *in-situ* polymerization, solution blending and melt blending. The solution blending method of preparation of the nanocomposites is generally not considered to be eco-friendly and cost-effective due to the use of excess amount of organic solvents. On the contrary, melt-mixing of polymers and polymers with nanofillers is the most effective way to obtain the blends/nanocomposites for commercial applications. The method is totally environmentally friendly and does not require any solvent. Table 6.1 (at the end of the chapter) provides details about the preparation of blends consisting of conducting polymers and elastomers such as NR, SBR, NBR, EVA and EPDM.

6.3 Morphology of Elastomer/Conducting Polymer Blends

6.3.1 Natural Rubber/Conducting Polymer Blends

Sukitpaneenit *et al.* [17] studied morphology of undoped and HCl-doped PANI/NR composite fibers with different compositions (0, 5, 10% w/w). It was noted that the roughness of the fiber surface increased with the increasing PANI content. Further scanning electron microscopy (SEM) exhibited increase in surface roughness of the nanocomposite fiber upon doping with HCl. SEM of cross-sections of freeze fractured surfaces also exhibited the presence of PANI particles in contrast to homogeneous morphology observed in neat NR. Camillo *et al.* [18] carried out optical microscopy analysis for samples obtained by adding PANI-EB (emeraldine base) dissolved in N-methyl-2-pyrrolidone (NMP) to the latex (NRL), PANI-EB dissolved in m-cresol to NR dissolved in xylol (NRD) and overlaying the surface of a pure NR cast film with a PANI layer grown by *in-situ* polymerization (NRO). It was qualitatively inferred that doped PANI was more homogeneously distributed in NRL followed by NRO and NRD. SEM images of NR containing 10 and 30 wt% PANI showed good dispersibility of PANI in the NR matrix, which contributed to an increase in the interfacial adhesion between filler and matrix of the blend [19]. Yong [20] used electron beam irradiation technique in order to crosslink the natural rubber-polyaniline dodecylbenzenesulphonate (PANI.DBSA) blends. The presence of the elongated rod-like structures of the PANI.DBSA dispersed particles was observed within the NR matrix. Further, extent of network formation enhanced with increasing PANI.DBSA loading.

In another study, the microscopy analysis of synthetic rubber (cis-1,4-polyisoprene) (56%):PANI (69%) composite films exhibited the presence of highly structured grains [32]. Yong [33] also examined the morphology of ENR-PANI.DBSA blends consisting 10 and 40 wt% PANI.DBSA through transmission electron microscopy (TEM). The analysis of ENR/PANI.DBSA (10 wt%) established the presence of regions corresponding to well blended ENR and PANI.DBSA. However, the presence of phase separated PANI.DBSA particles in ENR/PANI.DBSA (40 wt%) was observed, along with the formation of cluster/conductive pathways of PANI.DBSA particles.

Yong [34] studied the morphology of 2.5 and 5.0 wt% of PANI.DBSA loaded in ENR prepared by using internal mixer. The author observed better dispersion and presence of some large isolated PANI.DBSA particles in ENR/PANI.DBSA (2.5 wt%) unstrained blend exhibiting relatively low electric conductivity. TEM of unstrained and strained blends with 5.0 wt% of PANI.DBSA indicated formation of conductive network with elongated DBSA particles exhibiting high conductivity. Yong *et al.* [35] also studied morphology of ENR/PANI.DBSA (2.5 and 5 wt%) blends prepared by solution mixing. It was inferred that networks with higher loading of PANI.DBSA in ENR resulted in conductive networks due to better dispersion of PANI.DBSA particles in ENR matrix.

6.3.2 SBR/Conducting Polymer Blends

Almazan *et al.* [43] studied morphology of PANI/SBR:20/80 (PA80), PANI.DBSA/SBR:20/80 (PD80) and PANI.DBSA/SBR:10/90 (PD90) composites prepared by solution casting and noted larger amount of isolated polymer particles in PANI-DBSA/SBR compared to PANI/SBR composites. The improved dispersion of the PANI.DBSA could be explained due to its greater compatibility with SBR. Park *et al.* [44] studied fractured surface morphology of SBR/PANI.DBSA blends. The authors noted better dispersion for more than 2 wt% PANI.DBSA in latex SBR blends. Further, it was noted that reducing the PANI content in SBR/DBSA doped PANI composites decreased the extent of agglomeration and enhanced dispersion. The microscopy analysis of SBR/PANI and SBR/PPy (PANI and PPY: 2.9-4.9 wt%) indicated uniform coating of PANI and PPy respectively [48].

Liu *et al.* [45] studied the dispersion behavior of PPy@HNTs in XSBR and compared it with neat XSBR. The increasing loading of PPy@HNTs in XSBR resulted in uniform dispersion. It was concluded that PPy@HNTs formed continuous conductive network in the rubber matrix, especially at high loading.

6.3.3 NBR/Conducting Polymer Blends

The morphology of NBR-PANI.DBSA (40 wt%) blends has been studied with and without compatibilizer (hydroquinone) [50]. The corresponding TEM of blends with added compatibilizer in Figure 6.1 showed the lowest level of phase separation. These investigations also suggested that PANI.DBSA particles (50-1000 nm) in NBR matrix might be expected to make a contribution to the electrical conductivity.

In another study, the microscopy studies (both optical and transmission electron micrographs) also confirmed reduced amount of phase separation of NBR-PANI.DBSA blends prepared by thermomechanical mixing method [53]. The morphological studies also pointed out that the small PANI.DBSA colloidal particles contributed to the electrical conductivity of the blends.

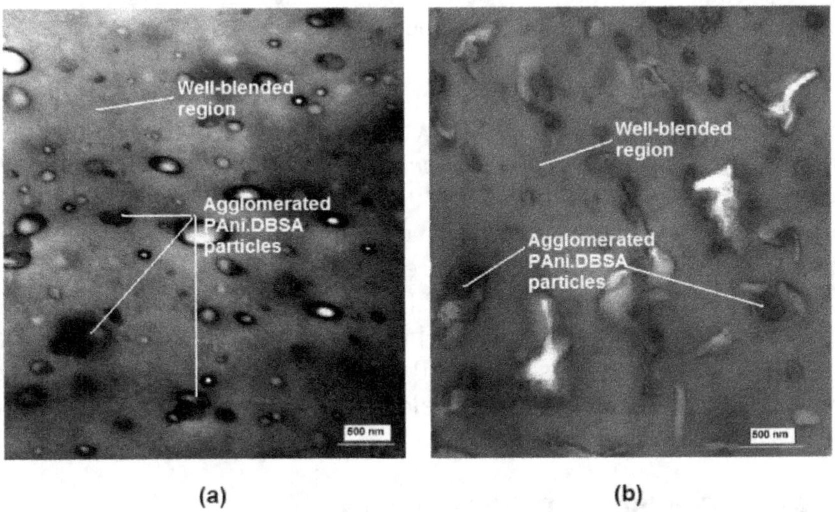

Figure 6.1 Transmission electron micrographs of NBR-PANI.DBSA blend with 40 wt% of PANI.DBSA (a) blend without hydroquinone and (b) blend with added hydroquinone. Reproduced from Reference 50 with permission from Malaysian Rubber Board.

Hanif and Yong [56] studied the morphology of 5.0 wt% and 50.0 wt% PANI.DBSA filled NBR blends. TEM images in Figure 6.2 showed the presence of scattered and isolated PANI.DBSA particles at low concentration, but connected to each other as the concentration of PANI.DBSA increased, providing a conducting path. NBR-PANI.DBSA (20 wt%) non-irradiated and electron beam irradiated blends showed identical morphology [51]. The dispersion of PANI.DBSA particles in NBR formed electrically conductive networks, which enhanced with increasing PANI.DBSA proportion.

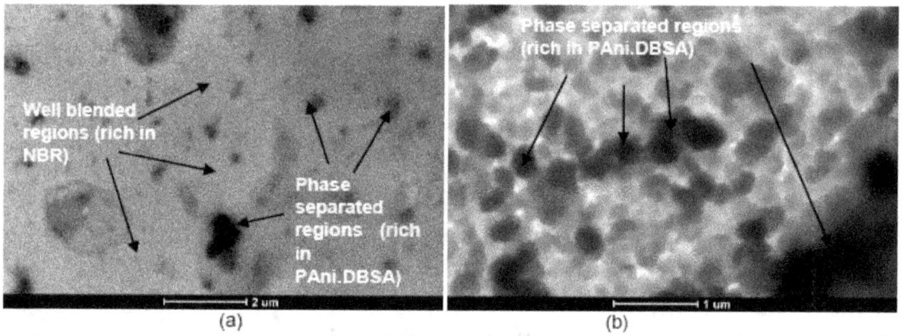

Figure 6.2 Transmission electron micrographs of NBR based blends; (a) at 5.0 wt% PAni.DBSA content (b) at 50.0 wt% PAni.DBSA content. Reproduced from Reference 56 with permission from AIDIC.

Yong [52] carried out microscopy studies on NBR-PP-PANI.DBSA TPV consisting 1.0, 10.0 and 30 wt% PANI.DBSA. The micrographs exhibited fine and elongated dispersion of PANI.DBSA particles. In addition, the analysis of NBR-PP-PANI.DBSA (20 and 30 wt%) showed the formation of conductive pathways by DBSA particles. In another study, transmission electron microscopic analysis of PANI.DBSA (10 wt%) filled NBR (CAN: 41.4 wt%) also showed the formation of cluster/conductive pathways [60]. It was suggested that PANI.DBSA particles (50-1000 nm) in the NBR-rich regions are likely to make a contribution to electrical conductivity even below the main percolation threshold.

6.3.4 EVA/Conducting Polymer Blends

The effect of the PANI-DBSA (33 wt%) addition on the morphology of the EVA/PANI blends was investigated [5]. The scanning microscopy analysis of the blend, prepared in a two-roll mill, showed the presence of uniformly dispersed PANI-DBSA domains in the EVA matrix. However, the similar blend prepared by internal mixer showed better dispersibility of PANI particles within the EVA matrix. In another study, Jeevananda and Siddaramaiah [67] observed uniform distribution of PANI in EVA-PANI (90/10) blends prepared by roll mill and internal mixing methods. It was inferred that the blend prepared in the internal mixer exhibited better dispersion of PANI particles within the EVA matrix. SEM micrographs of EVA-PANI (90/10) blend showed homogeneous morphology involving uniform distribution of PANI in EVA matrix. However, phase separation occurred in EVA-PANI (80/20) blends due to the aggregation of PANI in EVA matrix. EVA-PANI blend system up to 93/07 prepared by emulsion polymerization showed a homogeneous morphology with fine distribution of PANI granules in the EVA matrix [68].

EVA/PANI (30, 50, 60, 70 and 80 wt%) blends synthesized by inverse emulsion revealed both granular and fibrillar morphology [69]. EVA/PANI-DBSA (95/5 and 70/30 wt%) blend films were also analyzed by optical microscopy [70]. EVA41/PANI–DBSA (RP) displayed a well-defined conducting pathway with a PANI–DBSA content as low as 5 wt%. Paul and Pillai [71] studied the morphology of solution-mixed EVA-PANI (20 wt%) blend film and emulsion-polymerized PANI (18.5 wt%)-SPDPAA (3-pentadecylphenoxyacetic acid)-EVA blends. Some micro-voids were observed in the emulsion-polymerized blend (due to PANI aggregation) in contrast to solution-mixed blend. Yang and Ruckenstein [72] compared the morphology of cryogenically fractured surfaces of EVA blends with 30 wt.% PANI-DBSA without compatibilizer and in presence of 20 wt.% compatibilizer. The findings showed that addition of compatibilizer increased the dispersion of the PANI-DBSA throughout the EVA matrix. In another study, scanning electron images of tensile fracture surface of different EVA/PANI composites clearly indicated that ductile fracture surface of neat EVA changed to brittle fracture with progressive addition of PANI [61]. Such weak interfacial bonding between ductile EVA with rigid PANI could not effectively cause interfacial stress transfer when subjected to increased load, and as a result, the failure of interface took place.

6.3.5 EPDM/Conducting Polymer Blends

The optical micrographs of EPDM/PANI.DBSA-S blends prepared by solution doping showed homogeneous dispersion of PANI in EPDM

[76]. Further, the authors noted the formation of non-miscible blends of PANI.DBSA doped by grinding in a mortar and reactive processing with EPDM. Faez *et al.* [77] studied the morphology of NBR:(EPDM/PANI-DBSA) blends and observed the presence of PANI-DBSA agglomerates of larger size in NBR:PANI:PANI-DBSA (50:20:30 wt%) and NBR:PANI:PANI-DBSA (40:30:30 wt%) than NBR:PANI:PAni-DBSA (50:25:25 wt%) and NBR:PANI:PANI-DBSA (60:16:24 wt%) blends. The optical micrographs of PANI/EPDM rubber blends prepared in presence of maleic anhydride as a compatibilizer showed good distribution of the PANI complex in the matrix [79].

Srivastava and coworkers studied the morphology of rubber blends of EPDM and NBR loaded with 14 wt% of PS@PANI [82]. The presence of PANI shell (including few deformed shell) was clearly visible in the rubber matrix in EPDM/PS@PANI composites following homogeneous dispersion and interconnectivity. In contrast, the NBR/PS@PANI blend exhibited homogeneously dispersed leopard skin-like morphology of PS@PANI in NBR. Optical microscopy images also showed homogeneous dispersion of PS@PANI in the rubber matrix.

6.4 Properties of Elastomer/Conducting Polymer Blends

6.4.1 Electrical Conductivity of Rubber/Conducting Polymer Blends

Natural Rubber/Conducting Polymer Blends

The electrical conductivity was observed to increase with enhancing PANI loading level into NR elastomer matrix [19]. The dc electrical conductivity of the blend films (free-standing) containing 20 vol% of polyaniline in 80% of natural rubber, referred earlier as NRL, NRD, NRO fabricated using three different ways by casting, followed the order: NR<NRD<NRO<NRL [18]. This was in all probability associated with the compatibility between the conducting polymer and rubber. Sukitpaneenit *et al.* [17] measured dc electrical conductivity of undoped and HCl doped NR/PANI composites as a function of PANI content. The findings suggested that the electrical conductivity of the NR fiber bundle (10^{-10}S/cm) increased with increasing PANI content and attained a value of 10^{-3} S/cm for NR/PANI (5% w/w). Further increase of the PANI concentration to 10% w/w increased

the conductivity only marginally suggesting percolation threshold had already reached at a PANI content of 5% w/w. Yong [20] studied the electrical conductivity of NR-PANI.DBSA blends before and after electron beam irradiation. It was noted that the blends exhibited identical electrical conductivity values and very close percolation threshold.

Silva *et al.* [6] studied the electrical conductivity of NR/PANI-DBSA composite films as a function of the PANI-DBSA content. The authors found that conductivity increased with an increase in the PANI-DBSA content in the NR matrix and attained a value of 10^{-6} S/cm for 10 wt% of PANI-DBSA. The percolation threshold was found to be ~3.1 wt% and was found to be seven orders of magnitude higher in case of NR/PANI.DBSA (10 wt%) compared to neat NR. Hanif and Yong [56] also measured the electrical conductivity of NBR blends with different composition of fillers. The conductivity of NBR (1 x 10^{-8} S.cm^{-1}) improved on addition of either carbon black or PANI.DBSA by several orders of magnitude. It was also inferred that the conductivity of PANI.DBSA blends was greater than the corresponding carbon black blends. The estimated percolation threshold for NBR-PANI.DBSA and NBR-carbon black blends corresponded to 2.5 wt% and 30.0 wt% respectively.

SBR/Conducting Polymer Blends

Mathew and Neena [39] noticed extremely low percolation threshold in SBR/PANI composite synthesized via *ex-situ* technique. Martins *et al.* [47] studied the variation of the logarithm of the electrical conductivity as a function of PANI-DBSA-PSS (poly(styrene sulfonic acid)) complex concentration in SBR blends. It was noted that the conductivity of SBR/PANI-DBSA-PSS blends increased with increasing content of the conducting complex. The authors observed the percolation threshold at values lower than 10% (w/w) for the PANI-DBSA-PSS complex because the 90/10 blend was already conductive (ca. 10^{-7} S/cm). In case of 50% (w/w) of PANI-DBSA-PSS complex, the conductivity corresponded closely to intrinsic conductivity of the pure PANI-DBSA-PSS complex (approx. 10^{-2} S/cm). In another study, the electrical conductivity of the SBR/PANI-DBSA blends was also observed to gradually increase with PANI-DBSA contents [44]. At 0.5 wt% PANI-DBSA, the conductivity was already above 10^{-1} S/cm. It tend to level off above 1 wt% PANI-DBSA and appeared to reach a plateau at 4 wt% PANI-DBSA.

The onset of conductivity achieved with only 1 wt% of PANI-DBSA in SBR indicated percolation threshold to be ~ 13 times smaller than the calculated by theoretical predictions.

Liu and Liu [45] studied the extent of dispersion state of PPy@HNTs in XSBR matrix. Microscopy analysis demonstrated that PPy@HNTs were successfully incorporated into XSBR rubber matrix with good interfacial interactions with the polymer. Liu *et al.* [48] observed that the conductivity of SBR blends formed with PANI and PPy were in the range of 0.02-0.92 S cm^{-1} [48].

NBR/Conducting Polymer Blends

Yong [52] studied the variation of electrical conductivities for NBR-PP as a function of PANI.DBSA loading (wt%) (Figure 6.3). The electrical conductivity of blends increased, in general, with PANI.DBSA loading. It was also noted that added compatibilizer resulted in very high electrical conductivities (up to ~1.4 S/cm). The percolation threshold corresponded to ~1.5 wt%/0.85 volume% of PANI.DBSA.

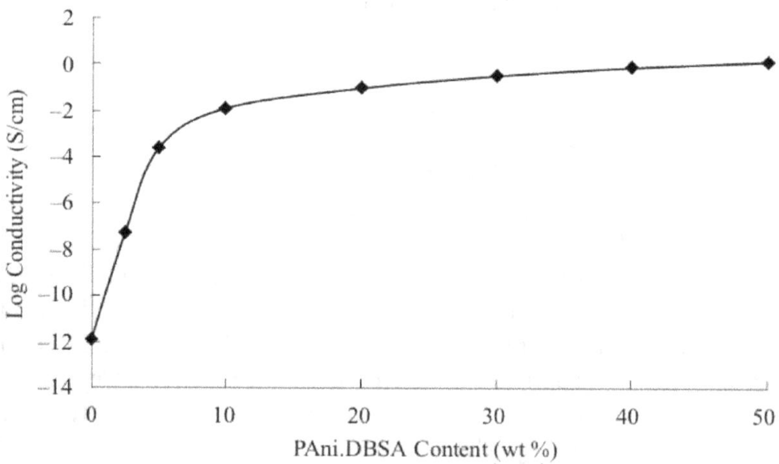

Figure 6.3 Electrical conductivities for NBR-PP TPV as a function of PAni.DBSA loading (wt%). Reproduced Reference 52 with permission from Malaysian Rubber Board.

Electrical conductivity measurements were also carried out on NBR-PANI.DBSA blends prepared by high temperature-mechanical mixing [54]. The investigations revealed no effect of adding vulcanizing agent (sulfur) on the electrical properties of vulcanized blends. The electrical conductivity of both peroxide vulcanized and non-vulcanized blends of NBR-PANI.DBSA increased with the proportion of PANI [55]. These blends attained plateau electrical conductivity of the order of 10^{-2} S/cm with about 40 wt% PANI.DBSA loading. Yong *et al.* [60] measured the electrical conductivity of NBR (ACN contents: 25.7, 33.1, 41.4 and 48.2 wt%)/PANI.DBSA blends by solution mixing and casting. The conductivity values up to 10^{-2} S cm^{-1} were achieved. The corresponding values of percolation threshold were found to be 7.4/6.4, 7.5/6.4, 6.6/5.8 and 6.0/5.4 wt%/vol% respectively.

EVA/Conducting Polymer Blends

Different theoretical models were also applied to predict dc conductivity of EVA/PANI composites prepared by *in-situ* polymerization [66]. According to this, Scarisbrick model showed similarity between experimental and theoretically predicted conductivity. The conductivity of EVA-PANI blends enhanced with on increasing PANI content [68]. The conductivity of EVA/PANI blends varied from 10^{-2} S cm^{-1} to $<10^{-6}$ S cm^{-1} depending on the EVA content [69]. Paul and Pillai [71] studied the variation of conductivity with weight percent of PANI for emulsion-polymerized melt-processable and solution-mixed PANI-SPDPAA-EVA films. The percolation threshold in the solution-mixed blend of PANI-SPDPAA-EVA was achieved at 10 wt% of PANI, (5×10^{-8} S cm^{-1}). In contrast, the percolation threshold in emulsion-polymerized PANI-SPDPAA-EVA appeared at a very low fraction of PANI. A maximum value of conductivity (0.85 S cm^{-1}) was obtained corresponding to 28.5 wt% of PANI. In another study, polypyrrole-poly(ethylene-co-vinyl acetate) composite exhibited a percolation threshold in the range 5-20 wt% and reached values as high as 5-7 S/cm [72]. The role of alkyl gallate compatibilizers in enhancing the conductivity of extruded PANI-EVA has also been studied [73].

EPDM/Conducting Polymer Blends

Faez *et al.* [76] measured the electrical conductivity of blends of EPDM and PANI-DBSA prepared by solution, grinding in a mortar and reactive processing as a function of PANI-DBSA weight fraction. It was noted that the conductivity values were higher for blends prepared with PANI-DBSA doped by reactive processing. Zn sulfonated EPDM/PANI exhibited increase in conductivity with enhancing PANI content, but it decreased with increasing sulfonate content [86]. The percolation threshold of the composite corresponded to ~13 wt% PANI. Room temperature dc conductivity measurements exhibited higher conductivity of PS@PANI core-shell incorporated NR and EPDM than the corresponding PS/PANI loaded rubber samples [82]. Also, it was noted that 25 wt% PS@PANI loaded rubber blends showed abrupt increase in conductivity. Conductivity measurements were also reported on flexible, free-standing and stretchable film of EPDM/PANI.DBSA blends prepared by mechanical process [87]. It was noted that the conductivity increased with PANI.DBSA content. The onset of the conductivity at low conducting polymer content indicated a low percolation threshold of the blend.

6.4.2 Mechanical Properties of Elastomer/Conducting Polymer Blends

NR/PANI Blends

Silva *et al.* [6] measured mechanical proprieties of NR/PANI-DBSA with proportions of 95/5 and 90/10. Young's modulus and tensile strength of NR/PANI-DBSA-90/10 are maximum enhanced to 1.50±0.20 and 2.20±0.10 MPa compared to the respective values of 0.68 ± 0.05 MPa and 1.36 ± 0.04 MPa corresponding to neat NR and was ascribed it to the rigidity of PANI. It was also inferred that the elongation at the break of the NR/PANI-DBSA-95/05 (699 ± 2 %) and NR/PANI-DBSA-90/10 (695 ± 7%) also improved compared to neat NR (766 ± 9 %). Tensile profile of NR/PANI-DBSA (90:10) remained more or less identical to NR/PANI-DBSA (95:05).This behavior could be related to the amount of water uptake in the composite owing to the hygroscopic characteristics of DBSA-doped PANI.

Tensile strength and hardness were observed to enhance with increasing PANI loading level into NR/PANI composites [19]. Helaly *et al.* [21] studied the effect of PANI on the physico-mechanical properties of NR vulcanizates before and after aging. The authors noted that vulcanizates containing polyaniline showed tensile values less than that of conventional antioxidant PBN. The mechanical properties of the bundles of PANI/NR composite fibers fabricated by a wet spinning process were also determined [17]. It was noted that the elongation at break (820%) of neat NR fiber bundle was reduced on the introduction of undoped PANI particles. Camillo *et al.* [18]

carried out tensile measurements for blends consisting of polyaniline and latex NR (NRL), NR dissolved in xylol (NRD), PANI layer grown by *in-situ* polymerization on NR (NRO) and pure NR cast film. It was concluded that NRL blend had optimal performance. The tensile stress and elongation remained almost same for NRL and NR up to 700%. Further, the stress-strain plots pointed out rupture point of NR and NRL corresponding to 1.4 ± 0.5 MPa and 1.1± 0.5 MP respectively.

The variation of tensile strength and elongation % at break with loading of electron beam irradiated NR-PANI.DBSA was also studied and corresponding findings for elongation are presented in Figure 6.4 [20]. It was observed that both tensile strength and EB % of NR enhanced with the PANI.DBSA proportion up to an optimum level (20.0 wt%) in electron beam irradiated PANI.DBSA. However, tensile strength and EB % decreased for blends with ≥30.0 wt% of PANI.DBSA due to increased phase separation between rubber host and PANI.DBSA. This study also showed superior tensile properties of NR/PANI.DBSA blends irradiated at higher doses (150 and 200 kGy) compared to irradiation at lower doses (50 and 75 kGy) and attributed the enhanced perfromance to higher crosslink densities.

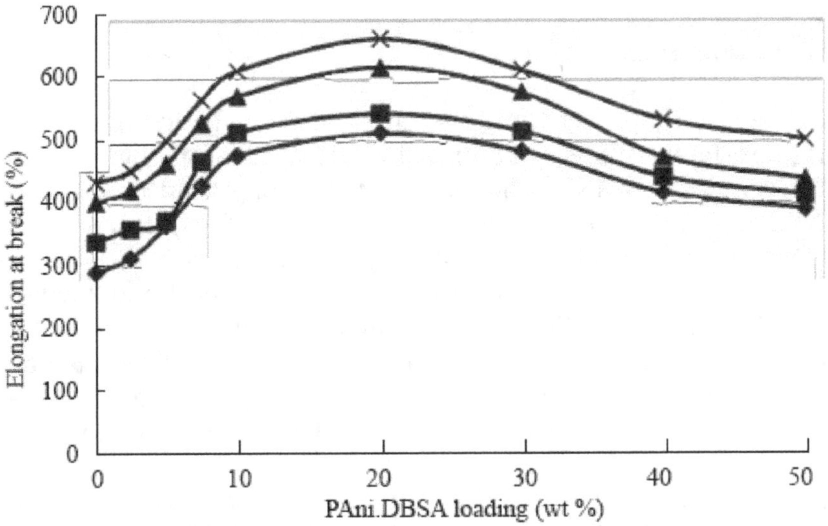

Figure 6.4 Elongation at break (EB %, error ±20%) of electron beam irradiated NR-PANI.DBSA blends. ◆-irradiated at 50 kGy; ■-irradiated at 75 kGy; ▲-irradiated at 150 kGy; ×-irradiated at 200kGy. Reproduced from Reference 20 with permission from Malaysian Rubber Board.

Baboo *et al.* [31] studied variation of Young's modulus, tensile strength and toughness of cis-polyisoprene (CPI) and trans-polyisoprene (TPI) blends (CPI-100/TPI-0, CPI-75/TPI-25, CPI-50/TPI-50, CPI-25/TPI-75 CPI-0/TPI-100) without PANI and with 0.1, 1 and 5 wt% of PANI. Mechanical properties in all CPI/TPI/PANI blends improved as compared to blends in absence of PANI. The maximum enhancement in the mechanical properties was achieved for CPI-75/TPI-25/PANI-5 wt% blend. In another study, Yong [33] observed that reinforcement of PANI.DBSA (2.5-30.0 wt%) in epoxidized natural rubber enhanced the tensile strength and elongation at break (%). At higher PANI.DBSA loadings (≥40.0 wt. %), tensile strength and EB % were observed to decrease due to the phase-separated PANI.DBSA particle regions.

Panigrahi and Srivastava [82] reported tensile strength and elongation at break of NR, NBR and EPDM loaded with different weight percentage of PS@PANI and PS/PANI blends prepared by core-shell (CS) and simple mixing (SM) methods. It was concluded that mechanical properties of the blends prepared by CS as well as SM were always higher as compared to individual rubbers. However, rubber blends prepared by the CS method exhibited superior tensile strength compared to the SM method.

SBR/PANI Blends

Park *et al.* [44] studied tensile strength and elongation at break of SBR latex/PANI-DBSA at the breaking point as a function of PANI-DBSA contents. An initial increase in the strength up to 1 wt% of PANI-DBSA was followed by sharp decrease till 6 wt% of PANI-DBSA. Elongation at break slightly increased till 1 wt% of PANI.DBSA, followed by an abrupt decrease between 1 and 2 wt% of PANI-DBSA, further followed by gradual decrease till 6 wt% PANI-DBSA. In another study, mechanical properties of 1.25, 2.5, 5.0 and 10 wt% of PPy-wrapped halloysite nanotubes (PPy@HNTs) loaded carbxylated SBR were investigated [45]. The corresponding values of elongation at break (%) were ~611, 661,609, 573 and 492. Tensile strength of XSBR/PPy@HNTs increased with PPy@HNTs loadings and attained enhancement of 1.1 time for XSBR/PPy@HNTs (10 wt%) blend as compared to neat XSBR. The tensile modulus at 100%, 300% and 500% strain of the XSBR/PPy@HNTs with PPy@HNT content was also studied. It was noted that the tensile modulus of XSBR/PPy@HNTs (5 wt%) at 500% enhanced by 2.07

times with respect to the neat xSBR. The observed improvement in the tensile properties was attributed to the uniform dispersion of PPy@HNT in the matrix. The possibility of good interfacial interactions between PPy@HNTs and rubber was also not ruled out.

The mechanical properties SBR and DBSA doped PANI, along with poly(styrene sulfonic acid) (PSS) acting as compatibilizer between PANI and SBR, were reported [47]. The Young's modulus and elongation were found to be strongly dependent on PANI concentration and the ratio between PANI-DBSA and PSS. It was also noted that elongation at break decreased considerably at all compositions due to incompatibility of PANI-DBSA in SBR. Overall, Young's modulus for the blend with 50% (w/w) PANI-DSSA-PSS showed a significant improvement.

NBR/PANI Blends

Yong [52] studied the tensile strength and elongation at break (%) of NBR-PP-PANI.DBSA thermoplastic elastomer vulcanizates (TPVs) prepared by using internal mixer. It was observed that the strength was maximum at PANI.DBSA 30.0 wt% loading in NBR-PP blend. However, the strength decreased for blends with ≥40.0 wt% of PANI loaded NBR-PP. Hanif and Yong [56] also measured tensile strength and elongation at break (%) of PANI.DBSA as well as carbon black modified NBR (up 20.0 wt%) and findings are displayed in Figure 6.5. It was also noted that tensile strength of NBR-PANI.DBSA and NBR-carbon black increased by 28 and 56% respectively. In contrast, elongation at break of NBR-PANI.DBSA compounds was found to be higher than carbon-filled compound. This is an indication that the addition of carbon black induced stiffness to the compound.

NBR blended with emulsion prepared polyaniline (doped with DBSA) showed hardness values of 4874 and 3272 corresponding to 5 and 6.25 wt% of PANI respectively [57]. Yong [51] also measured tensile properties of NBR-PANI.DBSA blends prepared by electron beam irradiation technique. Both tensile strength and EB % of blends enhanced with the PANI.DBSA proportion (up to an optimal level of 20.0 wt%). This was followed by decrease with 30.0 wt% PANI.DBSA in the blend due to the increase of phase-separated PANI.DBSA regions. The findings also demonstrated better tensile properties for the samples irradiated at higher doses (150 and 200 kGy) compared to irradiation at lower doses (50 and 75 kGy) due to the higher cross-linking densities. Soares *et al.* [58] recorded stress-

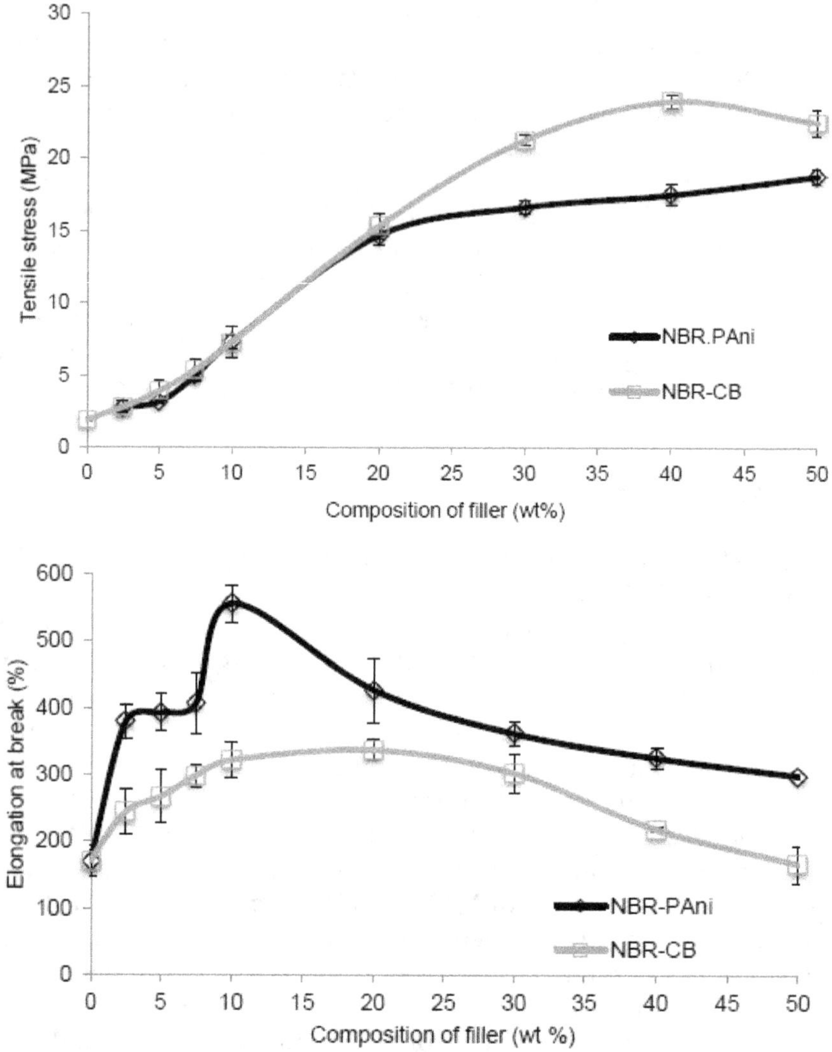

Figure 6.5 (a) Tensile strength and (b) elongation at break of NBR blends with different fillers Reproduced from Reference 56 with permission from AIDIC.

strain curves of NBR and physical/*in-situ* polymerized blend with DBSA (15% and 27%). The blends prepared by physical mixing resulted in an increase of both ultimate tensile strength and elongation at break, as compared to pure NBR. On the other hand, substantial increase in tensile strength and a decrease on elongation at break was observed in the blends prepared by *in-situ* polymerization. According to Luan *et al.* [59], NBR/PANI composite showed a higher elongation at break of 250% even after 7 cycles of immersion and relatively smaller strain dependent resistance change (190% at a 210% strain) [59].

EVA/PANI Blends

Mechanical properties investigations of EVA/PANI-DBSA blends showed a decrease in tensile strength and elongation at break with the addition of small amount of PANI-DBSA [5]. This was attributed to the blend incompatibility and plasticizing effect of DBSA. Tensile strength of EVA/PANI protonated with dopant (sulfonic acid of 3-pentadecylphenoxyacetic acid (SPDPAA), prepared by solution-mixing and emulsion-polymerization, decreased with the PANI content [71]. However, reduction in the tensile strength with PANI content was comparatively lower in the blend prepared by solution method as compared to emulsion method. The percent elongation at break in either of the blends decreased with increasing PANI content. But, decrease was found to be relatively much lower for the emulsion-polymerization method. Jeevananda and Siddaramaiah [68] studied mechanical properties of a series of EVA-PANI (98/02, 95/05, 93/07, 90/10, 85/15, 80/20) blends. The maximum increase in tensile strength was noticed corresponding to 7 wt% of PANI in EVA. It was attributed to the positive interactions between -NH (PANI) and -C=O group (EVA). The mechanical property analysis also showed reduction in elongation at break in the blends. In another study, tensile strength of the PANI-EVA (30, 50, 60, 70 and 80 wt.%) blends was reported to in the range 5.5-8.6 MPa [69]. A dramatic increase in elongation at break (218.3%) was observed in the case of EVA 80 blend. It was concluded that phase separation between the two components had a pronounced effect on the mechanical and electrical properties of the blends. Rahaman *et al.* [61] studied the mechanical properties of EVA/PANI composites and observed decrease in elongation at break and Young's modulus with increase in PANI concentration. The authors attributed the observed performance to relatively weak interfacial bonding between EVA/PANI phases and appreciable polarity differences.

EPDM/PANI Blends

Faez and De Paoli [76] studied the mechanical properties of blends such as EPDM/PANI-DBSA-S, EPDM/PANI-DBSA-M and EPDM/PANI–DBSA-R prepared by solution doping, grinding in a mortar and reactive processing respectively. It was noted that Young's modulus and tensile strength increased with PANI-DBSA concentration independently of the doping method. It was also noted that the Young's modulus of the blends at constant PANI.DBSA content followed the order: EPDM/PANI-DBSA (solution doping) < EPDM/PANI-DBSA (reactive processing) < EPDM/PANI-DBSA (grinding in mortar). It was also noted that EPDM/PANI-DBSA (reactive processing) had higher tensile strength in comparison to the other blends. Also, the tensile strength decreased for blends corresponding to 30% w/w of PANI-DBSA (reactive processing) and 20% w/w of PANI-DBSA-M (grinding in mortar). Faez *et al.* [77] also studied the mechanical properties of ternary blends consisting of nitrile rubber, EPDM rubber and polyaniline, doped with dodecylbenzenesulfonic acid. A reduction in soluble fraction (SF) as well as elongation at break was noticed on increasing the (EPDM/PANI-DBSA) content due to the higher extent of crosslinking. The maximum tension at break was attained in EPDM/PANI-DBSA (40 wt%) blend. The complete crosslinking was suggested at 50 wt% of EPDM/PANI-DBSA in the ternary blend.

Xie *et al.* [86] studied the tensile strength of PANI/zinc sulfonated EPDM ionomer based composites as a function of PANI content (sulfonate content). It was noted that the tensile strength of the composite containing ionomer with 10 and 24 mmol SO_3H/100 g increased quickly up to ~7 wt% PANI followed by a decrease until 12 wt% PANI content. The authors attributes this phenomenon to the reinforcing effect PANI-DBSA offered, whereas the excess PANI destroyed the physically crosslinked network formed by ionic domains. The tensile strength increased relatively slowly and subsequently decreased for the composites containing the ionomer with 40 mmol SO_3H/100 g.

6.4.3 Thermal Stability of Elastomer/Conducting Polymer Blends

NR/PANI Blends

Thermal stability of NR/PANI composites was analyzed from room temperature to 600 °C in nitrogen [19]. Loading of PANI in NR was observed to improve the thermal stability of the composites. It was also noted that NRP30 composite (NR: 100 phr, PANI: 30 phr) exhibited highest thermal stability. In another study, TGA of PANI (20 vol%)/NR (80 vol%) blend films, prepared by adding PANI-EB (emeraldine base) dissolved in *N*-methyl-2-pyrrolidone (NMP) to the latex (NRL) and overlaying the surface of a pure NR cast film with PANI layer grown by *in-situ* polymerization (NRO), showed no mass loss up to 250 °C in nitrogen indicating their stability [18]. In contrast, mass loss below 100 °C was observed in NR and PANI-EB dissolved in *m*-cresol to NR dissolved in xylol (NRD). Sukitpaneenit *et al.* [17] studied thermal stability behavior of NR/PANI blends and concluded that the degradation of NR in the composites took place before PANI. Further, no remarkable change was observed in the thermal stability of the PANI/NR composite.

Typical TG/differential TG (DTG) curves were obtained for the neat NR, PANI-DBSA and NR/PANI-DBSA composite with 10 wt% of PANI-DBSA in nitrogen atmosphere [6]. Thermograms of the neat NR and composite exhibited the same decomposition mechanism, showing good thermal stability up to ~200 °C. TG curve of DBSA-PANI showed three main weight loss stages: < 100 °C (water and other volatiles), 250-350 °C (DBSA and oxidation of PANI structure) and > 400-500 °C (degradation of the bound PANI-DBSA and decomposition of PANI). TG profile of NR showed remarkable weight loss in range of 300-450 °C corresponding to the structural decomposition of rubber. The NR/PANI-DBSA composite also showed a peak at ~370 °C (major peak) mainly attributed to rubber decomposition and weight losses, along with the peaks corresponding to the loss of water (60-100 °C) and the degradation of PANI and DBSA bound in the composite (400-500 °C).

Panigrahi and Srivastava [82] studied the thermal stability behavior of EPDM and EPDM/PS@PANI, NBR and NBR/PS@PANI and NR and NR/PS@PAN. It was inferred that the thermal stability was improved by 72 °C in EPDM/PS@PANI-14 blend compared to neat EPDM. In all probably, barrier effect caused by the PANI shell prevented the degradation of the polymer chains of EPDM and accounted for enhanced stability of EPDM/PS@PANI-14 blend.

SBR/PANI Blends

Liu and Liu [45] investigated the thermal stability behavior of XSBR/PPy@HNTs (1.25, 2.5, 5 and 10 wt%) composites. All thermograms appeared almost identical below < 450 °C. It was also noted that the thermal stability of XSBR slightly increased in high temperature region by the addition of PPy@HNTs.

PANI/NBR Blends

Tassi and De Paoli [63] investigated the thermal stability of pure NBR and PANI/NBR in the temperature range of 470-790 °C under nitrogen atmosphere. PANI/NBR showed a slow mass loss after 100 °C and attained 6% loss at 360 °C. This was in all probability associated with the expulsion of low-molecular weight oligomers adsorbed in the rubber. At further higher temperature, mass loss of 82% (360-500 °C) was observed followed by another mass loss of 12% (500-650 °C).

PANI/EVA Blends

PANI-EVA (30, 50, 60, 70 and 80 wt%) blends exhibited weight losses corresponding to loss of moisture, dopant and vinyl acetate, due to the degradation of both PANI and EVA [69]. The moisture content associated with the first step decreased with decreasing polyaniline content. The commencement of the second step occurred at a progressively lower temperature as the EVA content in the blend increased from 50 to 80 wt%.

TGA studied exhibited that EVA-PANI (90:10, 85:15, 80:20, 75:25 and 70:30) blends prepared by mechanical blending except EVA-PANI 90/10) underwent three step degradation process [67]. The degradation kinetic parameter was obtained for each stage of thermal degradation of the blends. The activation energy (E) of the blends for the thermal degradation process was in the range of 3.3-95.6 kJ/mol and of 9.6-135.0 kJ/mol for Broido and Coats-Redfern methods, respectively. Thermal stability behavior of EVA/PANI-DBSA (33 wt%) blends prepared by two roll mixing and internal mixing methods was studied under nitrogen in the temperature range of 40-600 °C [5]. The presence of two step decomposition was observed, i.e. thermal decomposition of PANI-DBSA including acetate groups of EVA (250-420 °C) and EVA backbone (> 420 °C). Paul and Pillai [71] carried out TGA analysis of PANI (different contents and protonated with SPDPAA) blended with EVA by emulsion-polymerization and solution mixing methods. The studies confirmed the thermal stability of EVA up to 300 °C, above which decomposition took place to liberate acetic acid (325 to 400 °C) and ethylene backbone (450 °C onwards). The decomposition process in PANI-SPDPAA-EVA took place in two-steps, particularly at a lower PANI content. The decomposition of emulsion-polymerized PANI-SPDPAA-EVA and pure EVA was almost similar at PANI content less than 11.42%. At higher PANI content (>11.42%), the system after 300 °C involved multiple-stage decomposition. Also, the findings revealed 5% weight loss at 250 °C in PANI-SPDPAA-EVA blend.

EPDM/PANI Blends

Domenech *et al.* [79] performed TGA analysis of EPDM rubber, EPDM-*g*-MAH0.5 rubber, PANI-DBSA cast from chloroform, PANI-DBSA/EPDM and PANI-DBSA/EPDM-*g*-MAH0.5 blends prepared with different concentrations of an oxidant agent. Increasing concentration of PANI.DBSA in the blends slightly decreased the thermal stability of the mixture. In another study, the TGA analysis of PANI(DBSA)/EPDM blends in nitrogen atmosphere showed more than one stage of degradation [80]. The mass loss between 380-520 °C was observed to be the most significant and corresponded to the simultaneous degradation of the bound dopant, PANI and elastomer.

6.4.4 Calorimetric Properties of Elastomer/Conducting Polymer Blends

NR/PANI Blends

The effect of the adding PANI-DBSA on the glass transition temperature (T_g) of NR was studied by Silva *et al.* [6]. T_g of NR (-63°C) showed no significant change up to 10 wt% PANI-DBSA filled in NR. According to Camillo *et al.* [18], (T_g) of NR (-67°C) appeared to be unaffected by PANI in PANI (20%)/NR (80 %) blend, irrespective of the methods used for its fabrication. The ring opening of epoxidized natural rubber (ENR) was also identified in ENR/PANI.DBSA blends by differential scanning calorimetry (DSC) [34].

Kalasad *et al.* [32] also studied DSC of composite films of cis-1,4 polyisoprene (rubber)/PANI of 56/44 and 31/59 vol% compositions. A broad endothermic peak observed in both blends was shifted to higher temperature (333 to 358 K) with increasing % of rubber content in the blends due to the hydrophobic nature of PANI. A broad exothermic peak at ~463 K was assigned to a series of chemical reactions.

SBR/PANI Blends

Liu and Liu [45] performed DSC analysis to investigate the influence of PPy@HNTs on the thermal properties of the XSBR. T_g of 0, 1.25, 2.5, 5 and 10 % PPy@HNTs filled XSBR was found to be 12.1, 10.3, 11.9, 11.0 and 11.2 °C respectively.

SBR and its doped PANI blends were reported by Martins *et al.* [47]. It was noted that SBR matrix exhibited T_g of -50 and 45°C, corresponding to the butadiene and styrene blocks, respectively. On blending SBR with increasing amount of doped PANI showed presence of only one T_g (between -50 and 45°C). This indicated that the presence of the conducting polymer contributed to a partial plasticization and produced a more homogeneous and compatible material.

NBR/PANI Blends

Sub-ambient temperature DSC analysis has been carried out on pure NBR and NBR/PANI.DBSA blends with and without added hydroquinone [50]. The T_g of SBR corresponded to -11 °C. Further, T_g of 90/10, 80/20, 70/30, 60/40 and 50/50 in NBR (wt%):PANI.DBSA (wt%) blends had values of -5, -5, -5, -5 and -7 °C respectively. In another study, NBR-PANI.DBSA blends prepared with NBR (48 wt% CAN) showed the largest temperature shifts as compared to pure polymers [53]. DSC analysis was also carried out on NBR, PANI and NBR/PANI blend using temperature ranges of - 80 to 20 °C and 20 to 600 °C [65]. Pure NBR and NBR/PANI showed an endothermic transition assigned to T_g at -60° and -45°C respectively. In contrast to pure NR, an exothermic transition was noticed in NBR/PANI at-20 °C. The high-temperature DSC of PANI/NBR showed thermal processes similar to pure PANI and pure NBR. However, in addition, it also exhibited heat variations at 260, 425, 440, 505 and 570 °C. These results strongly suggested that NBR/PANI was neither a composite nor a blend but a grafted copolymer. The effect of the addition of PANI-DBSA on the T_g of NR was also investigated by Hanif and Yong [56]. The results did not show a significant change in the T_g of NR (-63°C) with addition of up to 10 wt% of PANI-DBSA.

EVA/PANI Blends

DSC heating and cooling thermograms of pure EVA, pure PANI and their blend composite were studied by Rahaman *et al.* [61]. Distinct crystalline melting peak (~75 °C) appeared during heating for EVA and EVA/PANI composite. During cooling process, exothermic peak was also observed due to crystallization in EVA and EVA/PANI composite. However, no endothermic/exothermic peak appeared in neat PANI during heating/cooling process.

DSC analysis of PANI-SPDPAA-EVA with 7.18, 18.5 and 28.5 wt% of PANI was compared with pure EVA [71]. These study showed bimodal melting transition, identified as T_1 (52 °C) and T_2 (75 °C), in the first heating scans. DSC study also revealed the effect of doped PANI on the transitions associated with EVA. DSC study of PANI-DBSA/EVA composite films consisting 0, 0.25, 2,5, 7.5, 10 and 17.5 mass% of PANI content has also been carried out [75]. EVA copolymer had a bimodal melting transition in the first heating scan, denoted as T_1 (34.5 °C) and T_2 (48.9 °C). The lower temperature peak (at T_1) was assigned to the melting of fringed micellar crystals, whereas the higher temperature peak (at T_2) was attributed to the melting of more perfect chain-folded crystals. The two melting peaks of EVA copolymer appeared in the composites too, the T_2 peak being always predominant.

6.5 Application of Elastomer/Conducting Polymer Blend Nanocomposites

6.5.1 Sensing

Wang *et al.* [41] successfully coated a thin film of PANI onto flexible poly(styrene-butadiene-styrene) (SBS) fiber using *in-situ* dilute polymerization and evaluated the developed system as ammonia gas sensor. The SBS/PANI composite demonstrated overall superior performance for detection of ammonia gas.

Rahaman *et al.* [74] studied compressive stress (pressure) sensitivity of dielectric properties of EVA/PANI composites. The study revealed that EVA-PANI composites could be used as dielectric sensor. It was noted that dielectric constant and loss increased with an increment in applied pressure. The dielectric properties also increased with respect to time under constant pressure. Rahaman *et al.* [81] also studied the pressure sensitivity of EVA/PANI composites prepared through *in-situ* polymerization method. The investigation demonstrated the sensitivity of the developed composites to applied pressure through a decrease in resistivity with an increase in pressure.

6.5.2 Corrosion Inhibiting Behavior

High toxicity and environmental risks are involved in coatings of harmful chemicals in many commercial applications [43,83]. Therefore, new materials are needed to minimize the environmental and health impact arising from such coatings. In view of this, conducting polymers represent a promising alternative to achieve effective corrosion inhibition on carbon steel. Yong [33] investigated corrosion inhibiting behavior of ENR/PANI.DBSA (< 5.0 wt% and > 40.0 wt%) blends for carbon steel, based on total immersion and electrochemical corrosion tests. In another study, Almazan *et al.* [43] successfully synthesized PANI/SBR and PANI-DBSA/SBR composites. The authors found that the incorporation of DBSA impacted the nature of the nitrogen vibrations from PANI's amine group and also prevented rust formation from atmospheric exposure. The investigation on protective behavior of PANI on SBR coated stainless steel in different corrosion media (3.5% NaCl and 0.5 M HCl) revealed superior corrosion protection than PANI coating alone [42]. In another study, PANI/SBR composites have also been studied to determine the feasibility of their application as anti-corrosive coatings for carbon steel [81,84]

6.5.3 Electromagnetic Shielding Interference

Various conducting composites have been used as effective means of controlling electromagnetic interference (EMI) [4]. Several studies have also been reported on conductive blends/composites consisting of conducting polymers and rubber. Al-Ghamdi *et al.* [19] studied dielectric spectroscopy, absorption loss and reflection loss of NR/PANI composites in the frequency range from 1 to 12 GHz. The authors noted that the total electromagnetic interference shielding effectiveness (SE) by absorption and reflection loss depended on PANI content in the composites. SE values in NRP10, NRP20 and NRP30 samples corresponded to 30, 50 and 60 dB (12 GHz) respectively. These findings also suggested absorption as a dominant electromagnetic wave shielding mechanism.

Martins *et al.* [47] evaluated the attenuation of reflectivity for SBR blends with 10, 30 and 50% (w/w) of PANI-DBSA-PSS in the frequency range of 8-12 GHz. An increase in the attenuation values was noticed for SBR/PANI-DBSA-PSS blends with high PANI-DBSA-PSS concentrations and conductivity, following which EPDM/PANI-

DBSA blends were observed to produce high attenuation values of 20 dB. Yong [52] reported EMI SE measurements in the frequency range of 0-10 GHz on NBR-PP-PANI.DBSA thermoplastic elastomer vulcanizates (TPV) with 1.0, 2.5, 10, 20, 30, 40 and 50 wt%, as displayed in Figure 6.6. It was noted that EMI SE of all NBR-PP TPVs was enhanced with PANI.DBSA loading. This was attributed to the increment in electrical conductivity. EMI SE of 24-32 dB was recorded for NBR-PP TPVs blend with ≥30.0 wt % of PANI. DBSA.

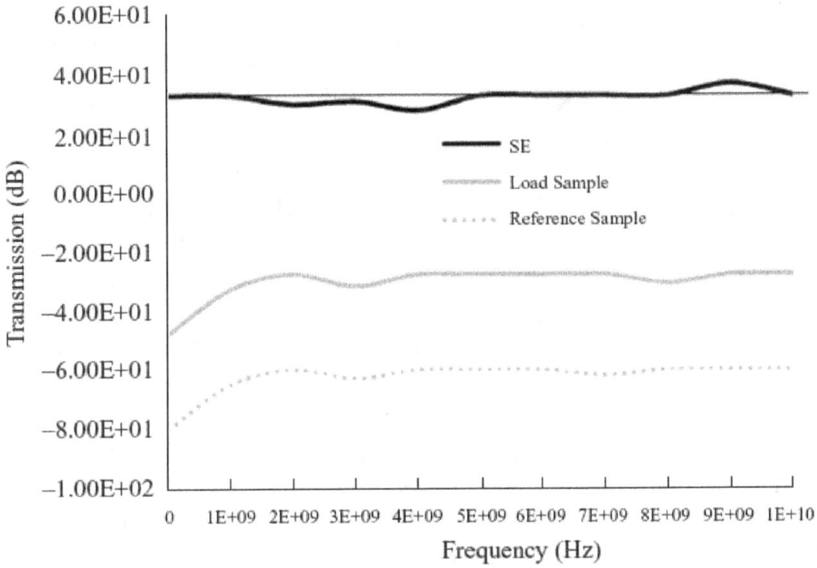

Figure 6.6 Raw results of electromagnetic interferences shielding effectiveness for NBR-PP TPV with 50.0 wt% PANI.DBSA. Reproduced from Reference 52 with permission from Malaysian Rubber Board.

Das et al. [16] studied EMI shielding of EVA with different concentrations of doped PANI in the X-band. The findings indicated that the composites having higher PANI loading (>23%) can be used for EMI shielding materials. Faez *et al.* [77] reported radiation absorbing properties of ternary blends based on nitrile rubber, EPDM rubber and polyaniline doped with dodecylbenzenesulfonic acid in the frequency range of 8-12 GHz. The blends exhibited good absorption performance between 11 and 12 GHz. Rahaman *et al.* [61] studied the effect of frequency on EMI shielding effectiveness of EVA/PANI composites. The effect was observed to be somewhat irregular and non-linear over the entire frequency range. EMI shielding effectiveness enhanced with the increasing PANI content in EVA matrix. It was suggested that PANI in EVA matrix formed conductive mesh, which absorbed microwave radiations by interfering with incident rays.

Recently, investigations were carried out to study EMI shielding as a function of frequency (1-8 GHz) for PS@PANI/rubber (EPDM, NBR, NR) blends consisting 6, 14, 25 and 40 wt% of filler [82]. SE was observed to be nearly independent of frequency in rubber/core@shell blends. Further, PS@PANI filled rubbers exhibited superior EMI SE compared to conventional PS/PANI/rubbers. The shielding efficiency of neat rubber, rubber blends of PS@PANI and PS/PANI followed the following order over the entire frequency range: rubber/PS@PANI (>30 dB) > rubber/PS/PANI (<15 dB) > neat rubber (<13 dB).

The reflectivity measurements were reported for conductive elastomeric blend of EPDM/PANI.DBSA in the range of 8-12 GHz [88]. The study showed that dopant and PANI concentrations as well as the thickness of the samples affected the properties. EPDM/PANI blends with wt/wt% of 50/50, 40/60 and 20/80 showed shielding values of ~15 dB. In another study, the investigations on understanding the influence of processing time and composition for the microwave absorption of EPDM/PANI blends prepared using EPDM and PANI.DBSA by mechanical mixing in an internal mixer were also carried out [89]. It was noted that processing time and composition affected microwave absorption behavior of the blends. The radar-absorbing materials prepared in this manner showed broadband behavior with microwave radiation absorption of up to 90%.

Table 6.1 presents the summary of the preparative methods reported in the recent studies for the development of conducting polymer/rubber blends and composites.

6.6 Conclusions

Conducting polymer/rubber blends are advanced materials with a superior set of properties which make them promising candidates for a wide variety of commercial applications. In this chapter, the synthesis, properties and applications of a diverse range of blends of natural rubber (NR), styrene butadiene rubber (SBR), nitrile rubber (NBR), ethylene-vinyl acetate (EVA) and ethylene propylene diene monomer rubber (EPDM) with conducting polymers have been reviewed in detail.

Table 6.1 Preparative methods of elastomer/conducting polymer blends

Elastomer	Conducting polymer	Blend preparation method	Ref.
NR	PANI-DBSA	Casting method	[6]
NR	PANI (up to 10% wt/wt) in its emeraldine base form	Wet spinning process followed by subjecting resulting fibers to doping by means of aqueous HCl solution to convert PANI (ES) to conducting PANI (Emeraldine form)	[17]
NR (80%)	20% in volume of PANI	Blend film fabricated adding PANI dissolved in NMP to the latex, adding PANI dissolved in m-cresol to NR dissolved in xylol, overlaying the surfaces of pure NR cast film with a PANI layer grown by *in-situ* polymerization	[18]
NR	0, 10, 20, 30 phr of PANI prepared by chemical redox polymerization of aniline	Open two-roll mill	[19]
NR	PANI.DBSA (2.5, 5.0, 7.5, 10.0, 20.0, 30.0, 40.0 and 50.0 wt%)	Internal mixer; electron beam irradiation	[20]
NR (and SBR)	o-aminophenol, poly(o-amino-hiophenol), poly(m-anisidine) and polyaniline	To roll mill	[21]
NR (and NBR)	Conducting PANI	Wet mixing process	[23]
NR	PANI	*In-situ* polymerization of aniline in the presence of NR latex	[25]
NR Latex (80%)	PANI (20%)	Subjected to casting PANI (undoped)/NR latex followed by treating with either HCl or corona discharge	[27]
TPNR (LLDPE:	PANI (10%)	Melt blending using	[28]

NR:LNR = 50:40:10) (90%)		internal mixer (mixing temperature:120-150 °C, mixing speed of rotation: 20-50 rpm, time of mixing: 13, 14, 15 minutes)	
TPNR (LLDPE:NR: LNR = 40:50:10)/TPNR (90 wt%)	PANI 10 wt %)	Melt blending	[29]
ENR (50.0-97.5 pphr)	PANI.DBSA (2.5-50.0 pphr)	Internal mixer/120 °C, 100 rpm	[32]
ENR	PANI.DBSA	Casting method using masterbatch, ENR (wt%):PAni.DBSA (wt%) = 50:50, 60:40, 70:30, 80:20, 90:10, 95:5, 97.5:2.5 and 99:1.=	[33]
ENR (50.0-97.5 pphr)	PANI.DBSA (2.5-50.0 pphr)	Two-roll mill (120 °C, 100 rpm); ENR (wt%): PAni.DBSA (wt%) = 50:50, 60:40, 70:30, 80:20, 90:10, 95:5, 97.5:2.5 and 99:1	[34]
SBR	PANI	Ex-situ technique	[39]
SBR	PANI	Electro-polymerization using potentiostatic method	[42]
SBR	PANI and PANI.DBSA	CP/SBR ratio = 05/95, 10:90, 15:85, 20:85)	[43]
SBR latex in	Aqueous PANI.DBSA	Solution blending	[44]
xSBR	PPy@HNTs	Solution mixing and casting drying method	[45]
SBR	PANI	Solution casting method	[46]
SBR	PANI.DBSA	Solution casting method	[85]
NBR	PANI	Electrochemical graft copolymerization of aniline with NBR	[63]
NBR	PANI.DBSA	Solution casting using wt% NBR: wt% PAni.DBSA as 50:50, 60:40, 70:30, 80:20, 90:10, 95:5, 97.5:2.5 and 99:1	[49]
NBR	PANI.DBSA	Internal mixer, 140 °C/100 rpm; NBR wt %: PAni.DBSA wt% (i.e., 99:1, 97.5:2.5, 95:5, 90:10, 80:20, 70:30,	[53]

		60:40, 50:50)	
NBR	PANI.DBSA	High temperature-mechanical mixing method using combination of internal mixer and two-roll mill	[54]
NBR	PANI.DBSA	Two roll-mill, NBR wt%:filler wt% blends (97.5:2.5, 95:5, 92.5:7.5, 90:10, 80:20, 70:30, 60:40, 50:50)	[56]
NBR	PANI.DBSA	Polymerization of aniline (*in-situ* emulsion) in presence of NBR	[58]
NBR	PANI.DBSA	Solution mixing and casting	[60]
NBR/PP (50:50 wt%)	PANI.DBSA	Internal mixer, 130 °C/100 rpm	[52]
NBR/EVA (50:50)	PANI	Internal mixer, 100 °C /60 rpm /4 min	[57]
NBR/EPDM	PANI.DBSA	Melt blending using an internal mixer at 80 °C, 30 and 60 rpm, with processing times of 12 or 3 min, respectively	[77]
EVA	PANI-DBSA	Two-roll mill (50 °C/10 min, 15 rpm) and in a Haake internal mixer (150 °C/10 min, 15 rpm)	[5]
EVA	PANI-DBSA	Brabander Plasticoder (at 50 °C, time of mixing: 6 min, rotor speed: 60 rev min^{-1})	[16]
EVA	PANI	Mechanical blending using plasticorder; EVA-PANI (%, w/w) = 90/10, 85:15, 80:20, 75:25, 70:30)	[67]
EVA	PANI	Emulsion polymerization technique, EVA-PANI (%wt/wt) = 98/02, 95/05, 93/07, ... 50/50	[68]
EVA	PANI	Polymerization of aniline monomer was carried out in a solution of EVA with camphor sulphonic acid as the dopant	[69]

EVA	PANI.DBSA	Solution-cast process	[70]
EVA	PANI protonated with dopant	Emulsion polymerization, solution-mixing method	[71]
EVA	PANI doped with DBSA		[73]
EVA	PANI-DBSA		[75]
EVA	PANI	*In-situ* polymerization	[81,83,84]
Poly(ethylene-co-vinyl acetate)	PPy	Emulsion pathway	[72]
EPDM	PANI.DBSA	Internal mixer (150 °C/50 rpm)	[76]
EPDM	PAni and zinc sulfonated ethylene	In-situ polymerization	[86]
EPDM	PANI.DBSA	Solution based method	[79]
EPDM	PANI.DBSA	Casting method	[80]
EPDM	curauá fibers coated with PAni	Mechanical mixing in a counter-rotating twin mixer	[78]

References

1. Ates, M., Karazehir, T., and Sarac, A. S. (2012) Conducting polymers and their applications. *Current Physical Chemistry*, **2**(3), 224-240.
2. Das, T. K., and Prusty, S. (2012), Review on conducting polymers and their applications. *Polymer-Plastics Technology and Engineering*, **51**(14), 1487-1500.
3. Gangopadhyay, R., and De, A. (2000) Conducting polymer nanocomposites: A brief overview. *Chemistry of Materials*, **12**, 608-622.
4. Dhawan, S. K., Singh, N., and Rodrigues, D. (2003) Electromagnetic shielding behaviour of conducting polyaniline composites. *Science and Technology of Advanced Materials*, **4**, 105-113.
5. Leyva, M. E., Soares, B. G., Mattoso, M. E., and Sens, M. (2001) Electrically conductive, melt-processed polyaniline/EVA blends. *Journal of Applied Polymer Science*, **82**, 114-123.
6. Silva, M. J., Sanches, A. O., Malmonge, L. F., and Malmonge J. A. (2014) Electrical, mechanical, and thermal analysis of natural rubber/polyaniline-DBSA composites. *Material Research*, **17**, 59-63.
7. Chen, C. H. (2002) Thermal studies of polyaniline doped with dodecyl benzene sulfonic acid directly prepared via aqueous dispersions. *Journal of Polymer Research*, **9**(3), 195-200.
8. Bhadra, S., Khastgir, D., Singha, N. K., and Lee, J. H. (2009) Progress in preparation, processing and applications of polyaniline. *Progress in Polymer Science*, **34**(8), 783-810.

9. Afzal, A. B., Akhtar, M. J., and Ahmad, M. (2010) Morphological studies of DBSA-doped polyaniline/PVC blends. *Journal of Electron Microscopy*, **59**(5), 339-344.
10. Yang, C., Cao, Y., Smith, P., and Heeger, A. J. (1993) Morphology of conductive, solution-processed blends of polyaniline and poly(methyl methacrylate). *Synthetic Metals*, **53**, 293-301.
11. Heeger, A. J. (1993) Polyaniline with surfactant counterions: conducting polymer materials which are processable in the conducting. *Synthetic Metals*, **55-57**, 3471-3482.
12. Laska, J., Zak, K., and Pron, A. (1997) Conducting blends of polyaniline with conventional polymers. *Synthetic Metals*, **84**(1-3), 117-118.
13. Kanazawa, K. K., Diaz, A. F., Krounbi, M. T., and Street, G. B. (1981) Electrical properties of pyrrole and its copolymers. *Synthetic Metals*, **4**, 119-130.
14. Ge, H., and Wallace, G. G. (1992) Properties of thermally treated polypyrroles. *Polymer*, **33**, 2348-2352.
15. Lee, B-L. (1992) Electrically conductive polymer composites and blends. *Polymer Engineering and Science*, **32**(1), 36-42.
16. Das, N. C., Shinichi, Y., Hikosaka, M., Chak, T. K., Khastgir, F., and Chakraborty, A. (2005) Electrical conductivity and electromagnetic interference shielding effectiveness of polyaniline-ethylene vinyl acetate composites. *Polymer International*, **54**, 256-259.
17. Sukitpaneenit, P., Thanpitcha, T., Sirivat, A., Weder, C., and Rujiravanit, R. (2007) Electrical conductivity and mechanical properties of polyaniline/natural rubber composite fibers. *Journal of Applied Polymer Science*, **106**, 4038-4046.
18. Camillo, E. C., Constantino, C. J. L., Teruya, M. Y., Alves, N., Mattoso, L. H. C., and Job, A. E. (2005) Dependence of the electrical conductivity and elastomeric properties on sample preparation of blends of polyaniline and natural rubber. *Journal of Applied Polymer Science*, **97**, 1498-1503.
19. Al-Ghamdi, A. A., Al-Hartomy O. A., Al-Solamy, F., Al-Hazmi, F., Al-Ghamdi, A. A., El-Mossalamy, E. H., and El-Tantawy F. (2014) On the prospects of conducting polyaniline/natural rubber composites for electromagnetic shielding effectiveness applications. *Journal of Thermoplastic Composite Materials*, **27**(6), 765-782.
20. Yong, K. C. (2013) Electron beam irradiated natural rubber-polyaniline dodecylbenzenesulphonate blends: Their preparation, electrical and physical properties. *Journal of Rubber Research*, **16**(1), 1-14.
21. Helaly, F. M. Darwich, W. M., Abd El-Gha€ar M. A. (1999) Effect of some polyaromatic amines on the properties of NR and SBR vulcanizates. *Polymer Degradation and Stability*, **64**, 251-257.
22. *Polyaniline Blends, Composites, and Nanocomposites*, Visakh, P. M.,

Pina, C. D., and Falletta, E. (eds.), Elsevier, The Netherlands (2017).

23. Sreeja, R., Sharma, P. D., Najidha, S. J., Jayen, S. R., and Predeep, P. (2006) Semi-interpenetrating conducting elastomer composite thin films from polyaniline. *Polymers & Polymer Composites*, **14**(3), 261-269.
24. Galiani, P. D., Malmonge, J. A., Santos, D. P., and Malmonge, L. F. (2007) Compósitos de borracha natural com polianilina. *Polímeros*, **17**(2), 93-97.
25. Honey, J., Joseph, R., Mathew, K. T. (2007) Dielectric behavior of natural rubber composites in microwave fields. *Journal of Applied Polymer Science*, **103**(4), 2682-2686.
26. Sun, Y., and Ruckenstein., A. (1995) Synthesis of polypyrrole nanoparticles in natural rubber-polystyrene blend via emulsion polymerization. *Synthetic Metals*, **72**, 261-267.
27. Job, A. E., Constantino, C. J. L., Mendes, T. S. G., Teruya M. Y., Alves N., and Mattoso, L. H. C. (2003) Effect of natural rubber latex on the conducting state of polyaniline blends determined by Raman spectroscopy. *Journal Raman Spectroscopy*, **34**, 831-836.
28. Zailan, F. D., Ahmad, S., Chen, R. S., and Shahdan, D. (2016) Effect of processing parameters on tensile properties of thermoplastic natural rubber (TPNR) blend with polyaniline. *AIP Conference Proceedings*, **1784**, 040029.
29. Zailan, F. D., Chen R. S., Ahmad S., and Shahdan, D., and Bahari, K. W. (2017) Influence of processing conditions on mechanical behaviours of thermoplastic natural rubber/polyaniline blend. *Journal of Mechanical Engineering*, **3**(1), 227-237.
30. Soares, B. G., Celestino, M. L., Magioli, M., Moreira, V., and Khastgir, D. (2010) Synthesis of conductive adhesives based on epoxy resin and polyaniline.DBSA using the in situ polymerization and physical mixing procedures. *Synthetic Metals*, **160**(17-18), 1981-1986.
31. Baboo, M., Sharma, K., and Saxena, N. S. (2011) Study on mechanical and thermal properties of composites of cis- and trans-Polyisoprene blends filled with polyaniline. *Polymer Composites*, **32**, 2028-2037.
32. Kalasad, M. N., Gadyal, M. A., Hiremath, R. K., Ikram, I. M., Mulimani, B. G., Khazi, I. M., Anantha Krishnan, S. K., and Rabinal, M. K. (2008) Synthesis and characterization of polyaniline rubber composites. *Composites Science and Technology*, **68**, 1787-1793.
33. Yong, K. C. (2014) Corrosion inhibiting behavior of epoxidized natural rubber-polyaniline dodecylbenzenesulfonate (ENR-PANI.DBSA) blends. *Journal of Rubber Research*, **87** (3), 526-537.
34. Yong, K. C. (2011) Epoxidized natural rubber-polyaniline dodecylbenzenesulfonate (ENR-PANI.DBSA) blends: Effect of strain. *Journal of Rubber Research*, **84**(4), 580-593.

35. Yong, K. C., Foot, P. J. S., Morgan, H., Cook, S., Tinker, A. J., and, Ahmad, M. A. (2008) Synthesis and characterization of polyaniline rubber composite. *Journal of Rubber Research,* **11**(2), 59-77.
36. John, H., Joseph, R., and Mathew, K. T. (2007) Dielectric behavior of natural rubber composites in microwave fields, *Journal of Applied Polymer Science,* **103**(4), 2682-2686.
37. Wang, H., Wen, H., Hu, B., Fei, G., Shen, Y., Sun, L., and Yang D. (2017) Facile approach to fabricate waterborne polyaniline nanocomposites with environmental benignity and high physical properties. *Scientific Reports,* **7**, 43694.
38. Mathew, A. M., and Neena, P. (2011) Effect of applied potential on the electrochemical deposition of styrene-butadiene co-polymer based conducting polymer composite. *AIP Conference Proceedings,* **1391**(1), 799.
39. Mathew, A. M., and Neena, P. (2011) Styrene-butadiene copolymer based highly conducting and flexible polymer composite film with low percolation threshold. *AIP Conference Proceedings,* **1391**(1), 793-795.
40. Mathew, A. M., Neena, P., Prasanth, S., Panwar, N. S., and Pradeep, P. (2008) Synthesis and characterization of polypyrrole coated SBR composites. *AIP Conference Proceedings,* **1004**, 264.
41. Wang, X., Meng, Si., Tebyetekerwa, M., Weng,, W., Pionteck J., Sun, B., Qin, Z., and Zhu, M. (2017) Nanostructured, polyaniline/poly(styrene-butadiene-styrene) composite fiber for use as highly sensitive and flexible ammonia sensor. *Synthetic Metals,* **233**, 86-93.
42. Mathew, A. M., and Predeep, P. (2012) Styrene butadiene copolymer based conducting polymer composite as an effective corrosion protective coating. *Progress in Organic Coatings,* **74**(1), 14-18.
43. Leon Almazan, C. M. D., Chavez-Cinco M. Y., Paramo Garcıa, U., Mendoza-Martınez, A. M., Estrada-Moreno, I. A., and Rivera-Armenta, J. L. (2016) Conducting polymer composites' characterization. *Polymer Bulletin,* **73**, 1595-1605.
44. Park, C-Y., Oh, Y-S, Baek, S-H., Kim, T-Y., and Suh S-H. (2006) Electrical and mechanical properties of polyaniline. DBSA/SBR latex blends. *Revue Roumaine de Chimie,* **51**(11), 1087-1094.
45. Liu, Y., and Liu, M. (2017) Conductive carboxylated styrene butadiene rubber composites by incorporation of polypyrrole-wrapped halloysite nanotubes. *Composites Science and Technology,* **143**, 56-66.
46. Almazan, C. M. D. L., Armenta, J. L. R., Cinco, M. Y. C., and Martinez, A. M. M. (2014) Synthesis and Characterization of PANI/SBR Composites as Corrosion Protection Coating for Carbon Steel Substrates. *2014 ECS and SMEQ Joint International Meeting,* Mexico.

47. Martins, C. R., Faez, R., Rezende, M. C., De Paoli, M. A. (2006) Reactive processing and evaluation of butadiene-styrene copolymer/Polyaniline conductive blends. *Journal of Applied Polymer Science*, **100**, 681-685.
48. Liu, C-F., Maruyama, T., and Yamamoto, T. (1993) Conductive blends of p-conjugated polymers and thermoplastic polymers in latex form. *Polymer Journal*, **25**, 363-372.
49. Chiu, H.-T., Chiang, T.-Y., Chang, C.-Y., and Kuo, M.-T. (2011) Blends of NBR, carbon black, and polypyrrole-Electrical conductivity and synergy effects. *GAK Gummi Fasern Kunststoffe*, **64**(10), 615-623.
50. Yong, K. C. (2003) Electrically conductive poly(butadieneco-acrylonitrile)-polyaniline dodecylbenzenesulphonate [NBR-PAni.DBSA] blends with hydroquinone as compatibiliser. *Journal of Rubber Research*, **15**(3), 153-166.
51. Yong, K. C. (2013) NBR-PANI.DBSA blends: Effect of electron beam irradiation. *Rubber Chemistry and Technology*, **86** (1), 68-85.
52. Yong, K. C. (2015) Electrically conductive thermoplastic elastomer vulcanisate based on NBR and PP blends with polyaniline: Preparation and characterization. *Journal of Rubber Research*, **18**(4), 189-202.
53. Yong, K. C., and Saad, C. S. M. (2009) Novel peroxide-vulcanized NBR-PAni.DBSA blends, Part 1: Preparation and characterization. *Journal of Applied Polymer Science*, **112**, 3199-3208.
54. Yong, K. C., and Saad, C. S. M. (2010) High temperature-mechanical mixing to prepare electrically conductive sulfur-vulcanised poly(butadiene-co-acrylonitrile)-polyaniline dodecylbenzenesulfonate blends. *Journal of Rubber Research*, **13**(1), 1-17.
55. Leyva, M. E., Barra, G. M. O., Gorelova, M. M., Soares, B. G., and Sens, M. (2001) Conducting SBS block copolymer–polyaniline blends prepared by mechanical mixing. *Journal of Applied Polymer Science*, **80**, 626-633.
56. Hanif, H. M., and Yong, K. C. (2017) Electrically conductive poly(butadiene-co-acrylonitrile) [NBR]- based blends: Optimization of formulation. *Chemical Engineering Transactions*, **56**, 973-978.
57. Souza, Jr., F. G., Almeida, M., Soares, B. G., and Pinto, J. C., (2007) Preparation of a semi-conductive thermoplastic elastomer vulcanizate based on EVA and NBR blends with polyaniline. *Polymer Testing*, **26**, 692-697.
58. Soares, B. G., Amorim, G. S., Oliveira, M. G., and Pereira da Silva, J. E. (2006) Conducting elastomer blends based on nitrile rubber and PANI.DBSA. *Macromolecular Symposia*, **233**, 95-101.
59. Luan, Y., Noh, J-S., Kim, S. H. (2017) Facile control of stretchability and electrical resistance of elastomer/ Polyaniline composites for stretchable. *Materials Chemistry and Physics*, **190**, 68-73.

60. Yong, K. C., Foot, P. J. S., Morgan, H., Cook, S., and Tinker A. J. (2006) Conductive poly(butadiene-co-acrylonitrile)-polyanilinedodecylbenzenesulfonate[NBR-PAni.DBSA] blends prepared in solution. *European Polymer Journal*, **42**, 1716-1727.
61. Rahaman, M., Nayak, L., Chaki, T. K., and Khastgir, D. (2012) Conductive composites made from in-situ polymerized polyaniline in ethylene vinyl acetate copolymer (EVA): Mechanical properties and electromagnetic interference shielding effectiveness. *Advanced Science Letters*, **18**, 54-61.
62. Cho, H. W., Hepowit, L. R., Nam, H-S., Kim, S. H., Lee, Y. M., Kim, J. H., Kim, K. M., Ko, J. M. (2012) Synthesis and supercapacitive properties of electrodeposited polyaniline composite electrode on acrylonitrile-butadiene rubber as a flexible current collector. *Synthetic Metals*, **162**, 410-413.
63. Tassi, E. L., and. De Paoli, M-. (1992) A An electroactive elastomer:polyaniline/nitrile rubber. *Polymer*, **33**, 2427-2430.
64. Prudencio, L., Camilo, F. F., and Faez, R. (2014) Ionic liquids as plasticizers in nitrile rubber/polyaniline blends. *Química Nova*, **37**, 618-623.
65. Soares, B. G., Amorim, G. S., and Souza Jr, F. G., Oliveira, M. G., and Pereira da Silva, J. E. (2006) The in situ polymerization of aniline in nitrile rubber. *Synthetic Metals*, **156**, 91-98.
66. Rahaman, M., Chaki, T. K., Khastgir, D (2012) Modeling of DC conductivity for ethylene vinyl acetate (EVA)/polyaniline conductive composites prepared through in-situ polymerization of aniline in EVA matrix. *Composites Science and Technology*, **72**(13), 1575-1580.
67. Jeevananda, T., and Siddaramaiah, J. T. (2001) Thermal and morphological studies on ethylene-vinyl acetate copolymer-polyaniline blends. *Thermochimica Acta*, **376**, 51-61.
68. Jeevananda, T., and Siddaramaiah, J. T. (2003) Synthesis and characterization of ethylene vinyl-acetate/Polyaniline blends prepared by emulsion polymerization technique. *Polymer Engineering and Science*, **43**, 1138-1150.
69. Rao, P. S., Subrahmanya, S., and Sathyanarayana, D. N. (2003) Synthesis by inverse emulsion pathway and characterization of conductive polyaniline-poly(ethylene-co-vinyl acetate) blends. *Synthetic Metals*, **139**, 397-404.
70. Guilherme, M. O., Barra, M. E. L., Soares, B. G., and Sens, M. (2002) Solution-cast blends of polyaniline-DBSA with EVA copolymers. *Synthetic Metals*, **130**, 239-245.
71. Paul, R. K., and Pillai, C. K. S. (2002) Melt/Solution processable conducting polyaniline: elastomeric blends with EVA, *Journal of Applied Polymer Science*, **84**, 1438-1447.
72. Yang, S., and Ruckenstein, E. (1993) Preparation and mechanical

properties of electrically conductive polypyrrole-poly(ethylene-co-vinyl acetate) composites, *Synthetic Metals*, **60**, 249-254.

73. Dopico-Garcia., M. S. Ares., A., Lasagabaster-Latorre., A., Garcia., X., Arboleda, L and Abad, M. J. (2014) Extruded polyaniline/EVA blends: Enhancing electrical conductivity using gallate compatibilizers. *Synthetic Metals*, **189**, 193-202.
74. Rahaman, M., Rout, T. K., and Khastgir, D. (2014) Polyaniline/Ethylene vinyl acetate composites as dielectric sensor. *Polymer Engineering and Science*, **54**(7), 1632-1639.
75. Tsocheva, T. T., and Terlemezyan, L. (2001) Structure of composite films containing polyaniline studied by DSC. *Journal of Thermal Analysis and Calorimetry*, **66**, 415-422.
76. Faez, R., and De Paoli, M-A. (2001) A conductive rubber based on EPDM and Polyaniline I: Doping method effect. *European Polymer Journal*, **37**, 1139-1143.
77. Faez, R., Reis, A. D., Soto-Oviedo, M. A., Rezende, M. C., and De Paoli, M-A. (2005) Microwave absorbing coatings based on a blend of nitrile rubber, EPDM rubber and polyaniline. *Polymer Bulletin*, **55**, 299-307.
78. Araujo, J. R., Adamo, C. B., Rocha, W. F. C., Costa e Silva, M. V., Carozo, V., Calil, V. L., and De Paoli M-A. (2014) Elastomer composite based on EPDM reinforced with polyaniline coated curaua fibers prepared by mechanical mixing. *Journal of Applied Polymer Science*, **131**, DOI: 10.1002/app.40056.
79. Domenech, S. C., Bortoluzzi, J. H., Soldi, V., Franco, C. V. (2003) Electroactive mixtures of polyaniline and EPDM rubber: Chemical, thermal, and morphological characterization, *Journal of Applied Polymer Science*, **87**: 535-547.
80. Schmidt, V., Domenech, S. C., Soldi, M. S., Pinheiro, E. A., and Soldi, V. (2004) Thermal stability of polyaniline/ethylene propylene diene rubber blends prepared by solvent casting. *Polymer Degradation and Stability*, **83**, 519-527.
81. Rahaman, M., Chaki, T. K., and Khastgir, D. (2013) Polyaniline, ethylene vinyl acetate semi-conductive composites as pressure sensitive sensor. *Journal of Applied Polymer Science*, **128**, 161-168.
82. Panigrahi R., Srivastava, S. K., and Pionteck, J. (2017) Fabrication of elastomeric blends involving core (polystyrene)@shell (polyaniline) approach, Their characterization and applications in electromagnetic shielding. *Rubber Chemistry and Technology*, **91**(1), 97-119.
83. Rahaman, M., Chaki, T. K., and Khastgir, D. (2012) Consideration of interface polarization in the modeling of dielectric property for EVA/PAni conductive composites prepared through in-situ polymerization of aniline in EVA matrix. *European Polymer Journal*, **48**, 1241-1248.

84. Tsanov, T., Mokreva, P., Tsocheva, D., and Terlemezyan, L. (2004) Changes during storage of electrically conductive blends polyaniline-poly(ethylene-co-vinyl acetate). *Journal of Macromolecular Science, Part C: Polymer Reviews*, **44**(4), 311-350.
85. Leon Almazan, C. M. D., Chavez-Cinco, M. Y., Paramo-Garcia, U., Mendoza-Martinez, A. M., Estrada-Moreno, I. A., and Rivera-Armenta, J. L. (2016) PANI/SBR composites as anticorrosive coatings for carbon steel I. Chemical, morphological and superficial characterization. *Polymer Bulletin*, **73**, 1595-1605.
86. Xie, H-Q., Pu, Q-L., and Xie, D. (2004) Preparation of conductive polyaniline-sulfonated EPDM ionomer composites from in situ emulsion polymerization and study of their properties. *Journal of Applied Polymer Science*, **93**, 2211-2217.
87. Alvares, D. A., Moreira, J. C., Soares, D. A. W., Machado, W. S., Scuracchio, C. H., and Onmori, R. K. (2006) Study of conductivity of EPDM rubber and PANI blends. *Macromolecular Symposia*, **245-246**, 343-346.
88. Faez, R., Martin, I. M., De Paoli, M.-A., and Rezende, M. C. (2001) Microwave properties of EPDM/PAni-DBSA blends. *Synthetic Metals*, **119**, 435-436.
89. Faez, R., Martin, I. M., De Paoli M-A., Rezende, M. C. (2002) Influence of processing time and composition in the microwave absorption of EPDM/PANI blends. *Journal of Applied Polymer Science*, **83**, 1568-1575.
90. Zoppi, R. A., De Paoli, M. A. (1997) Chemical preparation of conductive elastomeric blends: polypyrrole/EPDM III. Electrochemical characterization. *Journal of Electroanalytical Chemistry*, **437**, 175-182.
91. Zoppi, R. A., De Paoli, M-A. (1996) Chemical preparation of conductive elastomeric blends: polypyrrole /EPDM. II. Utilization of matrixes containing crosslinking agents, reinforcement fillers and stabilizers. *Polymer*, **37**, 1999-2009.
92. Zoppi, R. A., De Paoli, M-A. (1995) Conductive elastomer derived from polypyrrole and EPDM rubber: preparation and properties. *Polimeros: Ciencia e Tecnologia*, **5**, 19-31.

Chapter 7

Natural Fibers Reinforced Conductive Polymer Composites

A. Pangajam, S. Kumar and K. Dinakaran*

Department of Chemistry, Thiruvalluvar University, Vellore 632115, India
**Corresponding author*: kdinakaran.tvu@gmail.com

7.1 Introduction

Fiber reinforced polymer (FRP) composite materials show great promise for applications in aerospace, land and marine transportation, chemical process industries, electrical and electronic industries, etc., due to their light weight and excellent strength to weight characteristics. FRP exhibit "made-to-order" strength, stiffness and geometry, along with properties such as thermal, electrical, chemical and solvent resistance, dimensional stability, etc. At present, there is an increasing need for the development of polymeric materials with light weight structure, high fracture toughness, better impact strength, good thermal stability and improved mechanical properties in order to utilize them for high-performance engineering and aerospace applications and FRP fulfil most of these requirements.

Polymer composites of synthetic fibers such as glass, carbon, boron and aramid are non-biodegradable and expensive, along with reported health hazards. In addition, the depletion of petroleum resources as well as the increased awareness about environment and health have made it necessary to identify eco-friendly materials having effective design to meet the demands of our modern lifestyle. The limitations associated with synthetic fibers can be overcome by the introduction of natural fibers generated from renewable resources, which are becoming increasingly popular for use as reinforcements in polymer matrices. Natural fibers based reinforcements provide benefits to the environment with respect to degradability, renewability, minimal health hazards and utilization of natural materials. The successful use of these fibers is dependent on

Conducting Polymer Composites, edited by Vikas Mittal

their well-defined structural and mechanical properties. The properties of natural fibers are influenced by their origin, retting and treatment, climate conditions, age of the plants and extraction methods.

One of the factors which make plastics attractive for engineering utilization is the possibility of property enhancement through fiber reinforcement [1]. The use of natural fillers to reinforce polymers further promotes the degradability of the resulting product [2]. Fibers like oil palm empty fruit bunch and Kola nitida wood fiber as well as several fillers such as rice husk have been used as reinforcing agents for various polymeric resins. As mentioned earlier, there is an increasing interest in natural fiber reinforced polymers due to their ease of processing and low cost as some of these fibers are regarded as waste. In this chapter, the recent research studies specifically reporting the development of conducting composites using natural fibers as reinforcement have been summarized.

7.2 Classification of Natural Fibers

Natural plant fibers are classified based on their origin such as bast fibers, fruit, leaf fibers and seed fibers as shown in Figure 7.1. Such fibers include firewood, filament, coir, rice straw, bamboo, wheat straw, flax, hemp, rice peelings, bagasse, pineapple leaf, oil palm, date palm, ramie, jowar, curaua, kenaf, doum fruit, sisal, rapeseed waste, jute, etc. [3-6]. A general classification of natural fibers was reported by Al-Oqla and Sapuan [7], as shown in Figure 7.1. Such classification is helpful for the selection of natural fibers according to the required properties. Most of the natural fibers mentioned in this classification have been used to reinforce different polymer matrices. As mentioned earlier, several factors determine the mechanical, electrical and biological characteristics of natural fiber reinforced polymer composites.

7.3 Conducting Polymer Composites

The alternate single and double bond, known as conjugation, in the polymer backbone gives rise to the electronic conductivity in conjugated polymers (CPs) [8]. The mobile π electrons in the double bond of the conjugated polymer are delocalized to give rise to the conduction band and having uninterrupted π electrons in a linear chain may exhibit metallic behavior. However, such a system is unstable with respect to bond replacement, which causes the development of an energy gap in the electronic spectrum [9]. Dopant ions are introduced to the structure to overcome the energy gap and, hence, to impart conductivity to these polymers.

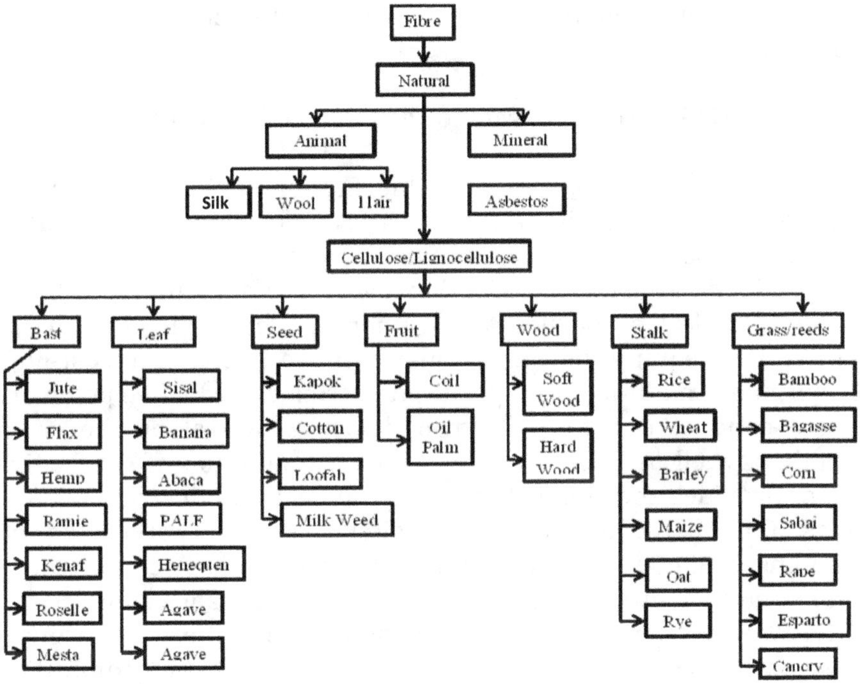

Figure 7.1 Classification of natural fibers.

CPs can be synthesized either chemically or electrochemically. Chemical methods of CP generation use either condensation or addition polymerization [10,11]. The advantages of electrochemical methods include ease of synthesis, immediate doping and trapping of molecules during synthesis, however, post-synthesis covalent changes in CPs is not easy [12]. On the other hand, chemical synthesis offers opportunities to adapt CP backbone covalently and makes the post-synthesis covalent modification possible, although this method is often more complicated [13]. Another large difference between electrochemical and chemical methods of CP synthesis is that electrochemical process can produce extremely thin CP films (of the order of 20 nm), while powders or extremely thick films are usually generated using chemical polymerization [14-17]. Furthermore, electrochemical synthesis is narrow to those systems in which the monomer can be oxidized upon use of potential to form reactive radical ion intermediates for polymerization [18]. The ordinary CPs (e.g. polypyrrole (PPy), polyaniline (PANI), etc.) can be generated chemically and electrochemically, however, several new CPs with customized monomers can only be synthesized by chemical polymerization [19,20].

7.4 Natural Fiber Reinforced Conductive Composites

In the recent past, textiles yarns and fabrics have been coated or blended with conductive fillers such as carbon, metal powder or conducting polymers. Incorporation of conductive fillers results in the production of smart fabrics, however, care must be taken to retain mechanical properties of the fibers. Recent technological advancement requires new components made out of low cost-high performance materials. Composites of intrinsic conductive polymers which conduct due to π-electron delocalization along the chain length, as well as composites in which the addition of conductive fillers provides the conductivity, have attracted industrial applications. Conductive polymer composites are easily processed by blending conductive fillers with polymers and more economical as compared to synthetic methodologies used to enhance conductivity [21,22]. Materials having a combination of good dielectric properties and mechanical strength are hard to develop, hence, conductive polymer composites encompassing electrical properties and mechanical strength of the filler and ease of fabrication are desirable. Conducting polymers incorporated with different types of metal nanoparticles are employed in switching devices, electromagnetic shielding of electronic equipment, conducting adhesives in electronics packaging, static charge dissipating materials, surge protection devices and underfill for flip chip [23-26]. In addition, the conducting polymer composites find applications in antennas, chemical detecting sensors, photo-thermal-optical recording, self-regulating heaters, tissue engineering, neural probes, drug delivery, bio-actuators, biosensors, electronic noses, electrochemical catalysts, and fuel cells [27-36].

In electronics applications, conductive composites are useful as adhesives and circuit elements. These materials normally consist of silver powder incorporated in a variety of polymers, including epoxy, silicone, polyurethane and polyimide. Their utility arises from very low electrical resistivity and high thermal conductivities. The positive temperature coefficient of resistance (PTCR) of conductive composites is used for self-limited heating systems and fault current limiters in circuit protection [37]. The later devices consist commonly of carbon black in a matrix of polyethylene. As the composite is heated by the passage of electrical current, the volume expansion near the melting point of the semi-crystalline polymer disrupts the conductive filler network, causing the resistance of the composite to increase tremendously.

The new hybrid materials of conducting polymers and natural fibers are expected to overcome the inherent processability difficulties of conducting polymers and open up new opportunities for the development of consumer products and applications. Some of the studies reporting such systems are discussed in the sections below.

7.4.1 Polyaniline-Natural Fiber Composites

A hybrid material using conducting polymers polyaniline and polypyrrole with cellulose substrates in the form of paper fibers and paper sheets were prepared by Johnston *et al.* [38]. The conducting polymers were observed to form spheres which fused into a continuous coating layer and encapsulated the natural fiber. The polymer interacted with cellulose fibers through the hydroxyl or amine groups present in the cellulose substrate surface by hydrogen bonding. The resulting materials exhibited the properties of both cellulose fiber (mechanical strength and integrity) and conducting polyaniline and polypyrrole (electrical conductivity, optical and chemical properties). The authors also deposited silver nanoparticles on the hybrid surface to impart anti-microbial activity to the composite [38]. In another study, nanocellulose as matrix material has been used to generate a composite with polyaniline [39]. The authors extracted nanocellulose through hydrolyzing bleached flax yarn using a strong acid at moderate temperatures. Thin films of nanocellulose dispersed PANI were prepared through *in-situ* polymerization of aniline-HCl using ammonium peroxydisulfate (APS) as an oxidant. The flexible composite films exhibited higher conductivity, which was significantly increased with enhancing the concentration of PANI from 10% to 30%. The PANI-nanocellulose composite with 30% PANI showed a conductivity value of 1.9×10^{-2} S/cm, which may represents its suitability for paper-based sensors as a flexible electrode.

Polyaniline and natural fiber were added in the epoxy (EP) matrix to prepare conductive thermoset material by Razak *et al.* [40]. Specifically, acetylated kenaf bast fiber (AKF) and conducting polyaniline (PANI) nanowires were used to produce the hybrid composites. The bast fiber composites displayed a significant improvement in tensile strength at 20 wt% AKF loading when compared to untreated kenaf fiber-epoxy composites. The ternary composites of EP/AKF/PANI were also produced by loading PANI coated AKF. The results indicated that the addition of PANI into epoxy-kenaf bast fiber enhanced the electrical properties without losing the mechanical strength. Further, the authors reported that the electrical percolation threshold was achieved at 11 wt% of PANI loading. Higher concentration of PANI in EP/AKF/PANI led to the reduction of tensile and flexural strength.

Conducting polymer composites containing polypyrrole and polyaniline encapsulated natural cellulose fibers have been prepared by Kelly *et al.* [41]. For this purpose, pyrrole and aniline monomers were polymerized in the presence of individual Kraft paper pulp fibers using ferric chloride and ammonium persulfate as oxidants. The nanocomposites of conducting polymers/cellulose fibers exhibited inherent properties such as electrical and chemical stabilities of polypyrrole and polyaniline as well as strength, flexibility and surface area characteristic of the cellulose fibers. In another study, polyaniline was electrochemically polymerized on conducting textiles of polyester covered with PPy)/anthraquinone sulphonic acid (AQSA) [42]. Electrochemical syntheses were performed by means of cyclic voltammetry (CV) and potentiostatic methods. Xia and Lu [43] reported highly conductive polymers (polypyrrole, polyaniline or poly3,4-ethylene-dioxythiophene)/silk fibroin composites prepared through *in-situ* polymerization. No modification of sill fibroin surface was carried out for composite synthesis. Polypyrrole/silk fibroin, polyaniline/silk fibroin and poly3,4-ethylene-dioxythiophene/silk fibroin composite fibers exhibited varied conductivity in the range of $3.8\text{-}4.2 \times 10^{-1}$, $0.9\text{-}1.2 \times 10^{-2}$ and $4.9\text{-}5.2 \times 10^{-3}$ S cm^{-1}, respectively. Improved thermal stability of silk fibroin fiber via coating process was also observed by the authors from the thermogravimetric analysis. Structural analysis revealed that hydrogen-bonding and electrostatic attraction existed between silk fibroin and conductive polymers. The peptide linkages (CO-NH) on silk fibroin attracted conjugated cation radicals during polymerization.

7.4.2 Polypyrrole-Natural Fiber Composites

Recent research reveals that polypyrrole can be blended with natural and man-made cellulose fibers such as cotton, viscose, cupro and lyocell, by *in-situ* polymerization. For instance, an aqueous solution of pyrrole monomer was applied on fibers where the monomer penetrated into the fiber and subsequent polymerization resulted in interpenetrating polypyrrole networks and fiber yarns, leading to high levels of blending [44]. Natural fiber reinforced polypyrrole could find application in the development of technical apparel such as antistatic fabrics, sportswear, heating and cooling equipment, etc. The authors also prepared polypyrrole in natural and man-made cellulose-based fibers by means of *in-situ* polymerization using chemical vapor phase deposition method [45]. The fabric was first impregnated in an aqueous solution of oxidant and dopant followed by exposure to pyrrole vapors and polymerization. The composite materials had significant differences in the structural, calorimetric and electrical properties. Vapor phase prepared fabrics showed a high uniform polypyrrole coating on the fiber surface and its partial penetration inside the amorphous zones of the fiber bulk. Polypyrrole-cellulose conducting composite textiles exhibited good performance towards light exposure and washing fastness tests.

Similarly, pyrrole monomer was polymerized in the presence of silk fabric by *in-situ* oxidative polymerization using ferric chloride as a catalyst by Cucchi *et al.* [46]. The amount of conductive polymer on the fabric could be tuned by changing the reaction time and concentration of pyrrole monomer in the reaction medium. The polypyrrole was found to be uniformly coated throughout the surface of the fabric, however, the polymerization occurred only at the surface of the fiber and not in the bulk. The polymerization process did not affect the intrinsic crystalline structure of silk. Polypyrrole-coated silk fabrics showed significant electrical properties. Particularly at higher polymer content, a linear fit was observed for the current versus voltage curve. The electrical resistance of polypyrrole coated silk fabrics decreased exponentially with longer reaction time or the concentration of pyrrole in the reaction system. The polypyrrole coated silk fabrics may be used for interactive, smart textiles and bio-based conductive composites for biomedical applications. Polypyrrole was also employed on a cellulose fiber (e.g., cotton) and a protein fiber (e.g., silk), after undergoing pre-treatment by two different methods, namely vapor phase method and liquid phase method [47]. The authors proved that vapor coating was more desired than the liquid phase method, as better gas penetration into the fibers was realized. The conductivity of polypyrrole coated fibers for cotton and silk fibers were reported to be 6.4×10^{-4} S/cm and 3.2×10^{-4} S/cm, respectively.

Chatterjee and Maity [48] prepared electrically conductive composites of natural and synthetic fibers by coating with polypyrrole. The authors chose common textile fibers such as cotton, wool, silk, polyester, nylon, acrylic and polypropylene for this purpose. The polymerization of pyrrole using ferric chloride as oxidant and p-toluene sulphonic acid as a dopant in an aqueous medium in the presence of natural fibers was carried out. Effect of the concentration of pyrrole, oxidant and dopant as well as substrate textile fibers on reaction kinetics of the *in-situ* chemical polymerization were investigated. The synthetic fibers had lower polypyrrole on the surface than that of the natural fibers, probably due to the electrostatic interactions between the fibers with polypyrrole, which led to improved fixation of polypyrrole on the natural fiber surfaces. Specifically, the hydrogen bonding facilitated significant chemical interaction in case of wool, silk, cotton and polyester fibers with polypyrrole.

Electrochemical capacitance behavior of polypyrrole composites with carbohydrate polymers like cotton, linen, modified cellulosic fiber-viscose rayon and synthetic polymer polyester fabrics were studied by Babu *et al.* [49]. PPy doped with anion was coated on fibers through a mild, room temperature wet *in-situ* chemical polymerization method using pyrrole monomer and ferric chloride as an oxidant. The authors fabricated a unit cell by assembling two symmetric textile electrodes separated by a solid polymer electrolyte membrane to study the capacitance behavior. The textile electrodes prepared with PPy-cotton and PPy-viscose showed the highest specific capacitance value of 268 F g^{-1} and 244 F g^{-1}, individually at a scan rate of 5 mV s^{-1}. The charge-discharge analysis also showed higher specific capacitance value for PPy-viscose and PPy-cotton. Thus, the authors demonstrated the simple and reproducible method for fabrication of textile-based supercapacitors and the chemistry of surface interaction of PPy molecules with natural and synthetic fabrics. In another study, electrically conductive paper composites were prepared using polypyrrole and pulp fibers (namely engineered fibers and a mixture of the modified and unmodified fibers) [50]. It was found that for the "mixture" method, low amount of conductive polymer was sufficient to achieve the same conductivity while a higher tensile strength in the paper was attained when compared with paper obtained exclusively from treated fibers. The long-time environmental aging stability of the composites was also analyzed, and the observed conductivity decay was interpreted through existing kinetic models. 3D nanostructured polypyrrole (3D PPy) composites reinforced with PPy coated nanocellulose (nanocellulose@PPy) were used to generate freestanding, mechanically strong and porosity optimized electrodes with large surface area by Wang *et al.* [51]. The nanocellulose@PPyelectrodes in symmetric energy storage devices exhibited capacitance of 46 F g^{-1} which corresponded to specific electrode capacitance of ~185 F g^{-1} based on the weight of the electrode and showed a value of 5.5 F cm^{-2} at a current density of 2 mA cm^{-2}. Nanocellulose@PPy electrodes showed a cell capacitance of 92% after 3000 charge/discharge cycles at 30 mA cm^{-2}.

Polymerization of pyrrole on cellulose fibers was reported by Beneventi *et al.* [52] by impregnating nanocellulose in $FeCl_3$ solution and re-dispersion in a pyrrole solution, followed by polymerization. The amount of adsorbed iron III on the $FeCl_3$-cellulose system was studied for its capability of polymerization of pyrrole on the fiber surface. The authors reported a lowering of electric resistance associated with a degradation of the cellulose fibers due to acid hydrolysis. In a similar study, Devi *et al.* [53] polymerized pyrrole in the presence of short nylon fibers using anhydrous ferric chloride as oxidant and p-toluenesulphonic acid as dopant. The synthesized polypyrrole and polypyrrole-coated nylon fibers (F-PPy) were used to generate composites with natural rubber, and the composites were studied for their cure pattern, filler dispersion, cure kinetics, DC conductivity, mechanical properties and surface morphology. The DC conductivity was observed to be enhanced for the F-PPy hybrid system compared to polypyrrole reinforced natural rubber composite. In addition, the F-PPy system exhibited better mechanical properties. Esfandiari [54] reported conductive filament yarns prepared by using cellulosic fibers (e.g. cotton) and a protein fiber (e.g. silk). The authors first pre-treated the fibers and subsequently coated them with polyaniline and polypyrrole by vapor and liquid phases deposition methods. The penetration of monomers into the fiber was suggested as the key to obtain inherent nanocomposites. Conducting properties of fibers, evaluated by four probe Method, revealed that the fibers had significant improvement in conductivity. From the microscopy and conductivity studies, it was also confirmed that the monomers deposited by gas phase method resulted in better adhesion between polymer and fiber, with a conductivity of 6.4×10^{-4} S/cm for cellulose fiber and 3.2×10^{-4} S/cm for silk fiber.

7.4.3 Polythiophene-Natural Fiber Composites

Mattana *et al.* [55] fabricated natural cotton fibers based conductive materials by coating cotton fiber with gold nanoparticles and poly(3,4-ethylene dioxythiophene) (PEDOT). The conductive cotton yarns may find applications in devices such as resistors, organic electrochemical transistors (OECTs) and organic field effect transistors (OFETs). The electrical and mechanical analysis performed on polythiophene and gold nanoparticles coated cotton yarns by the authors showed that the fabrics exhibited good conductivity and also maintained good flexibility. In another study, an electrically conductive material was fabricated by drop casting conductive poly(3,4-ethylene dioxythiophene) polystyrene sulfonate (PEDOT:PSS) films onto a cotton substrate by AlAmer [56]. The conductivity could be varied by changing the amount of PEDOT: PSS, which also controlled the sheet resistance of the cotton material.

Electrically conductive or photoluminescent recombinant silk fibers were developed by Muller *et al.* [57]. The natural spider silk fiber was stained with conjugated polyelectrolytes by a self-assembly process by controlling the pH and temperature. In addition, conductive polymer staining on natural *Bombyxmori* silk was also studied. Miniemulsified poly(3-octylthiophene) (P3OT) nanoparticles were developed to adsorb on softwood cellulose fibers by Sarrazin et al. [58]. Fiber functionalization by P3OT could be improved by pre-treatment of cellulose fibers with carboxymethyl cellulose. The iodine-doped pellets obtained from treated fibers showed a conductivity of $2\text{-}6\times10^{-4}$ S/cm. The conductivity of the hybrid material sharply decreased upon aging, probably due to the progressive de-doping and/or over-oxidation of P3OT by atmospheric O_2.

Wistrand *et al.* [59] reported electrically conductive nanocellulose fibers. For this, the authors developed polyelectrolyte multilayers on bleached softwood fibers and SiO_2 model surfaces by treating with PEDOT:PSS and poly(allylamine) (PAH). The layered sheets were tested for their tensile strength, adsorbed amount of polymer and electrical conductivity. The conductivity of the multilayer sheets exhibited 2-3 orders of magnitude enhancement. A difference in conductivity could also be noted depending on the PEDOT:PSS and PAH adsorbed in the outer layer, PAH lowered the conductivity as compared to PEDOT: PSS. However, PEDOT:PSS reduced the tensile strength of the paper.

7.4.4 Sisal Fiber Composites

In some studies reporting conducting composites using low- or non-conducting polymers, the electrical conducting properties of sisal fiber loaded polyester resin composites fabricated by resin transfer molding were studied with special reference to fiber content, frequency and temperature by Sreekumar *et al.* [60]. The dielectric constant (ε'), loss factor (ε''), dissipation factor and conductivity increased proportional to fiber loading for the entire range of frequencies. However, the increment was high at low frequencies, low at medium frequencies and very small at high frequencies. Also, the composites having a fiber content of 50 vol% showed higher values. The dielectric constant values increased with respect to temperature and decreased above glass transition temperature, which was attributed to the fiber content in the composites.

In another study, the electrical properties of short sisal fiber reinforced low-density polyethylene composites were also reported by Paul *et al.* [61]. Various surface treatments such as alkali, peroxide, stearic acid, permanganate and acetylation were performed on the sisal fibers to improve interfacial bonding. The dielectric constant increased significantly proportional to fiber loading and decreased with an increase in frequency. However, the dielectric constant values of the composites were found to decrease by chemical treatments due to the reduced hydrophilicity of treated natural fiber. The dielectric loss factors of treated sisal fiber reinforced low-density polyethylene composites were lower than that of untreated-sisal fiber low-density polyethylene composites. In another study, polypropylene matrix with sisal fibers having cylindrical pores and cenospheres having spherical pores was reported to exhibited low dielectric constant [62]. The authors presented the dielectric properties of sisal fiber reinforced polypropylene composites with and without cenospheres (both treated and untreated cenospheres). Higher loading of sisal fiber in polypropylene increased the dielectric constant ε' and enhanced the ac conductivity σ_{ac}. Overall, the sisal fiber-reinforced polypropylene composite having 20% sisal fiber with and without cenospheres exhibited enhanced dielectric constant, tan δ, and a.c. conductivity with an increase in temperature at different frequencies.

Ben Amor et al. [63] reported short palm tree lignocellulosic natural fiber-reinforced polyester composites. Electrical studies were performed in the temperature range of 40-200 °C and at a frequency range 1-100 kHz. The water molecules in palm fiber were observed to be responsible for orientation polarization. The relaxation process associated with conductivity occurring as a result of the carriers charges diffusion was also noted for temperature above glass transition and low frequencies. The authors also identified interfacial relaxation which was attributed to the charge accumulation at the Palm fibers/polyester interfaces. In another study, high-density polyethylene was reinforced with maleic anhydride treated natural fibers of banana, hemp and agave, and the composites were prepared in different ratios 60:40, 55:45, 50:50, and 45:55 (wt/wt) [64]. Overall, the increased crosslinking density between the polymer and fiber resulting from maleic anhydride treatment decreased the volume resistivity and surface resistivity of composites when compared to composites with untreated natural fibers.

7.4.5 Composites with Jute and Other Fibers

Red dye treated jute fabric reinforced polypropylene (PP) composites was prepared through compression molding process by Zaman *et al.* [65] Gamma radiation (250-1000 krad dose) was applied on both jute and polymer, and the composites were fabricated using non-irradiated and irradiated matrix and fibers. It was found that a irradiated jute and irradiated PP composite exhibited better results in terms of weathering and dielectric properties. In another study, Cabral *et al.* [66] studied mechanical strength, water uptake and dielectric properties of jute fiber-reinforced polypropylene (PP) composites. The authors reported that Young's modulus and tensile strength monotonically increased with increase in fiber volume fraction at the expense of impact resistance at higher fiber volume fraction. The authors also observed a change in the dependence of the water uptake and dielectric properties with fiber loading at a critical fiber content. Lai *et al.* [67] also reported compression molding process to prepare alkali, stearic acid, acetone, and potassium permanganate treated coconut coir reinforced polypropylene composites. The fiber treatment with various agents was carried out to improve the interfacial adhesion between the fibers and polypropylene matrix. The composites are tested for their mechanical strength, hardness and dielectric properties. Composites based on the modified fibers, in general, exhibited superior mechanical properties as compared to the composites containing unmodified fibers. The dielectric properties of the composites were influenced by the chemical treatment of the fiber. Also, the value of dielectric constant increased significantly with fiber loading.

7.5 Conclusions

Natural fibers present extensive opportunities for the development of eco-friendly conducting composites for a variety of commercial applications. The numerous sources, ability to control surface functionalization as well as excellent properties of these natural fibers make them suitable reinforcements for developing such polymeric systems.

References

1. Mastalygina, E. E., Popov, A. A., and Pantyukhov, P. V. (2017) Effect of biobased fillers nature on biodeterioration of hybrid polyethylene composites by mold fungi. *Materials Science and Engineering*, **213**, 012011.
2. Mohammed, L., Ansari, M. N. M., Pua, G., Jawaid, M., Islam, M. S. (2015) A Review on Natural Fiber Reinforced Polymer Composite and Its Applications. *International Journal of Polymer Science*, **2015**, Article ID 243947.
3. Al-Oqla, F. M., Sapuan, S. M., Ishak, M. R., and Nuraini, A. A (2015) Decision making model for optimal reinforcement condition of natural fiber composites. *Fibers and polymers*, **16**, 153-163.
4. Al-Oqla, F. M., Sapuan, S. M., Ishak, M. R., Nuraini, A. A. (2015) A decision-making model for selecting the most appropriate natural fiber-Polypropylene-based composites for automotive applications. *Journal of Composite Materials*, **50**(4), 543-556.
5. Majeed, K., Jawaid, M., Hassan, A., Abu Bakar, A., Khalil, H. P. S. A., Salema, A. A., and Inuwa, I. (2013) Potential materials for food packaging from nanoclay/natural fibres filled hybrid composites. *Materials & Design*, **46**, 391-410.
6. Jawaid, M. Khalil, H. P. S. A. (2011) Cellulosic/synthetic fibre reinforced polymer hybrid composites: A review. *Carbohydrate Polymers*, **86**(1), 1-18.

7. Al-Oqla, F. M., and Sapuan, S. M. (2014) Natural fiber reinforced polymer composites in industrial applications: feasibility of date palm fibers for sustainable automotive industry. *Journal of Cleaner Production*, **66**, 347-354.
8. Shirakawa, H., Louis, E. J., MacDiarmid, A. G., Chiang, C. K., and Heeger, A. J. (1977) Synthesis of electrically conducting organic polymers: halogen derivatives of polyacetylene, (CH)x. *Journal of the Chemical Society, Chemical Communications*, 578-580.
9. Nguyen, D., and Yoon, H. (2016) Recent advances in nanostructured conducting polymers: From synthesis to practical applications. *Polymers*, **8**(4), 118.
10. Naveen, M., Gurudatt, N., and Shim, Y. (2017) Applications of conducting polymer composites to electrochemical sensors: A review. *Applied Materials Today*, **9**, 419-433.
11. Kaur, G., Adhikari, R., Cass, P., Bown, M., and Gunatillake, P. (2015) Electrically conductive polymers and composites for biomedical applications. *RSC Advances*, **5**, 37553- 37567.
12. Bargon, J., Mohmand, S., and Waltman, R. J. (1983) Electrochemical synthesis of electrically conducting polymers from aromatic compounds. *IBM Journal of Research and Development*, **27**, 330-331.
13. Diaz, A. F., Kanazawa, K., and Gardini, G. P. (1979) Electrochemical polymerization of pyrrole. *Journal of the Chemical Society, Chemical Communications*, **14**, 635-636.
14. Dall'Olio, A., Dascola, G., Varacco, V., and Bocchi, V. (1968) Electron paramagnetic response and conductivity of an electrolytic oxypyrrole [(pyrrole polymer)] black. *C. R. S. Acad Sci Ser C*, **267**, 433-435.
15. Malhotra, B. D., Chaubey, A., and Singh S. P. (2006) Prospects of conducting polymers in biosensors. *Analytica Chimica Acta*, **578**(1), 59-74.
16. Ahuja, T., Mir, I., and Kumar, D. (2007) Biomolecular immobilization on conducting polymers for biosensing applications. *Biomaterials*, **28**(5), 791-805.
17. Guimard, N., Gomez, N., and Schmidt, C. (2007) Conducting polymers in biomedical engineering. *Progress in Polymer Science*, **32**(8-9), 876-921.
18. Waltman, R., and Bargon, J. (1986) Electrically conducting polymers: a review of the electropolymerization reaction, of the effects of chemical structure on polymer film properties, and of applications towards technology. *Canadian Journal of Chemistry*, **64**(1), 76-95.
19. Vert, M. (2007) Polymeric biomaterials: Strategies of the past vs. strategies of the future. *Progress in Polymer Science*, **32**, 755-761.
20. Hackett, A. J., Malmstrom, J., and Travas-Sejdic, J. (2017) Functionalization of conducting polymers for biointerface applications. *Progress in Polymer Science*, **70**, 18-33.

21. Deng, H., Lin, L., Ji, M., Zhang, S., Yang, M., and Fu, Q. (2013) Progress on the morphological control of conductive network in conductive polymer composites and the use as electroactive multifunctional materials. *Progress in Polymer Science*, **39**(4), 627-655.
22. Balint, R., Cassidy, N. J, and Cartmell, S. H. (2014) Conductive polymers: Towards a smart biomaterial for tissue engineering. *Acta Biomaterialia*, **10**, 2341-2353.
23. Long, Y., Li, M. M., Gu, C., Wan, M., Duvail, J. L., Liu, Z., and Fan, Z. (2011) Recent advances in synthesis, physical properties and applications of conducting polymer nanotubes and nanofibers. *Progress in Polymer Science*, **36**(10), 1415-1442.
24. Lasalu, C., Williams, D. E., and Travas-Sejdic, J. (2012) The application of nanopipettes to conducting polymer fabrication, imaging and electrochemical characterization. *Progress in Polymer Science*, **37**(9), 1177-1191.
25. Guo, B., Glavas, L., and Albertsson, A. (2013) Biodegradable and electrically conducting polymers for biomedical applications. *Progress in Polymer Science*, **38**(9), 1263-1286.
26. Smith, J. (1998) Intrinsically electrically conducting polymers. Synthesis, characterization, and their applications. *Progress in Polymer Science*, **23**(1), 57-79.
27. Guimar, N., Gomez, N., and Schmidt, C. (2007) Conducting polymers in biomedical engineering. *Progress in Polymer Science*, **32** (8-9), 876-921.
28. Gospodinova, N., and Terlemezyan, L. (1998) Conducting polymers prepared by oxidative polymerization: polyaniline. *Progress in Polymer Science*, **23**(8), 1443-1484.
29. Dutta, K., Kumar, P., Das, S., and Kundu, P. P. (2014) Utilization of conducting polymers in fabricating polymer electrolyte membranes for application in direct methanol fuel cells. *Polymer Reviews*, **54**, 1-32.
30. *Handbook of Conducting Polymers*, Skotheim, T. A., and Reynolds, J. R. (eds.), 2nd edition, CRC Press, USA (2007).
31. Borole, D. D., Kapadi, U. R., Mahulikar, P. P., and Hundiwale, D. G. (2006) Conducting polymers: an emerging field of biosensors. *Designed Monomers and Polymers*, **9**(1), 1-11.
32. *Nanostructured Polymer Blends*, Thomas, S., Shanks, R., and Chandran, S. (eds.), 1st edition, Elsevier, USA (2013).
33. Poblete, V., Alvarez, M., and Fuenzalida, V. (2009) Conductive copper-PMMA nanocomposites: microstructure, electrical behavior, and percolation threshold as a function of metal filler concentration. *Polymer Composites*, **30**, 328-333.
34. Sham, M.-L., and Kim, J.-K. (2004) Evolution of residual stresses in modified epoxy resins for electronic packaging applications. *Composites, Part A: Applied Science and Manufacturing*, **35**, 537-546.

35. Thakur, V. K., Thakur M. K., and Gupta, R. K. (2013) Rapid synthesis of graft copolymers from natural cellulose fibers. *Carbohydrate Polymers*, **98**, 820-828.
36. Al-Oqla, F. M., and Omar, A. A. (2012) A decision-making model for selecting the GSM mobile phone antenna in the design phase to increase overall performance. *Progress in Electromagnetics Research C*, **25**, 249-269.
37. Nan, C.-W., Shen, Y., and Ma, J. (2010) *Annual Reviews of Materials Research*, **40**, 131-151.
38. Johnston, J. H., Kelly, F. M., Burridge, K. A., and Borrmann, T. (2009) Hybrid materials of conducting polymers with natural fibres and silicates. *International Journal of Nanotechnology*, **6**, 312-328.
39. Liu, D. Y., Sui, G. X., amd Bhattacharyya, D. (2014) Synthesis and characterisation of nanocellulose-based polyaniline conducting films. *Composites Science and Technology*, **99**, 31-36.
40. Razak, S. I. A., Rahman, W. A. W. A., and Yahya, M. Y. (2013) Hybrid composites of short acetylated kenaf bast fiber and conducting polyaniline nanowires in epoxy resin. *Journal of Composite Materials*, **48**, 667-676.
41. Kelly, F. M., Johnston, J. H., Borrmann, T., and Richardson, M. J. (2007) Functionalised hybrid materials of conducting polymers with individual fibres of cellulose. *European Journal of Inorganic Chemistry*, **35**, 5571-5577.
42. Molina, J., del Rio, A. I., Bonastre, J., and Cases, F. (2009) Electrochemical polymerisation of aniline on conducting textiles of polyester covered with polypyrrole/AQSA. *European Polymer Journal*, **45**, 1302-1315.
43. Xia, Y., and Lu, Y. (2008) Fabrication and properties of conductive conjugated polymers/silk fibroin composite fibers. *Composites Science and Technology*, **68**, 1471-1479.
44. Dall'Acqua, L., Tonin, C., Peila, R., Ferrero, F., and Catellani, M. (2004) Performances and properties of intrinsic conductive cellulose–polypyrrole textiles. *Synthetic Metals*, **146**, 213-221.
45. Dall'Acqua, L., Tonin, C., Varesano, A., Canetti, M., Porzio, W., and Catellani, M. (2006) Vapour phase polymerisation of pyrrole on cellulose-based textile substrates. *Synthetic Metals*, **156**, 379-386.
46. Cucchi, I., Boschi, A., Arosio, C., Bertini, F., Freddi, G., and Catellani, M. (2009) Bio-based conductive composites: Preparation and properties of polypyrrole (PPy)-coated silk fabrics. *Synthetic Metals*, **159**, 246-253.
47. Hosseini, S. H., and Pairovi, A. (2005) Preparation of conducting fibres from cellulose and silk by polypyrrole coating. *Iranian Polymer Journal*, **14**, 934-940.
48. Chatterjee, A., and Maity, S. (2017) A comparative study of reaction kinetics of in-situ chemical polymerization of polypyrrole onto

various textile fibres. *Surface and Coatings Technology*, **324**, 569-576.

49. Babu, K. F., Subramanian, S. P. S., and Kulandainathan, M. A. (2013) Functionalisation of fabrics with conducting polymer for tuning capacitance and fabrication of supercapacitor. *Carbohydrate Polymers*, **94**, 487-495.
50. Huang, B., Kang, G., and Ni, Y. (2005) Electrically conductive fibre composites prepared from polypyrrole-engineered pulp fibres. *The Canadian Journal of Chemical Engineering*, **83**, 896-903.
51. Wang, Z., Tammela, P., Zhang, P., Huo, J., Ericson, F., Stromme, M., and Nyholm, L. (2014) Freestanding nanocellulose-composite fibre reinforced 3D polypyrrole electrodes for energy storage applications. *Nanoscale*, **6**, 13068-13075.
52. Beneventi, D., Alila, S., Boufi, S., Chaussy, D., and Nortier, P. (2006) Polymerization of pyrrole on cellulose fibres using a $FeCl_3$ impregnation-pyrrole polymerization sequence. *Cellulose*, **13**(6), 725-734.
53. Devi, D. S. P., Thekkedath, J., and Narayanan, S. K. N. (2012) Synthesis and characterization of conducting composites of polypyrrole/polypyrrole-coated short nylon fiber and natural rubber. *Polymer-Plastics Technology and Engineering*, **51**, 823-831.
54. Esfandiari, A. (2008) PPy covered cellulosic and protein fibres using novel covering methods to improve the electrical property. *World Applied Sciences Journal*, **3**(3), 470-475.
55. Mattana, G., Cosseddu, P., Fraboni, B., Malliaras, G. G., Hinestroza, J. P., and Bonfiglio, A. (2011) Organic electronics on natural cotton fibres. *Organic Electronics*, **12**, 2033-2039.
56. AlAmer, F. A. A. (2016) Method of Making Conductive Cotton Using Organic Conductive Polymer, US patent US20160258110A1.
57. Muller, C., Jansson, R., Elfwing, A., Askarieh, G., Karlsson, R., Hamedi, M., Rising, A., Johansson, J., Inganas, O., and Hedhammar, M. (2011) Functionalisation of recombinant spider silk with conjugated polyelectrolytes. *Journal of Materials Chemistry*, **21**, 2909-2915.
58. Sarrazin, P., Chaussy, D., Stephan, O., Vurth, L., and Beneventi, L. (2009) Adsorption of poly(3-octylthiophene) nanoparticles on cellulose fibres: Effect of dispersion stability and fibre pre-treatment with carboxymethyl cellulose. *Colloids and Surfaces A: Physicochemical and Engineering Aspects*, **349**, 83-89.
59. Wistrand, I., Lingstrom, R., and Wagberg, L. (2007) Preparation of electrically conducting cellulose fibres utilizing polyelectrolyte multilayers of poly(3,4-ethylenedioxythiophene):poly(styrene sulphonate) and poly(allyl amine). *European Polymer Journal*, **43**, 4075-4091.
60. Sreekumar, P. A., Saiter, J. M., Joseph, K., Unnikrishnan, G., and Thomas, S. (2012) Electrical properties of short sisal fiber rein-

forced polyester composites fabricated by resin transfer molding. *Composites, Part A: Applied Science and Manufacturing*, **43**(3), 507-511.

61. Paul, A., Joseph, K., and Thomas, S. (1997) Effect of surface treatments on the electrical properties of low-density polyethylene composites reinforced with short sisal fibers. *Composites Science and Technology*, **57**, 67-79.
62. Sharma, J., and Chand, N. (2013) Role of cenosphere addition on dielectric properties of sisal fiber-polypropylene composites. *Polymer-Plastics Technology and Engineering*, **52**, 743-753.
63. Ben Amor, I., Rekik, H., Kaddami, H., Raihane, M., Arous, M., and Kallel, A. (2009) Studies of dielectric relaxation in natural fiber–polymer composites. *Journal of Electrostatics*, **67**, 717-722.
64. Naik, J., and Mishra, S. (2005) Studies on electrical properties of natural fiber:HDPE composites. *Polymer-Plastics Technology and Engineering*, **44**(4), 687-693.
65. Zaman, H. U., Khan, A. H., Hossain, M. A., Khan, M. A., and Khan, R. A. (2012) Physico-mechanical, interfacial, degradation, and dielectric properties of jute/PP composites: Effect of dye and gamma radiation. *International Journal of Polymeric Materials and Polymeric Biomaterials*, **61**(8), 596-610.
66. Cabral, H., Cisneros, M., Kenny, J. M., Vazquez, A., and Bernal, C. (2005) Structure-properties relationship of short jute fiber-reinforced polypropylene composites. *Journal of Composite Materials*, **39**(1), 51-65.
67. Lai, C. Y., Sapuan, S. M., Ahmad, M., Yahya, N., and Dahlan K. Z. H. M. (2005) Mechanical and electrical properties of coconut coir fiber-reinforced polypropylene composites. *Polymer-Plastics Technology and Engineering*, **44**, 619-632.

Chapter 8

Conductive Polymer Composites for Biomedical Applications

Gagan Kaur[a,*] and Pathiraja Gunatillake[b,*]

[a]*Monash University, Materials Science and Engineering, Clayton, Victoria 3800, Australia*
[b]*CSIRO Manufacturing, Bayview Avenue, Clayton, Victoria 3168, Australia*
**Corresponding authors*: gagan.kaur@monash.edu; thilak.gunatillake@csiro.au

8.1 Introduction

Electrically conductive polymeric materials have recently received a great deal of attention to explore their potential in biomedical applications such as biosensors, biomedical implants and tissue engineering (Figure 8.1) [1]. Traditionally, metals such as platinum (Pt) and gold (Au) have been used in biomedical applications along with alloys of Pt-Au, diamond and iridium oxide. Some of the issues posed by these materials include mechanical property mismatch with biological tissue, limitations in applications requiring higher charge densities and difficulty in fabricating multi-electrode sensors as these often require other supporting substrates. Also, thin coatings of many polymers and other materials do not survive (e.g. in long term implants) on metal substrates/electrodes. Conductive polymeric biomaterials can provide solution to many of the above issues as these offer advantages such as a much larger surface area which can drastically reduce impedance compared to metal electrodes and ability to incorporate biological molecules that improves tissue-material interaction. Unlike metals, these polymeric materials also offer a number of processing options and their mechanical properties can be tailored to suit specific requirements.

Conjugated conducting polymers (CPs) such as polypyrrole (PPy), polyaniline (PANi) and poly(3,4-ethylenedioxythiophene) (PEDOT) exhibit electrical and optical properties similar to those of metals and

Conducting Polymer Composites, edited by Vikas Mittal

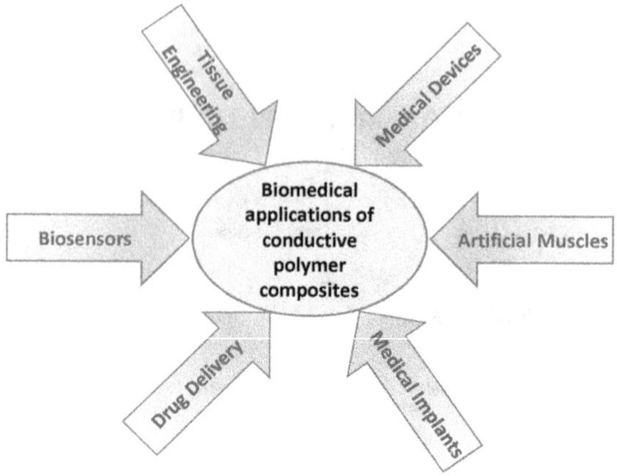

Figure 8.1 Biomedical applications of conductive polymer composites.

semiconductors; and offer advantages of conventional polymers such as ease of synthesis, good stability, ability to entrap and release biomolecules, etc. [2-4]. Furthermore, CPs can be potentially modified for electrical, chemical, physical and biocompatibility properties to better suit a specific application [3,4]. However, their use in biomedical applications is limited because CPs are often brittle and difficult to handle, and the use of larger dopants can further aggravate this effect [3,5]. Tailoring conducting polymers to have appropriate mechanical properties, electrical conductivity, processability as well as acceptable biocompatibility has been the major challenge in application of this class of polymers for clinically useful biomedical implants and devices. One way to overcome the shortcomings of CPs is to use these together with other polymers in the form of blends or composites in order to combine the useful properties of both materials. Conducting polymers like PPy and PANi have also been explored as conductive fillers, especially with natural polymers, in order to overcome the poor processability of these conducting polymers as well as to impart conductivity to otherwise insulating polymers [5-9]. The composites of CPs can provide the increased solubility and better mechanical properties necessary for various biomedical applications without significant compromise of their conductivity and other properties as discussed in later sections.

The demand for polymeric conductive materials has also led to more recent attention being directed towards conductive polymeric composites comprised of biocompatible insulating polymers with dispersed conductive fillers. Various biocompatible polymers such as polyurethanes are suitable for generating conductive composites for biomedical applications because of their biocompatibility, biostability, processability and good mechanical properties. Electrical conductivity can be imparted to insulating polymers by the incorporation of conductive fillers while still maintaining their polymeric characteristics [10]. Recently, nanosized conductive fillers including carbon nanotubes (CNTs), graphene and metal nanoparticles have generated a considerable interest and have been explored extensively in the development of polymer based conductive composites for biomedical applications [11-16]. The major focus of this chapter is to provide a comprehensive review of the investigations conducted over the last decade for the development of conducting composites for biomedical applications.

8.2 Conductive Polymer Composites

8.2.1 Electrical Conductivity of Polymeric Composites

A conductive composite is formed when an electrically conducting dispersed phase is added in sufficient quantities to a polymer matrix [10]. Various conductive fillers such as carbon black, carbon fiber, silver and other metallic particles have been reported in literature for making conductive composites. The more recent focus has been on the use of nanosized conductive fillers including graphene, CNTs and metal nanoparticles (such as gold and silver nanoparticles) for the development of polymer based conductive composites [11-14,16,17]. The conductivity of such composites arises due to the formation of conductive paths of filler particles within the polymer matrix [10,18]. The electrical conductivity of a composite is generally dependent on the weight/volume fraction of conducting filler, as described by the percolation theory [18,19]. At low filler loadings, the conductivity of the composite is still very close to that of the pure polymer matrix. At some critical loading known as percolation threshold, the conductivity starts increasing by many orders of magnitude with a very small increase in the filler loading. After the percolation threshold, the increase in conductivity levels off and approaches that of the filler material. It is at the percolation threshold that the concentration of filler is enough to form a continuous conductive network through the composite. It is now well established that the value of the percolation threshold decreases with increasing aspect ratio (ratio of length to diameter) of the filler which makes graphene, CNTs as well as metal nanowires more effective as fillers [10,19]. The formation of conductive paths is also governed by many other factors such as the state of dispersion, filler-matrix interactions, the geometry of filler and the intrinsic properties of the nanofillers [20]. For these reasons, choosing a composite preparation method that provides the desired level of filler distribution is of prime importance.

One of the main challenges in the fabrication of carbon based conductive polymer composites is that the carbon fillers are usually difficult to be homogenously dispersed within polymer materials [21,22]. Another challenge in fabrication of conductive composites for biomedical applications is to achieve both high conductivity and mechanical toughness at the same time. The limiting conductivity is as important as the percolation transition [10]. The high conductivity is often achievable at the cost of mechanical strength. It also seems fairly clear that the direct use of nanosized materials does not provide a way to improve the fabrication of conductive composite materials. However, if filler contact density can be reduced by sintering or using high-aspect ratio fillers, high conductivity can be achieved [10,23,24].

8.2.2 Methods of Preparation of Conductive Polymeric Composites

There are various methods of preparation of conductive composites. The most commonly used method of preparation of conductive polymer composites is the solution mixing method. By this method, a solution of polymer is prepared and the filler is separately dispersed in a suitable solvent by sonication. Once the filler is dispersed in the solvent, the polymer, which was previously dissolved in the same solvent, is added to the dispersion so that the polymer adsorbs on to the filler. The final step is removing the solvent by evaporation. Both organic solvents and water have been used to produce composites using this method [25].

The *in-situ* polymerization method is extensively used for the preparation of polymer/CNTs composites due to an advantage of formation of a covalent bond between the CNTs and matrix. In *in-situ* polymerization method, the filler is first dispersed in the liquid monomer. A suitable initiator is then diffused in the monomer/filler dispersion and polymerization is initiated either by heat or radiation.

Melt blending or melt processing method is preferred for industrial-scale processes because of its speed and simplicity. It also offers advantage of being free of solvents. In this method, graphene or other nanofiller is mixed with the polymer matrices in molten state. The thermoplastic polymers are mixed mechanically with filler at elevated temperatures using conventional methods such as extrusion and injection molding. The polymer chains get intercalated between the filler particles to form nanocomposites. The polymer chains experience a significant loss of conformational entropy during this process. This is also a preferred method for processing of polymers which are unsuitable for solution mixing or for *in-situ* polymerization.

More recently, the conductive composites have been explored for their use in 3D printing of electrodes and scaffolds. Melt processing method is particularly suitable for making carbon-based polymer composites for these applications because these can be easily manufactured into filament form for direct printing on desktop-sized low cost fused deposition modelling (FDM) printer. Carbon-based polymer composites have also advantages over metal-based or metal-alloy ink for 3D printed devices as these typically do not need additional processing step after printing such as thermal annealing or evaporation of solvent, as required by 3D printing of metal-ion and metal-colloid inks [26]. These materials also have significantly longer shelf life (years) than that of metallic ink (e.g. silver), which is typically stable for few weeks to months [26].

Recently, electrospinning has gained large attention probably due to the remarkable accessibility of fabrication of conducting polymer based composites and incorporation of drugs; capability to prepare controlled and oriented nanofibers; and feasibility to render scaffolds with the porosity required for effective tissue regeneration applications. Specifically, biopolymer-based conducting fibrous mats are of special interest for tissue engineering because they are able to stimulate specific cell functions or trigger cell responses in addition to the expected ability to physically support tissue growth [8]. Electrospinning CP-based composites is a simple way to prepare electrically conductive nanofibers especially for tissue scaffolds. For example, nanofibers composed of PANi and gelatin were shown to support the adhesion and proliferation of cardiomyocytes in *in vitro* studies [44]. It is an effective approach to fabricate long polymer composite fibers with diameters from micrometers down to 100 nm or even a few nanometers by using strong electrostatic forces [27]. In the usual electrospinning process, conducting polymer blend or polymer/filler dispersion is extruded from an orifice to form a small droplet in the presence of an electric field. Once the electric field reaches a critical value at which the repulsive electric force overcomes the surface tension of the polymer solution, the solution is ejected from the tip to a collector (Figure 8.2). While traveling to the collector, the solution jet solidifies due to the fast evaporation of the solvent and is deposited on a collector. Generally the fluid extension occurs first in uniform, and subsequently the straight flow lines undergo vigorous whipping and/or splitting motion due to fluid instability and electrically driven bending instability. Finally, the spun fibers are deposited commonly as a nonwoven web on a collector. Through an improved or modified electrospinning device, nanofibers with partial or good orientation could be fabricated.

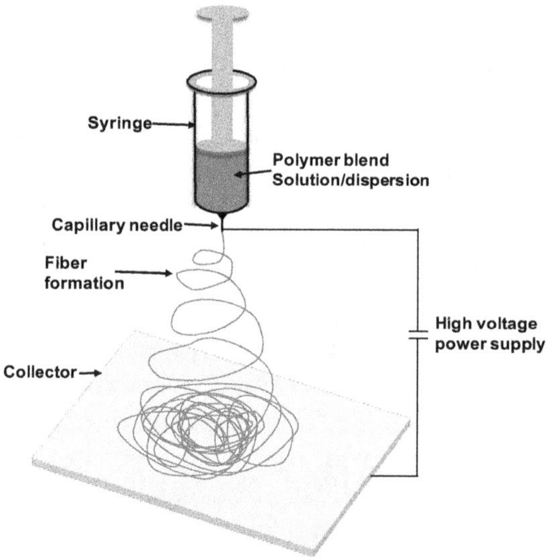

Figure 8.2 Schematic diagram of electrospinning process.

Some problems have been observed when electrospinning blends of two chemically different polymers, in which the polymer blend solutions are not homogenous owing to the different solvent system of the two polymers. One way to overcome this issue is to electrospin two polymer solutions simultaneously using a coaxial electrospinning technique [28]. Electrospun fibers made using the coaxial electrospinning technique have a bicomponent system that has properties from each of the polymeric components, e.g. one of the polymers could contribute to the mechanical strength while the other could provide the electrical conductivity [28]. Compared with other synthetic approaches, the electrospinning process seems to be the only method that can mass-produce continuous long nanofibers [28].

8.2.3 Composites of Conductive Nanoparticles/Fillers with Non-conjugated Polymers

Most of the literature available on conducting composites of non-conjugated polymers is very recent with initial work on polymers such as polystyrene, poly(ethylene terephthalate), nylon and polyurethanes for non-biomedical applications [4,19,29-33]. The work for biomedical applications is still in its infancy as researchers are still focusing more on fundamental work rather than on applications. The following sections present the current progress in developing conductive polymer composites of non-conjugated polymers with carbon as well as non-carbon conductive fillers such as graphene, CNTs and metal nanoparticles.

Polymer Composites of Non-conjugated Polymers with Carbon Fillers

Graphene and CNTs are the conductive carbon nanofillers which have been used for making conductive composites of polymers. Graphene is a two dimensional monolayer of sp^2 hybridized carbon arranged in honeycomb lattice and exhibits high mechanical strength, electrical conductivity and ultra-high specific surface area [34]. The graphene based polymer composites show superior mechanical, thermal and electrical properties as compared to the neat polymers [29,35]. However, stable dispersions of graphene in polar solvents can only be obtained using suitable surfactants due to its hydrophobic nature [36,37]. To overcome solubility issues associated with graphene, graphene oxide (GO) has also been explored for making conductive composites. It is similar to graphene but has the oxygen containing polar functional groups which enhance its biocompatibility, compatibility with polar solvents or with a polymer matrix, etc. [17,38-40]. Incorporation of hydrophilic graphene based fillers like GO also improves cell adhesion at the biomaterials surface [39].

Furthermore, there are various reports found in literature on the use of CNTs as conductive filler to make conductive polymer composites for various biomedical applications such as scaffolds for bone regeneration, tissue engineering and nerve regeneration [41-47]. Carbon nanotubes exhibit very good electrical conductivity of $>10^3$ S cm^{-1}, with a high aspect ratio reaching 100-1000 for μm long single-walled and multi-walled CNTs (SWCNTs and MWCNTs) [19,48,49]. Both single-walled and multi-walled CNTs have been extensively used as functional fillers for the preparation of highly conductive polymer composites.

In a study, conductive composite nanofibers of chitosan (CS) and MWCNTs were produced using conventional electrospinning technique for potential applications as conductive scaffolds and biomedical implants [50]. Several solvents including acetic acid and trifluoroacetic acid/dichloromethane (TFA/DCM) were investigated for the electrospinning of CS/MWNTs dispersion. The results showed that the TFA/DCM (volume ratio = 70:30) was the most preferred solvent system for nanofiber formation with acceptable electro-spinnability. The homogeneous nanofibers with an average diameter of 275 nm could be prepared with a conductivity of 9×10^{-5} S cm^{-1}. In another study, Gupta *et al.* [51] reported fabrication of MWCNT reinforced polyvinylidene fluoride (PVDF) composite with uniform dispersion and alignment of CNTs by solution casting technique. CNTs were aligned with the application of alternating voltage as well as pulsed current. The voltage-assisted films showed better dispersion and alignment of CNTs inside PVDF matrix while the current-assisted alignment led to the formation of shortest and continuous path for the flow of electrons, thus, resulting in the formation of highly anisotropic conductive film. Directly current passage assisted alignment of CNTs in the composite film recorded a 360% improvement in conductivity in the direction parallel to the alignment as compared to the structure with randomly aligned CNTs. At the same time, the composite in transverse direction to the alignment was totally insulating, indicating the efficiency of alignment. With the addition of only 0.5 wt% CNTs to PVDF matrix, the films showed improvement in elastic modulus and tensile strength by 180% and 150%, respectively, compared to pure PVDF film. However, the films behaved mostly isotropic in terms of mechanical properties, showing improvement in all directions with CNT reinforcement. This study provided an effective way to align CNTs in a thermoplastic matrix to tailor the directional electrical conductivity significantly for potential applications in biomedical devices.

Polyurethanes (PUs) are of particular interest to make conducting polymer composites by way of incorporating conductive particles because of their biocompatibility, biostability, processability and good mechanical properties [52]. Polyurethanes have found applications in numerous medical devices, including vascular grafts and pacemaker lead insulators [52-54]. The main focus of the initial attempts to generate conductive composites of PUs was their conductivity and shape memory properties rather than a wider exploration of properties relevant to biomedical applications [32,55,56]. Yousefi *et al.* [57] reported the preparation and conductivity of graphene reinforced polymeric composites from water-based colloidal dispersions of graphene oxide and PU latex. Choi *et al.* [55] examined the role of graphene as reinforcing filler in shape memory polyurethane nanocomposites. Rana *et al.* [56] also reported the preparation of caprolactone based PU nanocomposites for potential applications as shape memory materials for actuators, electronics and artificial muscles. The composites were prepared using both graphene and CNTs as conductive fillers and their effect on electrical and thermal conductivity of the composite was examined. CNTs and functionalized graphene sheets were incorporated as crosslinkers in the prepolymer. In comparison to pristine PU and CNT-crosslinked PU composite, the graphene-crosslinked PU composite exhibited better mechanical properties. The graphene-crosslinked PU composite also showed a higher electrical conductivity (1.67×10^{-3} S cm^{-1}) than the CNT-crosslinked PU composite (2.30×10^{-4} S cm^{-1}). Although these composites exhibited good conductivity and flexibility, their low melting point (36.3 °C) limited their applications in implantable biomedical devices.

In a more recent study by Gunatillake and co-workers, conductive composites of a biostable siloxane PU (Elast-Eon™) and graphene were prepared using different methods in order to compare the effect of the processing method on the conductivity of resulting composites (Figure 8.3) [58]. The solution mixing method was found to be best as it resulted in the composites with higher electrical conductivity. Also, it was suitable for preparing composites with better mechanical properties. A conductivity of 1.12×10^{-3} S cm^{-1} was achieved with 5 wt% loading of graphene. The *in vitro* cytotoxicity assay tests found that composites were not cytotoxic to living cells and potentially useful in biomedical applications.

In another study, Kim *et al.* [59] developed a nanocomposite of CNTs and polydimethylsiloxane (PDMS) elastomer by solution mixing method for its potential use in strain sensors and flexible electric circuits, along with bio-signal monitoring of brain, heart and muscle signals [59]. The scanning electron microscopy and turbiscan results demonstrated that the CNTs were homogeneously distributed in the PDMS polymer matrix. The CNT/PDMS nanohybrids possessed sheet resistance of $<20\ \Omega$/sq, tensile stress of ~3.65 MPa and a yield strain of >45%.

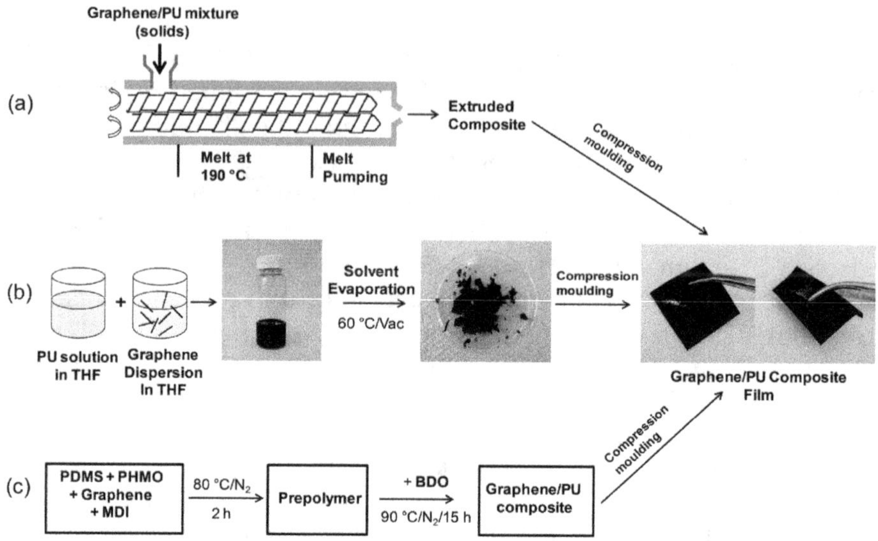

Figure 8.3 Preparation of graphene/Elast-Eon™ composites using melt mixing (a), solution mixing (b) and *in situ* polymerization (c). Reproduced from Reference 58 with permission from the Royal Society of Chemistry.

Polymer Composites of Non-conjugated Polymers with Non-carbon Conductive Fillers

Apart from carbon based conductive fillers, metal nanoparticles have also been explored to impart conductivity to non-conjugated insulating polymers. Kim *et al.* [60] demonstrated the fabrication of stretchable conductors of PU containing spherical gold nanoparticles deposited by either layer-by-layer assembly or vacuum-assisted flocculation. High conductivity and stretchability were observed in both composites despite the minimal aspect ratio of the nanoparticles. These materials achieved electrical conductivities of up to 11000 S cm^{-1} and also demonstrated the electronic tunability of mechanical properties resulting from dynamic self-organization of the nanoparticles under stress. In another study, Hu *et al.* [61] fabricated an elastomeric transparent composite electrode comprising of copper nanowires (CuNWs) embedded in the surface layer of an elastomeric PU matrix [61]. The composite electrode was fabricated by first forming a highly conductive CuNW network on glass, then overcoating with a layer of a liquid polyurethane precursor which was subsequently polymerized, and finally peeling off the resulting PU sheet. The composite was observed to retain elastomeric stretchability of the polymer matrix.

Pre-treatment of the CuNW network with 6-aminohexanoic acid enhanced the bonding between nanowires and PU matrix, and significantly improved the reversibility of the surface conductance of the composite electrode during repeated stretching at room temperature.

In a more recent study by Gunatillake and coworkers, conductive composites of a siloxane PU (Elast-Eon™), graphene and gold nanoparticles (AuNPs) were prepared using the solution mixing method [62]. The results showed that the addition of AuNPs significantly enhances the conductivity of graphene/PU composites. The conductivity of graphene/PU composite films with 5 wt% loading of graphene was 2.5×10^{-2} S cm^{-1}, which increased to 0.42 S cm^{-1} with the addition of 5 wt% AuNPs. Furthermore, the highest conductivity achieved was 1.388 S cm^{-1} for composites containing 5 wt% loading of AuNPs and 15 wt% of graphene. The mechanical properties of AuNPs/graphene/PU composites were poor, showing a significant loss of tensile strength and elongation compared to the unmodified polyurethane. The reduction in tensile strength and elongation was attributed to the presence of a significant level of particle agglomerates creating defects within the matrix leading to poor mechanical strength. Furthermore, the *in vitro* cytotoxicity assay tests showed that AuNPs/graphene/polyurethane composites were not cytotoxic to living cells and, as a result, were potentially suitable for biomedical applications [62].

8.2.4 Composites and Blends of Conducting Polymers

Conducting polymers such as PPy, PANi, polythiophene and their derivatives have been investigated for application in the biomedical field, in particular as components in biosensors and substrates for tissue engineering constructs [63-65]. Many of the literature studies targeting tissue engineering applications have focused on the use of conducting polymers to effectively deliver electrical signals from an external source to the seeded cells to enhance growth and proliferation. Several cell types including cardiomyocytes, neurons and osteoblasts respond to electrical signals by improving their functional outcomes. Among handful of available conducting polymers, PANi has received considerable interest for such applications due to ease of synthesis, tunable conductivity and biocompatibility. Similarly, PEDOT and PPy polymers have also been investigated for tissue engineering and related applications. Non-degradability, along with poor processability and biocompatibility as well as limitations in mechanical properties are the major drawbacks of conducting polymers to be useful in fabricating biomedical implants, particularly for tissue engineering applications. Furthermore, the electrically conductivity deteriorates when exposed to aqueous environments, thus, limiting the application where electrical stimulation is required for long durations.

As mentioned above, an effective way to improve mechanical properties of CPs is to create their composites or blends with other polymers that have better mechanical properties for the intended application. Blending a conducting polymer with biodegradable synthetic or natural polymers is one of the main approaches reported to overcome some of the limitations. Doping with large molecules can also be used to prepare conducting polymer composites with improved mechanical properties. However, these processes may cause interference with electron conjugation within the CP due to the presence of insulating molecules [3]. Other approaches investigated to address the above limitations include development of copolymers containing oligomer segments of conducting polymers and biodegradable polymers and polymer composites incorporating conducting nanoparticles. The use of conducting polymers in tissue engineering is one area where significant efforts have been made and the readers are referred to read the recent reviews on the use of conducting polymers for tissue engineering application [65-68]. This section focuses on combining conducting polymers and non-conducting synthetic polymers to prepare conducting polymer blends or composites to achieve good processability, mechanical properties and acceptable electrical conductivity. Studies on incorporating conducting polymer segments as part of a non-conducting synthetic polymer are also reviewed as this approach may allow the polymer to biodegrade. Furthermore, conductive hydrogels of CPs have also been discussed.

Polymer Blends of Conducting Polymers and Biodegradable Polymers

Synthetic biodegradable polymers such as polylactide (PLA)/glycolide [69-72], polycaprolactone (PCL) [73-75], poly(lactide-co-polycaprolactone) [8] and PUs [76] as well as natural polymers such as chitosan [77-79], gelatin [80, 81], collagen [9] and heparin [82-84] have received significant research attention for blending with conducting polymers. Remarkable improvements in the mechanical properties, processing and biocompatibility have been achieved by this approach.

Although PPy is considered a promising material for tissue engineering and bioelectronics applications, its capacity to hold electrical conductivity in aqueous environment is rather low. To address this issue, PPy particles were blended with poly(l,l-lactide) (PLLA) in 5:95 weight ratio and the blend showed slow deterioration of conductivity after a period of stable electroactivity measured using cyclic voltammetry [85,86]. Heparin was used as a dopant to further improve the electrical stability as well as cell adhesion. The composite membranes, when examined by X-ray photoelectron spectroscopy and scanning electron microscopy (SEM), showed that heparin appeared on the surface with a 'filament'-like morphology. Significantly improved electrical stability compared to unmodified composite as well as increased cell adhesion and growth confirmed the benefits of incorporating heparin [86]. Covalent attachment of bioactive molecules to PPy is another approach to fabricate composite membranes for tissue engineering applications. Bioactive molecules fibronectin (FN) and bovine serum albumin (BSA) have been chemically coupled to PPy nanoparticles and blended with PLLA fibers to produce scaffolds for nerve tissue repair [87,88]. The FN and BSA were incorporated to PPy particles during synthesis using a water-in-oil emulsion polymerization process [87]. Membranes containing 5% and 10% PPy particles with PLLA showed comparable cell adhesion, spreading and morphology to FN membranes when PPy was modified with FN. On the other hand, modification with BSA resulted in decreased cell adhesion as was the case with BSA membranes. Chemical coupling of BSA onto PLLA fibers using carbodiimide chemistry allowed the incorporation of 31 $\mu g/cm^2$ of protein on PLLA using this technique [88].

Chemical oxidative polymerization has also been used to prepare blends of PPy with PLLA or PCL with electrical conductivities of 2.7 ± 0.8 S cm^{-1} and 7.8 ± 2.3 S cm^{-1}, respectively [89]. These values were obtained when the PPy content was 40% in the composite. However, the cytocompatibility with dermal human fibroblast cells was compromised for higher PPy loadings. Another study reported the surface polymerization of PPy on electrospun PLLA fibers to produce three-dimensional fibrous scaffolds for nerve tissue engineering [90]. Rat pheochromocytoma cells cultured in fabricated three dimensional (3D) scaffolds showed much higher number of cells than on conductive fibrous meshes, suggesting that the 3D scaffolds may be useful in nerve tissue engineering with electrical stimulation to enhance cell growth. Similarly, dodecylbenzene sulfonic acid doped PANi and PLLA have been used to produced smooth fibers with diameters in the range 75 to 100 nm by electrospinning [91].

In a study by Shi *et al.* [69], composites of PPy and poly(d,l-lactide) (PDLLA) were prepared by emulsion polymerization of pyrrole in PDLLA solution followed by precipitation. The polymer blend with 3% PPy was cast onto a poly(tetrafluoroethylene) plate to produce a membrane with resistivity of 1×10^3 Ω /square. The composition retained conductivity in cell culture environment for over 40 days, and cell viability (human cutaneous fibroblasts) was significantly higher with electrical stimulation. In another study by the authors, solvent casting of a solution containing PPy particles and PLLA was used successfully to produce conducting membranes [70]. Human cutaneous fibroblasts cultured in multi-well electrical cell culture wells exhibited 40-fold increase in cell viability with electrical stimulation.

In an attempt to improve the biocompatibility of PPy blends, chitosan, a naturally occurring biocompatible polymer with anti-bacterial and wound healing properties, was blended with PPy [78]. A blend of PPy (2.5 wt%) and chitosan (97.5 wt%) was prepared by solvent casting method. The PPy/chitosan membrane supported adhesion, spreading and proliferation of Schwann cells with or without electrical stimulation. Another important observation was that the membrane supported enhanced expression and secretion of nerve growth factor and brain derived neurotrophic factor compared to that of controlled cells without electrical stimulation.

Polyurethane has been used to prepare blends with PPy to enhance biocompatibility as well as to prepare membranes with good mechanical properties [76]. The *in-situ* polymerization of pyrrole at different weight ratios was conducted in a polyurethane emulsion. The resulting blend contained a network of PPy nanoparticles in PU matrix, and the electrical conductivity was proportional to the PPy particle loading, while the modulus increased and elongation at break decreased. The conductivity increased from 9.95×10^{-11} S cm^{-1} for a blend with 1% PPy to 2.32×10^{-7} S cm^{-1} for 20% PPy.

Electrospinning method was used to produce conducting polyaniline/chitosan (PANi/CS) defect-free nanofibers of different ratios (1:3; 3:5 and 1:1) [79]. Increase in PANi content resulted in membranes with higher hydrophobicity and higher electrical conductivity. None of the membranes showed any toxic effects on osteoblasts and fibroblasts, and the membranes showed improved cell attachment and growth compared to tissue culture plastic. The membranes containing PANi/CS ratio of 1:3 supported improved cell attachment and proliferation for both cell lines due to a synergistic effect of hydrophilicity retention due to the high chitosan content and the conductivity that PANi introduced to the membrane.

Copolymers of Conducting Oligomers

Copolymerization of oligomeric segments of conducting polymers such as PANi has been another approach to introduce biodegradability to conducting polymers [92-94] with the objective of use in tissue engineering applications. The development of PANi-based polymers for tissue engineering of electrically responsive tissues has been reviewed recently [65].

Carboxylic acid capped aniline trimer was coupled with different architectures of pol(L-lactide) to produce conducting copolymers with conductivity comparable to that of PANi (Scheme 8.1) [92]. Using a similar approach, biodegradable conducting polymer hydrogels

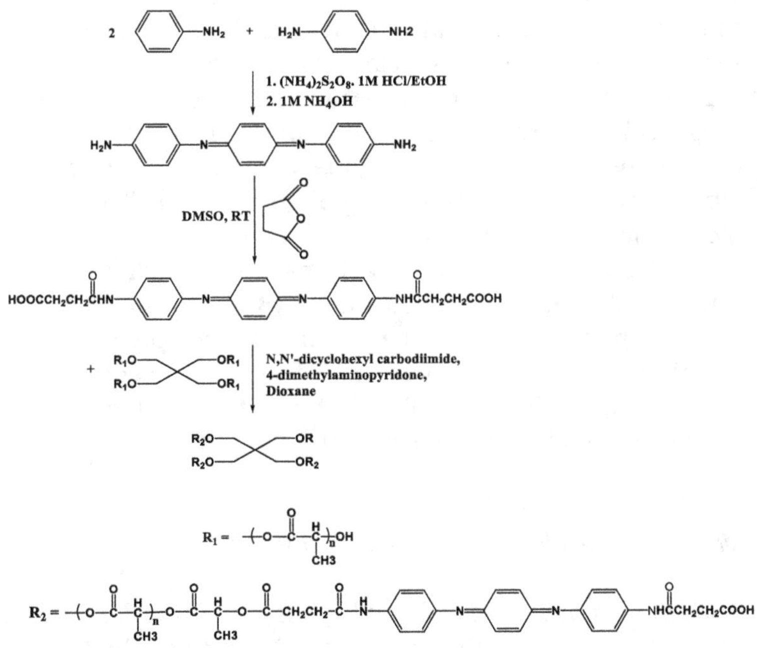

Scheme 8.1 Schematic synthesis of carboxyl-capped aniline trimer and the copolymers. Adapted from Reference 92 with permission from American Chemical Society.

were produced by coupling carboxy-capped aniline pentamer with the hydroxyl groups of hydroxyethylmethacrylate hydrogel [95]. Conductivity in the range 2.11×10^{-6} to 2.01×10^{-4} S cm^{-1} was achieved for hydrogels with different amount of aniline pentamer, and the highest conductivity was achieved when the pentamer content was 22%.

Acid functionalized PANi oligomers are self-doping due to the acid group acting as a dopant. Therefore, blending them with other polymers produces materials without compromising the conductivity of PANi. Several studies have focused on the preparation of blends of PANi with other polymers for tissue engineering applications. In a related study, the biodegradable natural polymer chitosan was crosslinked with conductive aniline pentamer and assessed for biocompatibility using C6 glioma cells [96]. The copolymer was found to be highly cytocompatible compared to chitosan and tissue culture plastic controls. In another study, glutaraldehyde was used as a crosslinker to produce a conducting hydrogel with conductivity in the range 1.14×10^{-4} to 4.23×10^{-4} S cm^{-1} [97]. The one-pot synthesis produced a material with both grafts and crosslinks.

Oligomers of pyrrole have also been considered to introduce conductivity to polymer networks via copolymerization. In one such study, a procedure to prepare biodegradable conducting copolymers containing trimeric pyrrole-thiophene-pyrrole segments was reported based on condensation polymerization of adipoyl chloride and propanol end-capped conducting oligomer [98]. The conductivity of the copolymer was in the order of 10^{-4} S cm^{-1} and showed no cytotoxicity. Furthermore, neuroblastoma cells attached and proliferated on the polymer surface. In a study by Zhang *et al.* [99], a copolymer based on poly(glycine ethyl ester) and aniline pentamer as side groups exhibited a conductivity of 2×10^{-5} S cm^{-1}. The copolymer lost 50% of its mass in 70 days in PBS buffer. The copolymer also had comparable cell viability to that of PDLLA when tested with RSC96 Schwann cells.

Conductive Hydrogels of Conducting polymers

Biocompatible conductive tough hydrogels represent a new class of advanced materials combining the properties of tough hydrogels and biocompatible conductors. Conducting polymers have been exploited for developing electroactive hydrogels which are polymeric blends combining the responsive properties of electroactive polymers and highly hydrated hydrogels within an aqueous milieu that is hospitable to biological molecules such as peptide sequences, enzymes antibodies and DNA [100]. The combination of hydrogels and inherently conductive electroactive polymers allows both materials to retain their unique responsive properties. In addition, the electroconductive hydrogels engender a new class of devices with low interfacial impedances suitable for neural prosthetic devices such as deep brain stimulation electrodes, electrically stimulated drug release devices (low voltage actuation) and implantable biosensors (potential for *in vivo* biocompatibility).

In a study by Javadi *et al.* [101], a self-assembled tough elastomeric, biocompatible and conductive composite structure was reported. The hydrogel comprised polyether-based linear PU, PEDOT doped with poly(4-styrenesulfonate) (PSS) and liquid crystal graphene oxide (LCGO). The polyurethane hybrid composite (PUHC) containing PEDOT:PSS, LCGO and PU exhibited a 10 fold increase in electrical conductivity along with a higher tensile modulus and yield strength compared to respective control samples. Furthermore, the PUHC was biocompatible and was found to support human neural stem cell (NSC) growth and differentiation to neurons and supporting neuroglia. The stimulation of PUHC also enhanced NSC differentiation with enhanced neuritogenesis compared to unstimulated cultures.

In another study, a highly swelling grafted hydrogel composed of poly(acrylic acid) (PAA) and polyvinyl alcohol (PVA) containing PANi nanoparticles was prepared by *in-situ* polymerization of aniline [102]. Impregnation of polyaniline into PVA-g-PAA resulted in a composite hydrogel which showed electroconductive and electroactive behavior. The electrical conductivity varied with varying content of PANi in the composite and was found in the range of 0.04-0.06 S cm^{-1} for 5% PANi content. The native and PANi impregnated matrix not only showed a moderate biocompatibility and good mechanical strength but also exhibited good swelling properties in both distilled water and electrolyte solution.

In another study by Shi *et al.* [103], synthesis of a bacterial cellulose (BC)/PANi nanofiber composite was reported, which is an electro-conductive hydrogel potentially suitable for biosensors and tissue engineering applications. The hydrogel was synthesized in ammonium persulphate solution by *in-situ* nano-assembly of BC nanofibers and PANi to enhance the electrical conductivity of BC nanofibers [103]. The electrical conductivity of composite hydrogels was observed to enhance from 10^{-8} to 10^{-2} S cm^{-1}.

8.3 Biocompatibility and Stability

There is a concern for safety risks of using metal nanoparticles and carbon based nanofillers for making conductive composites for biomedical applications. The risks are higher where composites are comprised of a biodegradable polymer which upon degrading may release the nanoparticles and nanofillers in the body. Hence, more research is now being focused on the toxicological studies on nano-systems for their improved characterization and reliable toxicity evaluations. The studies in this area have been predominantly based on *in vitro* assays, such as cell viability tests and cytokine release analyses for toxicity assessment of nanocomposites because deploying these materials for *in vivo* studies has even more challenges. However, the results from *in vitro* tests for cytotoxicity should be carefully applied to extrapolate the probable results in *in vivo* studies [104,105]. The mechanisms for adverse effects of metal nanoparticles have not been understood yet. It is also challenging to relate the reports on one type of nanoparticles to other types, due to the intrinsic differences in their physical properties (such as particle size and shape), chemical properties (such as surface chemistry and hydrophobicity) and methods of preparation as well as biological targets including cells, organs and tissues. The nanoparticles may not be filtered by the body's defense system due to their small size which suggests that they may cause inflammatory and/or toxic responses. The size and surface chemistry of nanoparticles also affect their cytotoxicity. Generally, smaller nanoparticles show greater toxicity because small nanoparticles are more easily taken up into the cell or even near the nucleus [105]. Furthermore, nanoparticles are less portable than the most chemical compounds. As a result, they require longer duration of analyses which would require the modification of the cytotoxicity protocols for nanoparticles and nanocomposites [104,105].

Toxicity of graphene based nanofillers is well reviewed in literature, but the research is still in infancy, and it is difficult to conclude the potential health hazards associated with their use [106-113]. However, the reports suggest that graphene based fillers can be used for various biomedical applications with consideration and controlled functionalization that can potentially reduce their toxicity [106,107,114].

The composites of conductive polymers with biodegradable polymers also pose safety risks upon degradation of the composite structure, since conductive polymers, in general, are not biodegradable.

Various reports can be found on evaluation of biocompatibility and cytotoxicity of PPy and PANi [115-118]. Biocompatibility of PPy is well established, e.g. an *in vitro* study was carried out to assess the biocompatibility of PPy mats in rats for intracranial triggered release of anti-epileptic drugs. The results showed that implanted PPy mats did not induce any obvious inflammation, trauma, gliosis and neuronal toxicity [116]. The studies on PANi have shown that both acid and base forms of PANI possess good biocompatibility properties, however at the same time, both polymers also showed considerable cytotoxicity, which being higher for PANi hydrochloride compared with PANi base [117]. Furthermore, the polymer purification via reprotonation/deprotonation cycle led to significant reduction in cytotoxicity [117]. These results indicated that the low molecular weight reaction residues or by-products, rather than PANi alone, are responsible for observed cytotoxicity of PANi [117,119].

Animal study data on conductive composites are rare. Subramanian *et al.* [72] have reported good biocompatibility and degradability of electrospun PLGA-polyhexylthiophene nanofibers based subcutaneous implantation in rats. Histology results demonstrated that implanted scaffolds showed decrease in inflammation over time up to eight weeks and well organized tissue structure and collagen deposition around the scaffold (Figure 8.4). In another *in vivo* study, Xu *et al.* [120] fabricated nerve conduits from PPy/PDLLA composite and used these to bridge a 10 mm defect in the sciatic nerve of a rat [120]. After six months, the rats with the PPy/PDLLA conduits showed functional recovery similar to that of the gold standard autologous nerve graft and significantly better than that of PDLLA conduits. These results suggest that such a conduit can potentially be used in nerve tissue regeneration, eliminating the drawbacks associated with using an autologous graft including limited donor source, donor site morbidity, multiple surgery sites and possible size mismatch.

8.4 Biomedical Applications of Conductive Composites

8.4.1 Tissue Engineering

Tissue engineering scaffolds play an important role in regenerative medicine. These provide a temporary three-dimensional support during tissue repair while also regulating the cell behavior such as cell adhesion, proliferation and differentiation [4,121-123]. Moreover, the modern biomedical nanotechnology requires 3-dimensional

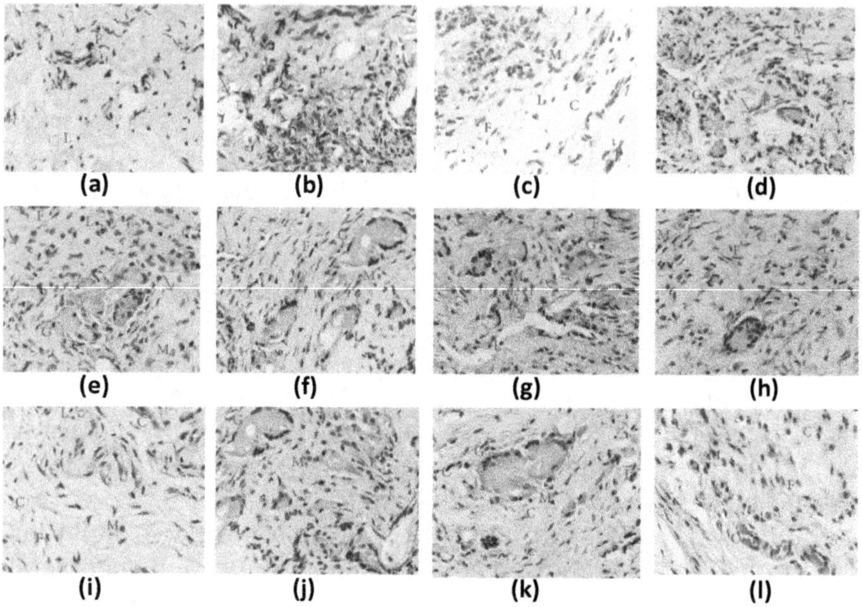

Figure 8.4 Micrographs of hematoxylin-eosin-stained tissue implanted with PLGA random nanofibers at (a) 2 weeks, (b) 4 weeks, and (c) 8 weeks; with PLGA-undoped polyhexylthiophene (PHT) random fibers at (d) 2 weeks, (e) 4 weeks, and (f) 8 weeks; with PLGA-doped PHT random fibers at (g) 2 weeks, (h) 4 weeks, and (i) 8 weeks; and with PLGA-doped PHT aligned nanofibers at (j) 2 weeks, (k) 4 weeks, and (l) 8 weeks. C: collagen; L: lymphocytes; M: macrophages; F: fibroblast; G: giant cell; and arrows indicate implant sites. Reproduced from Reference 72, Copyright © 2013 Anuradha Subramanian *et al.*

macrostructures with nano-textures to meet the requirements for practical applications in intricate biological systems and polymeric materials can fulfil these requirements. Researchers in the field of tissue engineering have investigated conducting polymers for the development of biomaterials that are responsive to various stimuli. The application of an electrical signal allows the stimulation of cells and tissues, hence, conducting polymers can provide the necessary platform enabling the development of tissue engineering scaffolds, biosensors, neural implants and drug delivery devices [121,122]. Recent reviews summarize the recent advances in the use of conducting polymers [65-67]. As mentioned earlier, common conducting polymers such as PPy, PANi and PEDOT have limitations in meeting property requirements for tissue engineering applications. In an attempt to overcome some of these limitations, researchers have developed conducting composites with improved biocompatibility, processability and mechanical properties. More recently, significant research effort has been directed towards development of conductive composites of biodegradable polymers, such as PCL and PLA, incorporating conductive fillers (e.g. PANi, PPy and graphene) for their potential use in tissue engineering [4,14,121,122,124]. This section provides a review of some of the fundamental studies in formulating various composites for these applications and illustrates a few examples where the focus was on the fabrication of tissue engineering constructs and their evaluations.

In a study, 3D composite nanofiber macrostructures of PCL with different concentrations of PANi were fabricated by electrospinning method using a specially designed collector [125]. The 3D structures possessed cap-like macrostructures with centimeter-scale thickness and interconnected pore nanotextures with nanometer-scale fibers. The biocompatibility of the 3D PANi/PCL composite nanofiber macrostructures was evaluated by culturing mouse myoblasts (C2C12 cells) as model cells. The initial responses of C2C12 cells to the 3D PANi/PCL composite macrostructures were significantly superior to those obtained for pure PCL, i.e. the cells exhibited typical myoblast-like morphologies with obvious pseudopodia and the moderate incorporation (<2.0 wt%) of conductive PANi facilitated cell proliferation showing that PANi possessed appreciable cell affinity. Moreover, the addition of conductive PANi to the 3D composite nanofiber macrostructures considerably enhanced myoblast differentiation and myotube maturation. In another study by Sayyar *et al.* [126], a composite of PCL and chemically reduced GO was prepared using covalent attachment as well as solution mixing. The obtained graphene nanocomposites possessed improved mechanical and electrical properties compared to neat PCL. The *in vitro* cytotoxicity testing results showed good adherence and proliferation of fibroblast cells on composite films confirming their biocompatibility for use as conducting substrates for the electrically stimulated growth of cells.

Gizdavic-Nikolaidis *et al.* [127,128] investigated conductive electrospun fibers produced by blending poly(aniline-co-3-aminobenzoic acid) (3ABAPANi) and PLA as wound dressings to allow cell growth and proliferation along with antibacterial properties [127,128]. The biocompatibility of 3D fiber mats produced by varying amounts of 3ABAPANi was demonstrated by seeding COS-1 fibroblasts. The proliferation rate was similar to that observed with tissue culture plate and glass substrates, while the fiber mat from 45:55 3ABAPANi:PLA composition had a significantly higher cell number compared to the two controls. Another study by Rahman *et al.* [129] reported electrospinning of PLA and poly(aniline-co-m-aminobenzoic acid) to produce nanofiber mats. The authors subsequently reported the suitability of these mats as scaffolds for culturing adipose derived stem cells (hASCs) [130]. The hASCs, when seeded on to these mats, attached and proliferated for seven days with similar proliferation rates to that of pure PANi.

Conductive nanofibrous PLA scaffolds with well-distributed PANi nanostructures were fabricated for bone tissue regeneration using *in-situ* polymerization/thermal induced phase separation method [131]. The *in-situ* polymerization led to the good distribution of PANi in the matrix, enhancing the conductivity of scaffolds to a level close to that of spongy bone. The live/dead staining and scanning electron microscopic observation of bone marrow derived mesenchymal stem cells on scaffolds indicated that these composite scaffolds exhibited good cell adhesion and non-cytotoxicity. Furthermore, compared to pure PLA scaffolds, conductive composite scaffolds provided better promotion of cell proliferation, osteogenic differentiation and calcium mineralization.

Electrically conductive polymeric hydrogels are also promising candidates as tissue engineering scaffolds due to their hydrated nature, good biocompatibility and their soft tissue-like mechanical properties [132-134].

8.4.2 Biosensors and Artificial Muscles

More recently, significant research work has been carried out to explore conductive polymer composites for their potential use in biosensors and actuators [135-139]. Graphene based composites have shown specific potential for application in a range of biosensors. These composites allow an accurate and selective detection of various biomarkers such as glucose, nicotinamine adenine dinucleotide, haemoglobin, adenosine triphosphate, cholesterol, catechol, ascorbic acid, uric acid, dopamine, heavy metal ions, amino acids and others [124,140]. A recent review summarizes the developments in the use of graphene based polymer composites as chemical sensors and biosensors [141]. For example, by combining the unique properties of graphene and ionic liquid (IL), with the dispersion and film-forming abilities of chitosan (CS) polymer, an electrochemical sensor based on graphene/IL/CS composite film was developed and evaluated for individual or simultaneous determination of guanine and adenine [142]. The graphene/IL/CS composite film sensor showed good sensitivity, stability and a low detection limit because of the unique electronic and catalytic properties of graphene and ionic liquid. The graphene/IL/CS composite did not only efficiently promote the electron transfer between the analytes and electrode surface but also enhanced the accumulation efficiency. Moreover, the fabricated biosensor exhibited good reproducibility and long-term stability making it potentially useful in the related physiology processes.

In another study, photolithographic technique was used to pattern the conducting polymer PEDOT:PSS on thin photocurable polydimethylsiloxane (PDMS) membranes via a conducting composite ink composed of a polymer and a biofriendly silk protein carrier [143]. The fabricated microelectrodes displayed detection of both electroactive (dopamine and ascorbic acid) and non-electroactive biomolecules (glucose) with high sensitivity. The electrodes did not delaminate or lose function under mechanical flexure due to the covalent linkage between the ink and photo-PDMS.

Conductive nanofibrous sheets composed of blends of PLA and PANi were fabricated for cardiac tissue engineering and cardiomyocytes-based 3D bio-actuators [144]. These PLA/PANi conductive nanofibrous sheets exhibited good cell viability and promoted differentiation of H9c2 cardiomyoblasts. The results also showed that PLA/PANi nanofibrous sheets enhanced the cell-cell interaction and maturation. Also, the cardiomyocytes on PLA/PANi conductive nanofibrous sheets could continue beating spontaneously with regular contraction patterns after 21 days of culturing.

8.4.3 Other Biomedical Applications

There is potential for use of conductive polymer materials as components in various medical devices and implants. For example, Williams *et al.* [145] aimed to produce an elastomeric polymer electrode for neural implants using a composite approach. A nanocomposite elastomeric material based on a biocompatible polyurethane (Pellethane80A) and multi-walled CNTs was prepared using solution mixing method. The percolation threshold of the PU/MWCNT films was high (between 7 and 10 wt%), however, the conductivity of the films was comparable to other similar systems reported in literature (up to $\sim 7\times10^{-2}$ S cm^{-1}). Although the stiffness of PU increased with increased loading of nanotubes, all composites were highly flexible and maintained elastomeric properties. In another study, a nanostructured composite of PPy and graphene nanosheets was obtained for potential use as a power source in implantable medical microdevices such as implantable diagnosis/monitoring capsules, microscale drug pumps and implantable micro-sensing devices [146]. The composite was fabricated via a simple one-step synthesis route and was applied as a cathode material coupled with zinc anode in simulated body fluid and 0.2 M phosphate-buffered saline. The battery with PPy/reduced GO cathode achieved an energy density of 264 mWh g^{-1} in PBS and 210 mWh g^{-1} in SBF, which suggested its feasibility for effectively powering a broad range of implantable medical microdevices.

Poly(propylene fumarate) (PPF) is a high-strength polymeric biomaterial which possesses cross-linkable functional groups, making it attractive for biomedical applications, such as bone defect filling or long-term drug delivery system [147]. Lalwani *et al.* [148] reported the preparation of PPF based biodegradable and biocompatible nanocomposites incorporating a low concentration (0.01-0.2 wt%) of graphene (single-walled and multi-walled GO nanoribbons and GO nanoplatelets) for the development of implants for bone tissue engineering applications. The resulting nanocomposites, especially GO nanoplatelets-PPF, exhibited significantly improved mechanical properties compared to neat PPF materials.

8.5 Conclusions and Future Direction

In summary, numerous literature studies have focused on developing electrically conducting composites to overcome the limitations of conventional conducting polymers for biomedical applications. Improving biocompatibility, processability and mechanical properties as well as retaining good electrical conductivity have been the major objectives. Composites of conducting polymers with non-conducting polymers and conducting nanoparticles, along with the synthesis of copolymers incorporating oligomer segments of conducting polymers are the main approaches investigated in the recent years. Achieving high electrical conductivity generally comes with a significant compromise in mechanical properties. Blends of conducting polymers and non-conducting polymers typically provide materials with improved cell compatibility but with poor conductivity, primarily due to the poor compatibility and lack of organized conducting pathways as a consequence. Incorporation of conducting nanoparticles into non-conducting polymers with good mechanical properties has paved the way to achieve good conductivity, but the need to have high loading of nanoparticles invariably leads to rather brittle materials. Among the conducting nanoparticles/nanofillers, graphene appears to be more attractive due to its high conductivity and ease of incorporation, resulting in composites with good electrical conductivity at a relatively low percolation threshold. A major challenge is to formulate composites so that the nanoparticles organize to form conducting pathways within the composite matrix at a relatively low loading. In more recent research, conductive composites of polyurethane and biocompatible natural polymers have been investigated for biomedical applications with some degree of success in achieving composites with reasonable mechanical performance, biocompatibility and conductivity.

Most studies have focused on investigating the use of conductive composites in biosensors, tissue engineering and related applications. Composites based on graphene have received the greatest research interest for this purpose and several reports provide proof-of-concept studies for their use as chemical and biosensors. A significant amount of further developmental work would be required to overcome many prevalent challenges, before advancing these for commercial application.

Considering the range of properties and the level of conductivity achieved with promising biocompatibility, the conductive composites are becoming attractive as substrates for a wide variety of tissue engineering applications. The ability to stimulate cells through the applications of an electrical signal and fabrication into fibers and membranes allow the use of these materials in neural implants, drug delivery devices and tissue engineering scaffolds. Although numerous literature reports on fabrication of such constructs are available, data on their performance in *in vivo* environments is lacking. However, the *in vitro* results provide strong support for their potential for these applications.

Investigations into the development and use of electrically conducting polymer composites for biomedical applications are relatively new. However, the level of understanding of the properties of these materials and their limitations are sufficient to realize their potential in various biomedical applications. Biosensors, tissue engineering substrates and related applications appear to be the main areas of focus for further research and development work.

References

1. Jagur-Grodzinski, J. (2012) Biomedical applications of electrically conductive polymeric systems. *e-Polymers*, **12**(1) 722-740.
2. Heeger, A. J. (2001) Semiconducting and metallic polymers: The fourth generation of polymeric materials (nobel lecture). *Angewandte Chemie, International Edition*, **40**(14), 2591-2611.
3. Guimard, N. K., Gomez, N., and Schmidt, C. E. (2007) Conducting polymers in biomedical engineering. *Progress in Polymer Science*, **32**(8-9), 876-921.
4. Guo, B., Glavas, L., and Albertsson, A. C. (2013) Biodegradable and electrically conducting polymers for biomedical applications. *Progress in Polymer Science* **38**(9), 1263-1286.
5. Collier, J. H., Camp, J. P., Hudson, T. W., and Schmidt, C. E. (2000) Synthesis and characterization of polypyrrole–hyaluronic acid composite biomaterials for tissue engineering applications. *Journal of Biomedical Materials Research*, **50**(4), 574-584.
6. Lee, R. J., Temmer, R., Tamm, T., Aabloo, A., and Kiefer, R. (2013) Renewable antioxidant properties of suspensible chitosan-polypyrrole composites. *Reactive and Functional Polymers*, **73**(8), 1072-1077.
7. Takano, T., Mikazuki, A., and Kobayashi, T. (2014) Conductive polypyrrole composite films prepared using wet cast technique with a pyrrole-cellulose acetate solution. *Polymer Engineering and Science*, **54**(1), 78-84.
8. Jeong, S. I., Jun, I. D., Choi, M. J., Nho, Y. C., Lee, Y. M., and Shin, H. (2008) Development of electroactive and elastic nanofibers that contain polyaniline and poly(L-lactide-co-epsilon-caprolactone) for the control of cell adhesion, *Macromolecular Bioscience*, **8**(7), 627-637.
9. Kim, H. S., Hobbs, H. L., Wang, L., Rutten, M. J., and Wamser, C. C. (2009) Biocompatible composites of polyaniline nanofibers and collagen. *Synthetic Metals*, **159**(13) 1313-1318.
10. Untereker, D., Lyu, S., Schley, J., Martinez, G., and Lohstreter, L. (2009) Maximum conductivity of packed nanoparticles and their polymer composites. *ACS Applied Materials and Interfaces*, **1**(1), 97-101.
11. Du, F., Scogna, R. C., Zhou, W., Brand, S., Fischer, J. E., and Winey, K. I. (2004) Nanotube networks in polymer nanocomposites: Rheology and electrical conductivity. *Macromolecules*, **37**(24), 9048-9055.
12. Winey, K. I., Kashiwagi, T., and Mu, M. (2007) Improving electrical conductivity and thermal properties of polymers by the addition of carbon nanotubes as fillers. *MRS Bulletin*, **32**(4), 348-353.
13. Antunes, M., and Velasco, J. I. (2014) Multifunctional polymer foams

with carbon nanoparticles. *Progress in Polymer Science*, **39**(3), 486-509.
14. Das, T. K., and Prusty, S. (2013) Graphene-based polymer composites and their applications. *Polymer - Plastics Technology and Engineering*, **52**(4), 319-331.
15. Kuilla, T., Bhadra, S., Yao, D., Kim, N. H., Bose, S., and Lee, J. H. (2010) Recent advances in graphene based polymer composites. *Progress in Polymer Science*, **35**(11), 1350-1375.
16. Stejskal, J. (2013) Conducting polymer-silver composites. *Chemical Papers*, **67**(8), 814-848.
17. Bai, H., Li, C., and Shi, G. (2011) Functional composite materials based on chemically converted graphene. *Advanced Materials*, **23**(9), 1089-115.
18. Brigandi, P. J., Cogen, J. M., and Pearson, R. A. (2014) Electrically conductive multiphase polymer blend carbon-based composites. *Polymer Engineering and Science*, **54**(1), 1-16.
19. Gurunathan, T., Rao, C. R. K., Narayan, R., and Raju, K. V. S. N. (2013) Polyurethane conductive blends and composites: synthesis and applications perspective. *Journal of Materials Science*, **48**(1), 67-80.
20. Clingerman, M. L., King, J. A., Schulz, K. H., and Meyers, J. D. (2002) Evaluation of electrical conductivity models for conductive polymer composites. *Journal of Applied Polymer Science*, **83**(6), 1341-1356.
21. Zhao, P., Luo, Y., Yang, J., He, D., Kong, L., Zheng, P., and Yang, Q. (2014) Electrically conductive graphene-filled polymer composites with well organized three-dimensional microstructure. *Materials Letters*, **121**(0), 74-77.
22. Stefanescu, E., Daranga, C., and Stefanescu, C. (2009) Insight into the broad field of polymer nanocomposites: from carbon nanotubes to clay nanoplatelets, via metal nanoparticles. *Materials*, **2**(4), 2095-2153.
23. Jiang, H., Moon, K. S., Li, Y., and Wong, C. P. (2006) Surface functionalized silver nanoparticles for ultrahigh conductive polymer composites. *Chemistry of Materials*, **18**(13), 2969-2973.
24. Reich, S., Mazur, S., Avakian, P., and Wilson, F. C. (1988) Sintering of silver interlayers in polyimide films. *Journal of Applied Physics*, **62**(1), 287-292.
25. G. Kaur, R. Adhikari, P. Cass, M. Bown, P. Gunatillake, (2015) Electrically conductive polymers and composites for biomedical applications, *RSC Advances*, **5**(47) 37553-37567.
26. Kwok, S. W., Goh, K. H. H., Tan, Z. D., Tan, S. T. M., Tjiu, W. W., Soh, J. Y., Ng, Z. J. G., Chan, Y. Z., Hui, H. K., Goh, K. E. J. (2017) Electrically conductive filament for 3D-printed circuits and sensors. *Applied Materials Today*, **9**, 167-175.
27. Long, Y.-Z., Li, M.-M., Gu, C., Wan, M., Duvail, J.-L., Liu, Z., and Fan, Z. (2011) Recent advances in synthesis, physical properties and appli-

cations of conducting polymer nanotubes and nanofibers. *Progress in Polymer Science*, **36**(10), 1415-1442.
28. Park, J. S. (2010) Electrospinning and its applications. *Advances in Natural Sciences: Nanoscience and Nanotechnology*, **1**(4), 043002.
29. Stankovich, S., Dikin, D. A., Dommett, G. H. B., Kohlhaas, K. M., Zimney, E. J. Stach, E. A. Piner, R. D. Nguyen, S. T., and Ruoff, R. S. (2006) Graphene-based composite materials. *Nature*, **442**(7100), 282-286.
30. Kim, C. I. Oh, S. M., Oh, K. M., Gansukh, E., Lee, H. I., and Jeong, H. M. (2014) Graphenes for low percolation threshold in electroconductive nylon 6 composites. *Polymer International*, **63**(6), 1003-1010.
31. Li, Z., Luo, G., Wei, F., and Huang, Y. (2006) Microstructure of carbon nanotubes/PET conductive composites fibers and their properties. *Composites Science and Technology*, **66**(7-8), 1022-1029.
32. Xinzhao, X., Guoming, L., Dongyan, L., Guoxin, S., and Rui, Y. (2018) Electrically conductive graphene-coated polyurethane foam and its epoxy composites. *Composites Communications*, **7**, 1-6.
33. Zhang, H.-B., Zheng, W.-G., Yan, Q., Yang, Y., Wang, J.-W., Lu, Z.-H., Ji, G.-Y., and Yu, Z.-Z. (2010) Electrically conductive polyethylene terephthalate/graphene nanocomposites prepared by melt compounding. *Polymer*, **51**(5), 1191-1196.
34. Sun, X. Sun, H. Li, H. Peng, H. (2013) Developing polymer composite materials: Carbon nanotubes or graphene? *Advanced Materials*, **25**(37), 5153-5176.
35. Grossiord, N., Hermant, M.-C., and Tkalya, E. (2012) Electrically conductive polymer-graphene composites prepared using latex technology. In: *Polymer-Graphene Nanocomposites*, Mittal, V. (ed.), Royal Society of Chemistry, UK, pp. 66-85.
36. Fernandez-Merino, M. J., Paredes, J. I., Villar-Rodil, S., Guardia, L., Solis-Fernandez, P., Salinas-Torres, D., Cazorla-Amoros, D., Morallon, E., Martínez-Alonso, A., and Tascon, J. M. D. (2012) Investigating the influence of surfactants on the stabilization of aqueous reduced graphene oxide dispersions and the characteristics of their composite films. *Carbon*, **50**(9), 3184-3194.
37. Pinto, A. M., Martins, J., Moreira, J. A., Mendes, A. M., and Magalhaes, F. D. (2013) Dispersion of graphene nanoplatelets in poly(vinyl acetate) latex and effect on adhesive bond strength. *Polymer International*, **62**(6), 928-935.
38. Du, J., and Cheng, H. M. (2012) The fabrication, properties, and uses of graphene/polymer composites. *Macromolecular Chemistry and Physics*, **213**(10-11), 1060-1077.
39. Pinto, A. M., Goncalves, I. C., and Magalhaes, F. D. (2013) Graphene-based materials biocompatibility: A review. *Colloids and Surfaces B: Biointerfaces*, **111**, 188-202.
40. Singh, V., Joung, D., Zhai, L., Das, S., Khondaker, S. I., and Seal, S. (2011) Graphene based materials : Past, present and future. *Progr-*

ess in Materials Science, **56**(8), 1178-1271.

41. Shao, S., Zhou, S., Li, L., Li, J., Luo, C., Wang, J., Li, X., and Weng, J. (2011) Osteoblast function on electrically conductive electrospun PLA/MWCNTs nanofibers. *Biomaterials*, **32**(11), 2821-2833.
42. Nardecchia, S., Serrano, M. C., Gutierrez, M. C., Ferrer, M. L., and Monte, F. d. (2013) Modulating the cytocompatibility of tridimensional carbon nanotube-based scaffolds. *Journal of Materials Chemistry*, **B1**(24), 3064-3072.
43. Edwards, S. L., Church, J. S., Werkmeister, J. A., and Ramshaw, J. A. M. (2009) Tubular micro-scale multiwalled carbon nanotube-based scaffolds for tissue engineering. *Biomaterials*, **30**(9), 1725-1731.
44. McKeon-Fischer, K. D., Flagg, D. H., and Freeman, J. W. (2011) Coaxial electrospun poly(ε-caprolactone), multiwalled carbon nanotubes, and polyacrylic acid/polyvinyl alcohol scaffold for skeletal muscle tissue engineering. *Journal of Biomedical Materials Research - Part A*, **99** A(3), 493-499.
45. Lewitus, D. Y., Landers, J., Branch, J. R., Smith, K. L., Callegari, G., Kohn, J., and Neimark, A. V. (2011) Biohybrid carbon nanotube/agarose fibers for neural tissue engineering. *Advanced Functional Materials*, **21**(14), 2624-2632.
46. Kabiri, M., Soleimani, M., Shabani, I., Futrega, K., Ghaemi, N., Ahvaz, H. H., Elahi, E., Doran, M. R. (2012) Neural differentiation of mouse embryonic stem cells on conductive nanofiber scaffolds. *Biotechnology Letters*, **34**(7), 1357-1365.
47. Behan, B. L., DeWitt, D. G., Bogdanowicz, D. R., Koppes, A. N., Bale, S. S., and Thompson, D. M. (2011) Single-walled carbon nanotubes alter Schwann cell behavior differentially within 2D and 3D environments. *Journal of Biomedical Materials Research, Part A*, **96** A(1), 46-57.
48. Kaneto, K., Tsuruta, M., Sakai, G., Cho, W. Y., Ando, Y. (1999) Electrical conductivities of multi-wall carbon nano tubes. *Synthetic Metals*, **103**(1–3), 2543-2546.
49. Badaire, S., Poulin, P., Maugey, M., and Zakri, C. (2004) In situ measurements of nanotube dimensions in suspensions by depolarized dynamic light scattering. *Langmuir*, **20**(24), 10367-10370.
50. Mahdieh, Z. M., Mottaghitalab, V., Piri, N., and Haghi, A. K. (2012) Conductive chitosan/multi walled carbon nanotubes electrospun nanofiber feasibility. *Korean Journal of Chemical Engineering*, **29**(1), 111-119.
51. Gupta, P., Rajput, M., Singla, N., Kumar, V., and Lahiri, D. (2016) Electric field and current assisted alignment of CNT inside polymer matrix and its effects on electrical and mechanical properties. *Polymer*, **89**, 119-127.
52. Gogolewski, S. (1989) Selected topics in biomedical polyurethanes. A review. *Colloid & Polymer Science*, **267**(9), 757-785.

53. Xue, L., and Greisler, H. P. (2003) Biomaterials in the development and future of vascular grafts. *Journal of Vascular Surgery*, **37**(2), 472-480.
54. Rahimi, A., and Mashak, A. (2013) Review on rubbers in medicine: Natural, silicone and polyurethane rubbers. *Plastics, Rubber and Composites*, **42**(6), 223-230.
55. Choi, J. T., Dao, T. D., Oh, K. M., Lee, H. I., Jeong, H. M., and Kim, B. K. (2012) Shape memory polyurethane nanocomposites with functionalized graphene. *Smart Materials and Structures*, **21**(7), 9.
56. Rana, S., Cho, J. W., and Tan, L. P. (2013) Graphene-crosslinked polyurethane block copolymer nanocomposites with enhanced mechanical, electrical, and shape memory properties. *RSC Advances*, **3**(33), 13796-13803.
57. Yousefi, N., Gudarzi, M. M., Zheng, Q., Aboutalebi, S. H., Sharif, F., and Kim, J.-K. (2012) Self-alignment and high electrical conductivity of ultralarge graphene oxide-polyurethane nanocomposites. *Journal of Materials Science*, **22**(25), 12709-12717.
58. Kaur, G., Adhikari, R., Cass, P., Bown, M., Evans, M. D. M., Vashi, A. V., and Gunatillake, P. (2015) Graphene/polyurethane composites: fabrication and evaluation of electrical conductivity, mechanical properties and cell viability. *RSC Advances*, **5**(120), 98762-98772.
59. Kim, J. H,. Hwang, J.-Y., Hwang, H. R., Kim, H. S., Lee, J. H., Seo, J.-W. Shin, U. S., and Lee, S.-H. (2018) Simple and cost-effective method of highly conductive and elastic carbon nanotube/polydimethylsiloxane composite for wearable electronics. *Scientific Reports*, **8**(1), 1375.
60. Kim, Y., Zhu, J., Yeom, B., Di Prima, M., Su, X., Kim, J.-G., Yoo, S. J., Uher, C., and Kotov, N. A. (2013) Stretchable nanoparticle conductors with self-organized conductive pathways. *Nature*, **500**(7460), 59-63.
61. Hu, W. L., Wang, R. R., Lu, Y. F., and Pei, Q. B. (2014) An elastomeric transparent composite electrode based on copper nanowires and polyurethane. *Journal of Material Chemistry*, C, **2**(7), 1298-1305.
62. Al-Attabi, N. Y., Kaur, G., Adhikari, R., Cass, P., Bown, M., Evans, M., Gunatillake, P., Malherbe, F., and Yu, A. (2017) Preparation and characterization of highly conductive polyurethane composites containing graphene and gold nanoparticles. *Journal of Materials Science*, **52**(19), 11774-11784.
63. Guo, B. L., and Ma, P. X. (2014) Synthetic biodegradable functional polymers for tissue engineering: a brief review. *Science China-Chemistry*, **57**(4), 490-500.
64. Guo, B. L., Glavas, L., and Albertsson, A. C. (2013) Biodegradable and electrically conducting polymers for biomedical applications. *Progress in Polymer Science*, **38**(9), 1263-1286.
65. Qazi, T. H., Rai, R., Boccaccini, A. R. (2014) Tissue engineering of electrically responsive tissues using polyaniline based polymers: A

review. *Biomaterials*, **35**(33), 9068-9086.
66. Balint, R., Cassidy, N. J., and Cartmell, S. H. (2014) Conductive polymers: Towards a smart biomaterial for tissue engineering. *Acta Biomaterialia*, **10**(6), 2341-2353.
67. Bendrea, A. D., Cianga, L., and Cianga, I. (2011) Progress in the field of conducting polymers for tissue engineering applications. *Journal of Biomaterials Applications*, **26**(1), 3-84.
68. Das, T. K., and Prusty, S. (2012) Review on conducting polymers and their applications. *Polymer-Plastics Technology and Engineering*, **51**(14), 1487-1500.
69. Shi, G. X., Rouabhia, M., Wang, Z. X., Dao, L. H., and Zhang, Z. (2004) A novel electrically conductive and biodegradable composite made of polypyrrole nanoparticles and polylactide. *Biomaterials*, **25**(13), 2477-2488.
70. Shi, G. X., Rouabhia, M., Meng, S. Y., Zhang, Z. (2008) Electrical stimulation enhances viability of human cutaneous fibroblasts on conductive biodegradable substrates. *Journal of Biomedical Materials Research, Part* A, **84A**(4), 1026-1037.
71. Forciniti, L., Guimard, N. K., Lee, S., and Schmidt, C. E. (2010) Unique electrochemically synthesized polypyrrole: poly(lactic-co-glycolic acid) blends for biomedical applications. *Journal of Material Chemistry*, **20**(40), 8865-8874.
72. Subramanian, A., Krishnan, U. M., and Sethuraman, S. (2013) In vivo biocompatibility of PLGA-Polyhexylthiophene nanofiberscaffolds in a rat model. *Biomed Research International*, **2013**, Article ID 390518.
73. Xie, J. W., MacEwan, M. R., Willerth, S. M., Li, X. R., Moran, D. W., Sakiyama-Elbert, S. E., and Xia, Y. N. (2009) Conductive core-sheath nanofibers and their potential application in neural tissue engineering. *Advanced Functional Materials*, **19**(14), 2312-2318.
74. Kim, Y. D., and Kim, J. H. (2008) Synthesis of polypyrrole-polycaprolactone composites by emulsion polymerization and the electrorheological behavior of their suspensions. *Colloid & Polymer Science*, **286**(6-7), 631-637.
75. Basavaraja, C., Kim, W. J., Kim, D. G., and Huh, D. S. (2011) Synthesis and characterization of soluble polypyrrole-poly(epsilon-caprolactone) polymer blends with improved electrical conductivities. *Materials Chemistry and Physics*, **129**(3), 787-793.
76. Broda, C. R., Lee, J. Y., Sirivisoot, S., Schmidt, C. E., and Harrison, B. S. (2011) A chemically polymerized electrically conducting composite of polypyrrole nanoparticles and polyurethane for tissue engineering. *Journal of Biomedical Materials Research Part A*, **98A**(4), 509-516.
77. Wan, Y., Wu, H., and Wen, D. J. (2004) Porous-conductive chitosan scaffolds for tissue engineering, 1.Preparation and characterization.

Macromolecular Bioscience, **4**(9), 882-890.

78. Huang, J. H., Hu, X. Y., Lu, L., Ye, Z., Zhang, Q. Y., and Luo, Z. J. (2010) Electrical regulation of Schwann cells using conductive polypyrrole/chitosan polymers. *Journal of Biomedical Materials Research, Part A*, **93A**(1), 164-174.
79. Moutsatsou, P., Coopman, K., and Georgiadou, S. (2017) Biocompatibility assessment of conducting PANI/chitosan nanofibers for wound healing applications. *Polymers*, **9**(12), 687.
80. Li, M. Y., Guo, Y., Wei, Y., MacDiarmid, A. G., and Lelkes, P. I. (2006) Electrospinning polyaniline-contained gelatin nanofibers for tissue engineering applications. *Biomaterials*, **27**(13), 2705-2715.
81. Ghasemi-Mobarakeh, L., Prabhakaran, M. P., Morshed, M., Nasr-Esfahani, M. H., and Ramakrishna, S. (2009) Electrical stimulation of nerve cells using conductive nanofibrous scaffolds for nerve tissue engineering. *Tissue Engineering Part A*, **15**(11), 3605-3619.
82. Garner, B., Georgevich, A., Hodgson, A. J., Liu, L., and Wallace, G. G. (1999) Polypyrrole-heparin composites as stimulus-responsive substrates for endothelial cell growth. *Journal of Biomedical Materials Research*, **44**(2), 121-129.
83. Stewart, E. M., Liu, X., Clark, G. M., Kapsa, R. M. I., and Wallace, G. G. (2012) Inhibition of smooth muscle cell adhesion and proliferation on heparin-doped polypyrrole. *Acta Biomaterialia*, **8**(1), 194-200.
84. Li, Y. L., Neoh, K. G., Cen, L., and Kang, E. T. (2003) Physicochemical and blood compatibility characterization of polypyrrole surface functionalized with heparin. *Biotechnology and Bioengineering*, **84**(3), 305-313.
85. Zhang, L. L., Meng, S. Y., and Zhang, Z. (2011) Electroactivity and Stability of Polylactide/Polypyrrole Composites. *Journal of Biomaterials Science, Polymer Edition*, **22**(14), 1931-1946.
86. Meng, S. Y., Rouabhia, M., Shi, G. X., and Zhang, Z. (2008) Heparin dopant increases the electrical stability, cell adhesion, and growth of conducting polypyrrole/poly(L,L-lactide) composites., *Journal of Biomedical Materials Research, Part A*, **87A**(2), 332-344.
87. Akkouch, A., Shi, G. X., Zhang, Z., and Rouabhia, M. (2010) Bioactivating electrically conducting polypyrrole with fibronectin and bovine serum albumin. *Journal of Biomedical Materials Research, Part A*, **92A**(1), 221-231.
88. Qin, J. B., Huang, Z. B., Yin, G. F., Yang, A. N., and Han, W. (2016) Fabrication and protein conjugation of aligned polypyrrole-poly(L-lactic acid) fibers film with the conductivity and stability. *Journal of Nanoscience and Nanotechnology*, **16**(3), 2371-2377.
89. Boutry, C. M., Gerber-Horler, I., and Hierold, C. (2013) Electrically conducting biodegradable polymer composites (polylactide-polypyrrole and polycaprolactone-polypyrrole) for passive resonant circuits. *Polymer Engineering and Science*, **53**(6), 1196-1208.

90. Jin, L., Feng, Z. Q., Zhu, M. L., Wang, T., Leach, M. K., and Jiang, Q. (2012) A Novel fluffy conductive polypyrrole nano-layer coated PLLA fibrous scaffold for nerve tissue engineering. *Journal of Biomedical Nanotechnology*, **8**(5),779-785.
91. Patra, S. N., Bhattacharyya, D., Ray, S., and Easteal, A. J. (2009) Electrospun poly(Lactic acid) based conducting nanofibrous networks. *IOP Conference Series: Materials Science and Engineering*, **4**, 012020.
92. Guo, B. L., Finne-Wistrand, A., and Albertsson, A. C. (2010) Molecular architecture of electroactive and biodegradable copolymers composed of polylactide and carboxyl-capped aniline trimer. *Biomacromolecules*, **11**(4), 855-863.
93. Liu, Y. D., Cui, H. T., Zhuang, X. L., Wei, Y., and Chen, X. S. (2014) Electrospinning of aniline pentamer-graft-gelatin/PLLA nanofibers for bone tissue engineering. *Acta Biomaterialia*, **10**(12), 5074-5080.
94. Chen, J., Guo, B. L., Eyster, T. W., and Ma, P. X. (2015) Super stretchable electroactive elastomer formation driven by aniline trimer self-assembly. *Chemistry of Materials*, **27**(16), 5668-5677.
95. Guo, B. L., Finne-Wistrand, A., and Albertsson, A. C. (2011) Versatile functionalization of polyester hydrogels with electroactive aniline oligomers, *Journal of Polymer Science, Part A: Polymer Chemistry*, **49**(9), 2097-2105.
96. Hu, J., Huang, L., Zhuang, X., Zhang, P., Lang, L., Chen, X., Wei, Y., and Jing, X. (2008) Electroactive aniline pentamer cross-linking chitosan for stimulation growth of electrically sensitive cells. *Biomacromolecules*, **9**(10), 2637-2644.
97. Zhang, L., Li, Y., Li, L. C., Guo, B. L., and Ma, P. X. (2014) Non-cytotoxic conductive carboxymethyl-chitosan/aniline pentamer hydrogels. *Reactive & Functional Polymers*, **82**, 81-88.
98. Rivers, T. J., Hudson, T. W., Schmidt, C. E. (2002) Synthesis of a novel, biodegradable electrically conducting polymer for biomedical applications. *Advanced Functional Materials*, **12**(1), 33-37.
99. Zhang, Q. S., Yan, Y. H., Li, S. P., and Feng, T. (2010) The synthesis and characterization of a novel biodegradable and electroactive polyphosphazene for nerve regeneration. *Materials Science & Engineering C*, **30**(1), 160-166.
100. Guiseppi-Elie, A. (2010) Electroconductive hydrogels: Synthesis, characterization and biomedical applications, *Biomaterials*, **31**(10), 2701-2716.
101. Javadi, M., Gu, Q., Naficy, S., Farajikhah, S., Crook, J. M., Wallace, G. G., Beirne, S., and Moulton, S. E. (2018) Conductive tough hydrogel for bioapplications. *Macromolecular Bioscience*, **18**(2), 1700270.
102. Bajpai, A. K., Bajpai, J., and Soni, S. N. (2009) Designing polyaniline (PANI) and polyvinyl alcohol (PVA) based electrically conductive nanocomposites: preparation, characterization and blood compatible study. *Journal of Macromolecular Science, Part A*, **46**(8), 774-

782.
103. Shi, Z., Zang, S., Jiang, F., Huang, L., Lu, D., Ma, Y., and Yang, G. (2012) In situ nano-assembly of bacterial cellulose-polyaniline composites. *RSC Advances*, **2**(3), 1040-1046.
104. Zare, Y., and Shabani, I. (2016) Polymer/metal nanocomposites for biomedical applications. *Materials Science and Engineering C*, **60**, 195-203.
105. Choi, J., and Wang, N. S. (2011) Nanoparticles in biomedical applications and their safety concerns. In: *Biomedical Engineering: From Theory to Applications*, Fazel-Rezai, R. (ed.), IntechOpen, UK, pp. 299-314.
106. Syama, S., and Mohanan, P. V. (2016) Safety and biocompatibility of graphene: A new generation nanomaterial for biomedical application. *International Journal of Biological Macromolecules*, **86**, 546-555.
107. Sharma, P., Mehra, N. K., Jain, K., and Jain, N. K. (2016) Biomedical applications of carbon nanotubes: A critical review. *Current Drug Delivery*, **13**(6), 796-817.
108. Guazzo, R., Gardin, C., Bellin, G., Sbricoli, L., Ferroni, L., Ludovichetti, F. S., Piattelli, A., Antoniac, I., Bressan, E., and Zavan, B. (2018) Graphene-based nanomaterials for tissue engineering in the dental field. *Nanomaterials*, **8**(5), E349.
109. Gurunathan, S., and Kim, J. H. (2016) Synthesis, toxicity, biocompatibility, and biomedical applications of graphene and graphene-related materials. *International Journal of Nanomedicine*, **11**, 1927-1945.
110. Ou, L., Song, B., Liang, H., Liu, J., Feng, X., Deng, B., Sun, T., and Shao, L. (2016) Toxicity of graphene-family nanoparticles: A general review of the origins and mechanisms. *Particle and Fibre Toxicology*, **13**(1), 57.
111. Mundra, R. V., Wu, X., Sauer, J., Dordick, J. S., and Kane, R. S. (2014) Nanotubes in biological applications. *Current Opinion in Biotechnology*, **28**, 25-32.
112. Wolfram, J., Zhu, M., Yang, Y., Shen, J., Gentile, E., Paolino, D., Fresta, M., Nie, G., Chen, C., Shen, H., Ferrari, M., and Zhao, Y. (2015) Safety of nanoparticles in medicine. *Current Drug Targets*, **16**(14), 1671-1681.
113. Orecchioni, M., Bedognetti, D., Sgarrella, F., Marincola, F. M., Bianco, A., and Delogu, L. G. (2014) Impact of carbon nanotubes and graphene on immune cells. *Journal of Translational Medicine*, **12**(1), 138.
114. Costa, P. M., Bourgognon, M., Wang, J. T.,W., and Al-Jamal, K. T. (2016) Functionalized carbon nanotubes: From intracellular uptake and cell-related toxicity to systemic brain delivery. *Journal of Controlled Release*, **241**, 200-219.

115. Wang, X., Gu, X., Yuan, C., Chen, S., Zhang, P., Zhang, T., Yao, J., Chen, F., and Chen, G. (2004) Evaluation of biocompatibility of polypyrrole in vitro and in vivo. *Journal of Biomedical Materials Research, Part A*, **68**(3), 411-422.
116. Bauquier, S. H., McLean, K. J., Jiang, J. L., Boston, R. C., Lai, A., Yue, Z., Moulton, S. E., Halliday, A. J., Wallace, G., and Cook, M. J. (2017) Evaluation of the biocompatibility of polypyrrole implanted subdurally in GAERS. *Macromolecular Bioscience*, **17**(5), doi: doi: 10.1002/mabi.201600334.
117. Humpolicek, P., Kasparkova, V., Saha, P., and Stejskal, J. (2012) Biocompatibility of polyaniline. *Synthetic Metals*, **162**(7), 722-727.
118. Almira, R., Asta, K., Stasys, T., and Arunas, R. (2007) Biocompatibility of polypyrrole particles: an in-vivo study in mice. *Journal of Pharmacy and Pharmacology*, **59**(2), 311-315.
119. Stejskal, J., Hajna, M., Kasparkova, V., Humpolicek, P., Zhigunov, A., and Trchova, M. (2014) Purification of a conducting polymer, polyaniline, for biomedical applications. *Synthetic Metals*, **195**, 286-293.
120. Xu, H., Holzwarth, J. M., Yan, Y., Xu, P., Zheng, H., Yin, Y., Li, S., and Ma, P. X. (2014) Conductive PPY/PDLLA conduit for peripheral nerve regeneration. *Biomaterials*, **35**(1), 225-235.
121. Ghasemi-Mobarakeh, L., Prabhakaran, M. P., Morshed, M., Nasr-Esfahani, M. H., Baharvand, H., Kiani, S., Al-Deyab, S. S., and Ramakrishna, S. (2011) Application of conductive polymers, scaffolds and electrical stimulation for nerve tissue engineering. *Journal of Tissue Engineering and Regenerative Medicine*, **5**(4), e17-e35.
122. Abd Razak, S. I., Wahab, I. F., Fadil, F., Dahli, F. N., Md Khudzari, A. Z., and Adeli, H. (2015) A review of electrospun conductive polyaniline based nanofiber composites and blends: Processing features, applications, and future directions. *Advances in Materials Science and Engineering*, **2015**, Article ID 356286.
123. Guo, B., Lei, B., Li, P., and Ma, P. X. (2015) Functionalized scaffolds to enhance tissue regeneration. *Regenerative Biomaterials*, **2**(1), 47-57.
124. Silva, M., Alves, N. M., Paiva, M. C. (2018) Graphene-polymer nanocomposites for biomedical applications. *Polymers for Advanced Technologies*, **29**(2), 687-700.
125. Song, J., Gao, H., Zhu, G., Cao, X., Shi, X., Wang, Y. (2016) The construction of three-dimensional composite fibrous macrostructures with nanotextures for biomedical applications. *Biofabrication*, **8**(3), 035009.
126. Sayyar, S., Murray, E., Thompson, B. C., Gambhir, S., Officer, D. L., and Wallace, G. G. (2013) Covalently linked biocompatible graphene/polycaprolactone composites for tissue engineering. *Carbon*, **52**, 296-304.
127. Gizdavic-Nikolaidis, M., Ray, S., Bennett, J. R., Easteal, A. J., and

Cooney, R. P. (2010) Electrospun functionalized polyaniline copolymer-based nanofibers with potential application in tissue engineering. *Macromolecular Bioscience*, **10**(12), 1424-1431.

128. Gizdavic-Nikolaidis, M., Ray, S. Bennett, J., Swift, S., Bowmaker, G., and Easteal, A. (2011) Electrospun Poly(aniline-co-ethyl 3-aminobenzoate)/Poly(lactic Acid) nanofibers and their potential in biomedical applications. *Journal of Polymer Science, Part A: Polymer Chemistry*, **49**(22), 4902-4910.
129. Rahman, N. A., Gizdavic-Nikolaidis, M., Ray, S., Easteal, A. J., and Travas-Sejdic, J. (2010) Functional electrospun nanofibres of poly(lactic acid) blends with polyaniline or poly(aniline-co-benzoic acid). *Synthetic Metals*, **160**(17-18), 2015-2022.
130. Rahman, N. A., Feisst, V., Dickinson, M. E., Malmstrom, J., Dunbar, P. R., and Travas-Sejdic, J. (2013) Functional polyaniline nanofibre mats for human adipose-derived stem cell proliferation and adhesion. *Materials Chemistry and Physics*, **138**(1), 333-341.
131. Chen, J., Yu, M., Guo, B., Ma, P. X., and Yin, Z. (2018) Conductive nanofibrous composite scaffolds based on in-situ formed polyaniline nanoparticle and polylactide for bone regeneration. *Journal of Colloid and Interface Science*, **514**, 517-527.
132. Khorshidi, S., and Karkhaneh, A. (2018) Hydrogel/fiber conductive scaffold for bone tissue engineering. *Journal of Biomedical Materials Research, Part A*, **106**(3), 718-724.
133. Sawyer, S. W., Dong, P., Venn, S., Ramos, A., Quinn, D., Horton, J. A., and Soman, P. (2018) Conductive gelatin methacrylate-poly(aniline) hydrogel for cell encapsulation. *Biomedical Physics and Engineering Express*, **4**(1), 015005.
134. Hosseinzadeh, S., Rezayat, S. M., Giaseddin, A., Aliyan, A., and Soleimani, M. (2018) Microfluidic system for synthesis of nanofibrous conductive hydrogel and muscle differentiation. *Journal of Biomaterials Applications*, **32**(7), 853-861.
135. Moon, J.-M., Thapliyal, N., Hussain, K. K., Goyal, R. N., and Shim, Y.-B. (2018) Conducting polymer-based electrochemical biosensors for neurotransmitters: A review, *Biosensors and Bioelectronics*, **102**, 540-552.
136. Jain, R., Jadon, N., and Pawaiya, A. (2017) Polypyrrole based next generation electrochemical sensors and biosensors: A review. *TrAC Trends in Analytical Chemistry*, **97**, 363-373.
137. Naveen, M. H., Gurudatt, N. G., and Shim, Y.-B. (2017) Applications of conducting polymer composites to electrochemical sensors: A review. *Applied Materials Today*, **9**, 419-433.
138. Nambiar, S., and Yeow, J. T. W. (2011) Conductive polymer-based sensors for biomedical applications. *Biosensors and Bioelectronics*, **26**(5),1825-1832.
139. Okuzaki, H. (2017) Electromechanically active polymer soft actua-

tors. *Kobunshi*, **66**(7), 346-348.

140. Lian, H., Sun, Z., Sun, X., and Liu, B. (2012) Graphene doped molecularly imprinted electrochemical sensor for uric acid. *Analytical Letters*, **45**(18), 2717-2727.
141. Lei, W., Si, W., Xu, Y., Gu, Z., and Hao, Q. (2014) Conducting polymer composites with graphene for use in chemical sensors and biosensors. *Microchimica Acta*, **181**(7), 707-722.
142. Niu, X., Yang, W., Ren, J., Guo, H., Long, S., Chen, J., and Gao, J. (2012) Electrochemical behaviors and simultaneous determination of guanine and adenine based on graphene–ionic liquid–chitosan composite film modified glassy carbon electrode. *Electrochimica Acta*, **80**, 346-353.
143. Pal, R. K., Pradhan, S., Narayanan, L., and Yadavalli, V. K. (2018) Micropatterned conductive polymer biosensors on flexible PDMS films. *Sensors and Actuators B: Chemica*l, **259**, 498-504.
144. Wang, L., Wu, Y., Hu, T., Guo, B., and Ma, P.X. (2017) Electrospun conductive nanofibrous scaffolds for engineering cardiac tissue and 3D bioactuators. *Acta Biomaterialia*, **59**, 68-81.
145. Williams, C. M., Nash, M. A., and Poole-Warren, L. A. (2005) Electrically Conductive Polyurethanes for Biomedical Applications. Smart Materials, Nano-, and Micro-Smart Systems, Australia, doi: 10.1117/12.585321.
146. Li, S., Shu, K., Zhao, C., Wang, C., Guo, Z., Wallace, G., and Liu, H. K. (2014) One-step synthesis of graphene/polypyrrole nanofiber composites as cathode material for a biocompatible zinc/polymer battery. *ACS Applied Materials & Interfaces*, **6**(19), 16679-16686.
147. Ulery, B. D., Nair, L. S., and Laurencin, C. T. (2011) Biomedical applications of biodegradable polymers. *Journal of Polymer Science, Part B: Polymer physics*, **49**(12), 832-864.
148. Lalwani, G., Henslee, A. M., Farshid, B., Lin, L., Kasper, F. K., Qin, Y.-X., Mikos, A. G., and Sitharaman, B. (2013) Two-dimensional nanostructure-reinforced biodegradable polymeric nanocomposites for bone tissue engineering. *Biomacromolecules*, **14**(3), 900-909.

Chapter 9

Conducting Nanocomposites of Biodegradable Poly(Butylene Succinate) with Thermally Reduced Graphene Nanosheets

Pramoda K. Pallathadka,[a,*] Koh Xue Qi,[a] Akhila Khatta,[b] Gisha E. Luckachan[c] and Vikas Mittal[c,*,**]

[a]*Institute of Materials Research and Engineering, Materials Analysis and Characterisation Group, A*STAR (Agency for Science, Technology and Research), 4 Funsionopolis Way 117602, Singapore*
[b]*School of Materials Science and Engineering, Nanyang Technological University, 639798, Singapore*
[c]*Department of Chemical Engineering, The Petroleum Institute (part of Khalifa University of Science and Technology), Abu Dhabi, UAE*
**Corresponding authors:* pramoda-kp@imre.a-star.edu.sg; vik.mittal@gmail.com
***Current address*: Bletchington, Wellington County, Australia

9.1 Introduction

Plastics have made a major contribution to our daily life, however, they also pose serious environmental issues. Conservation and protection of the global environment has emerged as the main challenge in order to further improve the quality of life [1-2]. In this light, biodegradable polymeric materials may be seen as a promising alternative to non-degradable plastics. In fact, the development of biodegradable polymeric materials with excellent properties has become an irreversible trend worldwide. Poly(butylene succinate) (PBS), a typical synthetic biodegradable aliphatic polyester, is generally synthesized from diols and dicarboxylic acids. It has potential applications as a packaging material that can alleviate environmental pollution caused by commodity polymers as it is not only biodegradable, but also has a good elongation at break, along with other physical properties similar to polyethylene. Among the industrially available biodegradable polymers, PBS is also well known for its reasonable production cost and good performance characteristics [1-3]. However, it still lacks the conductivity as well as mechanical

Conducting Polymer Composites, edited by Vikas Mittal

and thermal properties required for industrial applications. Therefore, incorporating other suitable materials into the PBS matrix may be an effective way to enhance the utility of PBS.

Preparation and characterization of polymer nanocomposite materials have aroused a great deal of interest among researchers over the past few years. This is mainly due to their enhanced properties including mechanical strength, barrier behavior, thermal stability and heat deflection temperature. This property enhancement is mainly due to the high aspect ratio of the nanoparticle reinforcements in the polymer matrices [1-4]. Common nano-reinforcements are organoclay, carbon nanotubes (CNTs), graphite, polyhedral oligomeric silsesquioxanes and SiO_2. Graphite has a layered structure similar to clay constituted by a large number of graphene nanosheets held together by van der Waals forces [5-11]. They form aggregates when dispersed in the polymer matrices due to their poor miscibility with organic polymers, therefore, pretreatment and modification of graphite is carried out to achieve a homogeneous dispersion of graphene nanosheets. The surface treatment of graphene makes it more compatible with the polymer matrices providing good dispersion, and resulting in composites with excellent mechanical and electrical properties [3-6].

There are a number of studies on PBS nanocomposites with fillers like TiO_2, silica, clay, carbon black and CNTs, but very limited studies with graphene or graphene oxide are available [9-17]. Findings from previous studies [15-18] are persuasive that graphene is able to make polymeric materials much stronger and more protective. For instance, in a recent study, melt-processed bio-nanocomposites of PBS:chitosan (CS) blends with silica, aluminosilicate and graphene were compared for their performance [17]. 10% fraction of graphene enhanced the tensile modulus of the nanocomposites by >225%, whereas it was 80% and 65% for corresponding silica and silicate containing composites. Also, graphene nanocomposites had much higher storage modulus due to the formation of network structures of high aspect ratio graphene platelets at much lower filler weight fraction. Figure 9.1 also shows the relative storage modulus of the nanocomposites at a frequency of 1 rad/s as a function of filler concentration. The impact of increasing graphene content beyond 5 wt% was less significant indicating its effectiveness at lower weight fractions. On the other hand, silica and silicate fillers were effective in modulus enhancement only at high filler fraction.

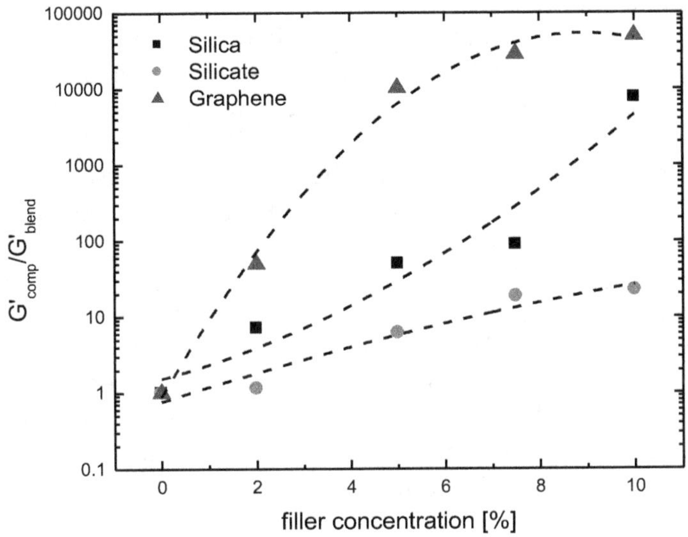

Figure 9.1 Relative storage modulus (G') of nanocomposites as a function of filler concentration (1 rad/s angular frequency). Reproduced from Reference 17.

The current study deals with the preparation of PBS/graphene nanosheets nanocomposites, especially for enhancing electrical conductivity. Investigations of the structure, morphology, thermal, electrical and mechanical properties of the obtained PBS nanocomposites have been carried out to evaluate the influence of filler reinforcement on the properties of neat PBS.

9.2 Experimental

9.2.1 Materials

PBS (Bionelle 1050 from Showa Highpolymer Co., Japan) was kindly donated by Prof. Shih from Chaoyang University, Taiwan. Thermally reduced graphene was produced by thermal exfoliation of precursor graphite oxide using modified Hummer's method. A short description for the preparation of graphene is as follows: 5 g of graphite powder was added with 125 ml concentrated H_2SO_4, to which 2.5 g of $NaNO_3$ was subsequently added. 15 g of $KMnO_4$ was added after stirring the mixture for 30 min while keeping in an ice-bath (5 °C), and the temperature was raised gradually to 35 °C. Deionized water was added after stirring for 2 h, and the temperature was raised to

100 °C. Subsequently, the mixture was quenched and diluted by pouring it into 1.5 l de-ionized water. This was followed by addition of 30% H_2O_2 until the evolution of bubbles of hydrogen stopped. The solution was then filtered, and the residues were dispersed in 2 l de-ionized water and were added with dilute HCl (6%) (2 l) to remove the SO_4^{2-} ions. The dispersion was filtered, and the filtrate was analyzed for SO_4^{2-} and Cl^- ions using $BaSO_4$ and $AgNO_3$ respectively. The washed graphite oxide was dried under vacuum at 60 °C for 24 h. Graphene was generated via thermal exfoliation by placing 1 g of graphite oxide in a long quartz tube with 25 mm internal diameter and sealed at one end. The sample was flushed with nitrogen, followed by insertion of the tube in a tube furnace preheated to 1050 °C. The tube was held in the furnace for 30 s. The confirmation of graphene formation was achieved through Raman spectroscopy, microscopy and X-ray diffraction.

9.2.2 Sample Preparation

Prior to the melt mixing process, the PBS resin was dried in a vacuum oven at 80 °C overnight in order to remove any traces of moisture which might cause its potential hydrolytic degradation during melt processing. The PBS/graphene nanocomposites were prepared using a co-rotating twin screw-extruder. The mixing temperature was kept around ~170 °C at a screw speed of 270 rpm. Disk-shaped (diameter of 25 mm and thickness of 1 mm) and rectangular samples (thickness of 1 mm) were prepared using a hot press operated at 150 °C and 10 kPa.

9.2.3 Characterization

The nanocomposites were characterized for their structure, thermal, electrical and mechanical properties. X-ray Diffraction (XRD) measurements were carried out on a Bruker GADDS powder X-ray diffractometer using a reflection geometry and CuKα radiation (wavelength λ= 0.154 nm) operated at 40 kV and 40 mA. Differential scanning calorimeter was used to determine enthalpy and peak temperatures for melting and crystallization. Dielectric analyzer (DEA) and dynamic mechanical analyzer (DMA) were used to determine the glass transition temperatures of the nanocomposites. MicroXact 8 A4P-200 4-point probe resistivity mapping system and nanoindentation measurements were performed to understand their conductivity and mechanical behavior. The effect of crystallization temperature on the crystalline morphology and spherulitic growth was studied via polarized optical microscopy (POM) coupled with a Linkam hot stage. The surface morphology was examined by transmission electron microscopy (TEM) using a JEOL 2000EX(II) (JEOL Ltd., Tokyo, Japan) equipment on the microtomed samples. Thermal properties of the samples were analyzed using Netzsch thermogravimetric analyzer (TGA). Nitrogen was used as a carrier gas and the scans were obtained from 50 to 700 °C at a heating rate of 20 °C/min.

FTIR-ATR measurements were conducted on PBS and PBS/graphene nanocomposites using a Bruker VERTEX 70 spectrometer equipped with Pike MIRacle ZnSe ATR accessory, DT-DLaTGS detector and KBr beam-splitter. ATR accessory had a ZnSe crystal of refractive index 2.4@1000 cm^{-1} with a depth penetration of 2 mm at 45° angle of incidence. IR spectra were acquired between 4000 cm^{-1} and 600 cm^{-1} using the OPUS software at 4 cm^{-1} resolution with an average of 60 scans. Normal ATR spectra of PBS was obtained by keeping a square piece of PBS on the ZnSe crystal surface. Stressed PBS spectra were collected by smashing the square piece of the polymer by applying mechanical stress using the piston on ATR accessory.

9.3 Results and Discussion

9.3.1 Structure of Neat PBS and PBS/Graphene Nanocomposites

XRD is an important tool for determining the crystal structure of the polymers in the nanocomposites. XRD analysis was performed at room temperature, and the patterns are shown in Figure 9.2. The diffraction patterns of the neat PBS and its graphene nanocomposite samples were quite similar. This indicated that the samples had the same crystal structure. The main characteristic diffraction peaks observed at 2θ values of 19.7, 21.9, and 22.8° were assigned to (020), (021) and (110) planes. In the case of nanocomposites, the diffraction peaks more or less were unchanged from neat PBS indicating that the incorporation of graphene nanosheets in the PBS matrix did not alter the crystal forms. Similar observation was made by Liang *et al.* [19] in the carbon fiber reinforced PBS composites. Additional insights into the crystal forms gained from the FTIR-ATR spectra are presented in Figures 9.3a and 9.3b. Pure PBS exhibited a

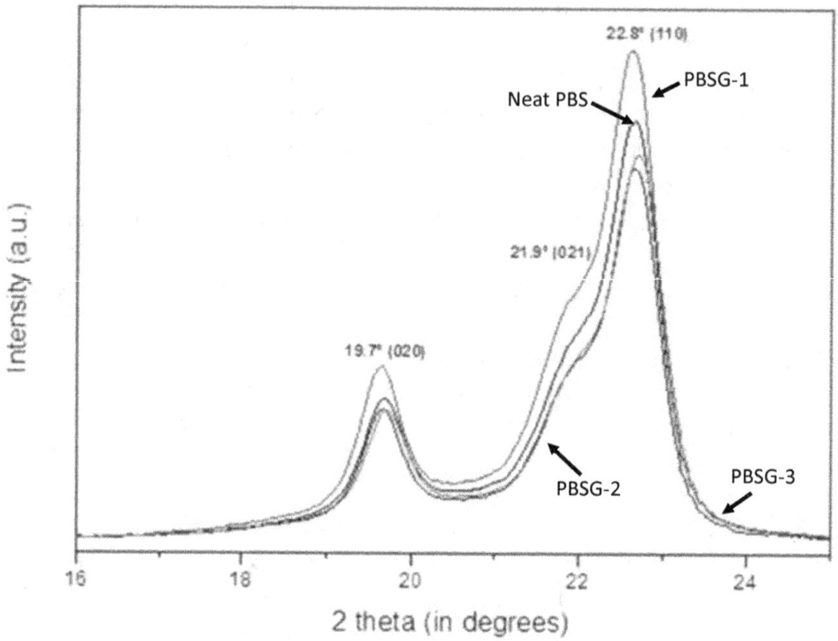

Figure 9.2 XRD patterns of PBS and PBS/graphene nanocomposites.

different spectrum on application of mechanical stress (as shown in Figure 9.3a), as compared to the normal PBS spectrum. ATR spectrum of PBS collected at normal conditions (spectrum (a) in Figure 9.3a) showed ester C=O stretching vibrations at 1714 cm^{-1}, CH_2 bending at 1470 cm^{-1} and 1414 cm^{-1} as well as CH_2 wagging at 1375 cm^{-1} and 1360 cm^{-1}. Multiple bands in the regions of 1300 cm^{-1} to 1210 cm^{-1} and 1200 cm^{-1} to 1100 cm^{-1} were attributed respectively to asymmetric and symmetric stretching vibrations of C-O-C linkage. O-C-C stretching vibrations were observed at 1044 cm^{-1}, and the region below 1000 cm^{-1} corresponded to CH_2 sketch deformations [20-23]. By the application of a mechanical stress, the intensity of C-O-C asymmetric vibrations decreased significantly and the peak maxima of symmetric vibration shifted from 1173 cm^{-1} to 1153 cm^{-1} (spectrum (b) in Figure 9.3a). A doublet corresponding to CH_2 twisting appeared at 1330 cm^{-1} to 1310 cm^{-1} [24]. Changes in the intensity and peak shift of CH_2 bending and wagging vibrations were also observed. Since IR is sensitive to conformational changes of polymer chains, it can be deduced from the ATR results that PBS chains underwent phase transitions with the applied stress.

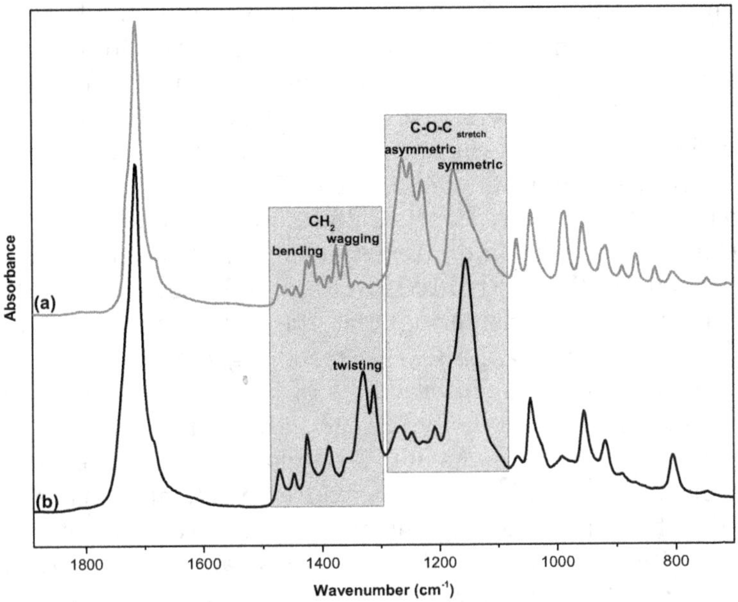

(a)

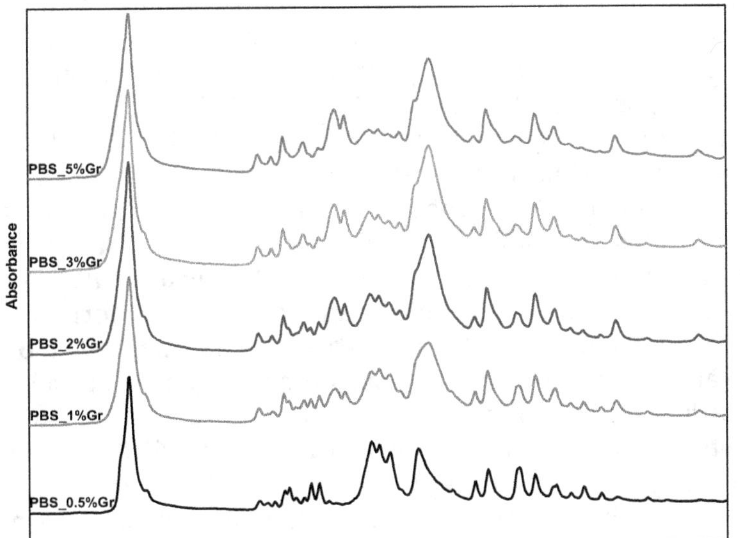

(b)

Figure 9.3 (a) FTIR-ATR spectra of PBS taken under normal and stressed conditions; (b) FTIR-ATR spectra of PBS/graphene nanocomposites with different percentages of graphene taken under normal conditions.

Ichikawa *et al.* [25,26] reported two crystalline modifications in PBS: α- and β- forms. The authors observed α-form in original fiber and β-form in a strained fiber. This solid-state crystal transition between α- and β-forms was reversible with respect to the successive application and release of strain. Therefore, the spectra obtained under stress and normal conditions in Figure 9.3a can be assigned respectively to β- and α-forms of PBS polymer. It is clear from Figure 9.3a that the shift and intensity changes of the IR bands in the region from 1400 cm^{-1} to 1100 cm^{-1} are highly sensitive to the α-β phase transitions. Considering that the bands at this region are strongly correlated to the C-O-C stretching modes resulting from different intermolecular interactions of ester groups, the relatively large peak shift and intensity change indicated that there were intense twists in the segments of C-O-C linkages during the α-β transition process of PBS [25,26]. Furthermore, the changes in the intensity of CH_2 bending and wagging mode and formation of new CH_2 twisting vibrations undoubtedly demonstrated a cooperative response of all PBS chain segments during the transition process [25, 26]. All spectra of PBS/G nanocomposites given in Figure 9.3b were collected under normal conditions. IR spectra similar to α form of PBS was observed at a low percentage of 0.5% filler content. On further increase of graphene content to 1%, new CH_2 twisting bands started to appear at 1330 cm^{-1} to 1310 cm^{-1}, followed by an intensity decrease of C-O-C asymmetric stretching at 1300 cm^{-1} to 1210 cm^{-1} and the peak maxima shift of C-O-C symmetric stretching from 1172 cm^{-1} to 1153 cm^{-1}. These changes became more prominent at high concentrations of graphene in the nanocomposites and finally a spectrum similar to β- form of PBS was observed for PBS/5% graphene nanocomposite. It should be noted that the β- form of PBS is unstable and is formed only under stressed conditions [27]. Gradual observation of this form in the nanocomposites under normal conditions as a function of graphene content indicated that the graphene platelets incorporated in the polymer matrix stabilized PBS in its β-form during the synthesis process.

To reveal the dispersion state of graphene in the PBS matrix, transmission electron microscopy analysis was carried out on the PBS/graphene nanocomposites with 2% and 5% filler content, as shown in Figure 9.4. As can be seen, the TEM images confirmed that the graphene nanosheets were homogeneously dispersed in the PBS matrix. Such dispersion is expected to enhance the mechanical and electrical properties of the composites significantly.

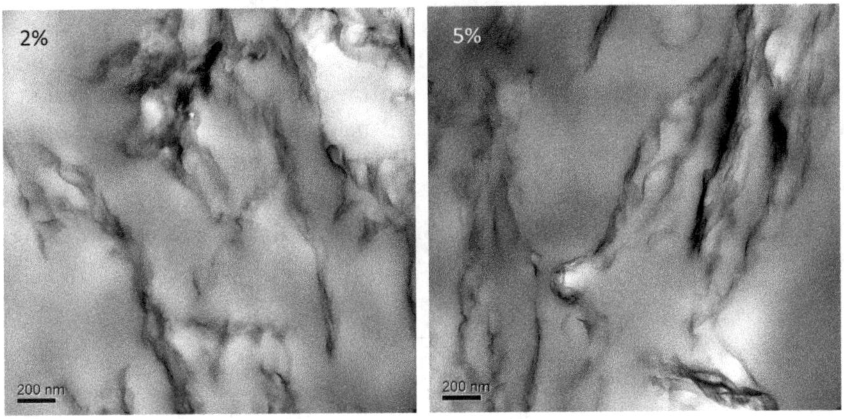

Figure 9.4 TEM micrographs of PBS/graphene nanocomposites with (a) 2% and (b) 5% filler content.

9.3.2 Thermal Properties of PBS/Graphene Nanocomposites

The 5% weight loss temperatures of PBS and PBS/graphene nanocomposites are tabulated in Table 9.1, and TGA curves under nitrogen atmosphere are shown in Figure 9.5. The 5% weight loss temperature for neat PBS was around 320 °C, and it progressively increased to 355 °C at 5 wt% graphene loading. This means that the T_d of the nanocomposite increased by 35 °C as compared to that of the pristine PBS sample indicating that the layered structure of graphene delayed the thermal degradation of the PBS matrix. It is, thus,

Table 9.1 Thermal properties of PBS/graphene nanocomposites

Sample	T_d (°C)	T_m (°C)	delta H (J/g)	% crystallinity
PBS	320	110.9	79.8	39.9
PBS/GNC0.5	330	111.0	64.3	32.5
PBS/GNC1	342	111.1	63.4	31.7
PBS/GNC2	347	111.1	62.6	31.3
PBS/GNC3	354	111.1	62.2	31.1
PBS/GNC5	355	111.2	61.3	30.7

clear that the incorporation of thermally reduced graphene nanosheets into PBS matrix enhanced the overall thermal degradation behavior of the neat PBS. This enhancement in the thermal stabilization might be attributed to the good dispersion of graphene nanosheets in the PBS matrix. The well dispersed nanosheets would result in an effective barrier to the volatile molecules produced from the PBS thermal decomposition. Similar results were also obtained in other filler reinforced PBS nanocomposites [19,28].

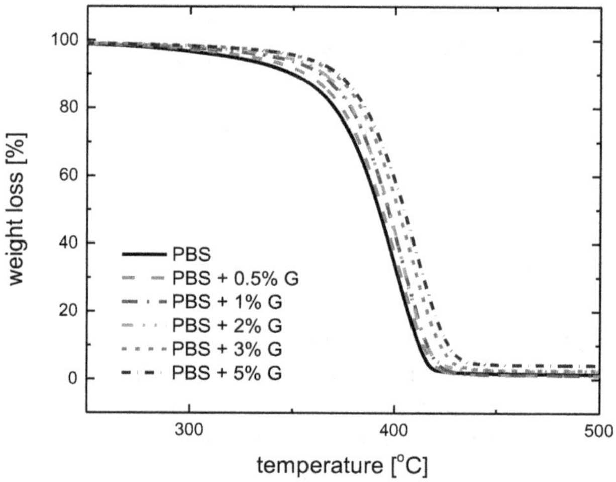

Figure 9.5 TGA thermograms of PBS/graphene nanocomposites as a function of filler content.

Differential scanning calorimetry (DSC) experiments on neat PBS and its graphene nanocomposite samples were carried out to understand the effect of incorporating graphene nanosheets on the crystallization behavior of the PBS matrix. The exothermic peak representing the peak crystallization temperature of PBS (73 °C) shifted to a higher temperature (77 °C) with the incorporation of graphene nanosheets implying that the crystallization of PBS was faster in the presence of graphene nanosheets. The graphene loading had heterogeneous nucleation effect on the crystallization of PBS. It is believed that the addition of graphene nanosheets induces the crystallization of PBS. However, compared to neat PBS, the T_m (~111 °C) was not much affected by the addition of the graphene nanosheets, probably due to their low content in the nanocomposites. In addition, both the melting and crystallization peaks in PBS/graphene nanocomposites were narrower than in neat PBS polymer, suggesting a narrower crystallite size distribution in PBS/graphene nanocomposites.

The crystallinity (χ_c) in all samples was calculated as follows:

$$\chi_c(\%) = \frac{\Delta H_m}{\Delta H_0} \times 100\%$$

where ΔH_m is the measured melting enthalpy (from DSC) and ΔH_o is the enthalpy of pure PBS crystal (200 J/g) [29]. As shown in Table 9.1, the crystallinity of PBS/graphene nanocomposites was reduced after the addition of the graphene. However, the change in PBS crystallinity was too small to induce any significant impact on the mechanical properties of the composites. Therefore, the enhancement of the strength and modulus of PBS/graphene nanocomposites can still be attributed to the homogeneous dispersion of graphene sheets in the polymer matrix and the strong interfacial interactions between the two components [30].

The nucleation effect of the nanofillers and the spherulite evolution during the melt crystallization for the neat PBS and its nanocomposites were monitored with the help of polarizing optical microscopy (POM). The samples were heated above their melting points and cooled to their peak crystallization temperature, as determined previously from DSC. Micrographs were taken of their crystal morphology, and the time taken for full crystallization within the field of view was recorded (Figure 9.6). The spherulites formed in neat PBS (Figure 9.6a) were regular and around 50 μm in size, which is typical for homogenous crystallization. On the other hand, the spherulites of the PBS/graphene nanocomposites were randomly spread out and non-uniform in size, as the filler platelets acted as nucleating sites for heterogeneous nucleated crystallization. Also, in agreement with the DSC data, crystallization took place within a shorter time in the composites as compared to neat PBS. It, thus, confirmed the fact that the graphene nanosheets dispersed in the nanocomposites served as nucleating agents for PBS crystals during their melt crystallization.

9.3.3 Electrical Properties of PBS/Graphene Nanocomposites

Table 9.2 presents the surface electrical conductivity of PBS as well as PBS/graphene nanocomposites. The electrical conductivity of the

(a)

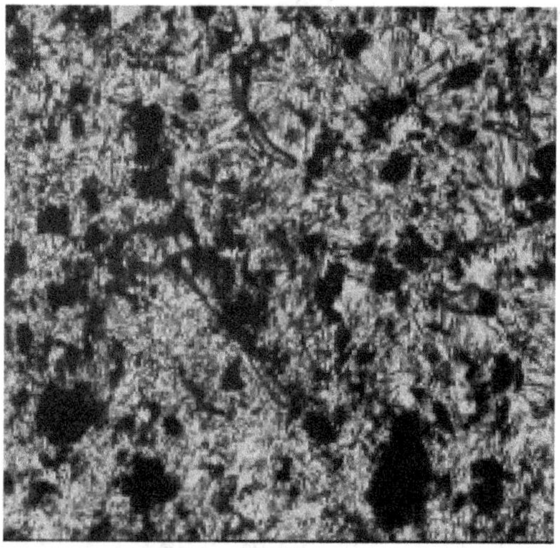

(b)

Figure 9.6 POM morphology of (a) neat PBS and (b) PBS-1 wt% graphene nanocomposite.

composites increased gradually with graphene loading. For instance. the electrical conductivity increased from 3.32×10^{-9} S/cm for neat PBS to 2.05×10^{-3} S/cm for 5 wt% PBS/graphene nanocomposite clearly indicating a 10^5-fold enhancement. This improvement can be attributed to the interconnected conducting networks formed by graphene nanosheets in the PBS matrix. Ali *et al.* [31] has also showed that the electrical conductivity of PBS/CNTs composites increased with addition of CNTs.

Table 9.2 Electrical conductivity of nanocomposites (average of 6 trials)

Sample	Sheet resistance	Thickness (cm)	Resistivity	Conductivity (S/cm)
PBS	2.01E+09	0.15	0.302	3.32 E-09
PBS/GNC1	1.38E+09	0.14	0.194	5.16 E-09
PBS/GNC2	1.37E+09	0.15	0.206	4.85 E-09
PBS/GNC5	3.25E+03	0.15	0.488	2.05 E-03

In many real time applications, a polymer nanocomposite will be subjected to different temperature and frequencies. Dielectric analyzer can provide reliable information about the relaxation behavior of these composites. In order to evaluate the effect of graphene nanosheets on temperature dependent relaxation behavior of PBS matrix, the thermo-mechanical properties of the materials were measured in the temperature range -100 to 100 °C, as shown in Figure 9.7. The glass transition temperature (T_g) of the nanocomposites shifted to a higher temperature with graphene content: the T_g values were ~2 to ~20 °C at the graphene nanosheets loading of 0, 1, and 2 wt% respectively. The increase in T_g with addition of graphene nanosheets was nearly 18 °C and correlated to the restricted movements of the PBS polymer chains due to their interaction with the graphene nanosheets. Figure 9.7a shows the variation in permittivity with temperature at 100 kHz. Inside the tested temperature range, graphene-reinforced nanocomposites showed substantially higher values of permittivity from 3.2 to 4.2 as compared to neat PBS (2.4 to 2.7). The increase in their permittivity up to 10 °C could be attributed to the immobilization of polymer chains onto the graphene surface, thus, increasing the effective volume of the filler.

9.3.4 Dynamic Mechanical Properties of PBS/Graphene Nanocomposites

Figure 9.8 shows the storage modulus (G') and tan δ plots with temperature for neat PBS and its graphene nanocomposites. G' is related

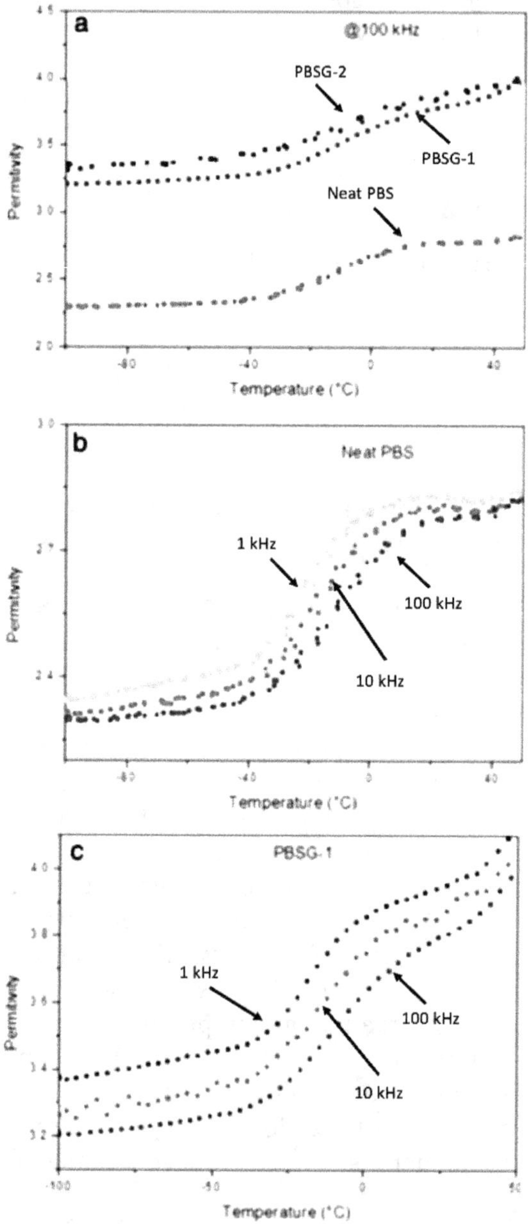

Figure 9.7 (a) Permittivity vs. temperature for neat PBS and its nanocomposites, (b) permittivity vs. temperature for neat PBS as a function of frequency and (c) permittivity vs. temperature for PBS/graphene nanocomposite (1 wt%) as a function of frequency.

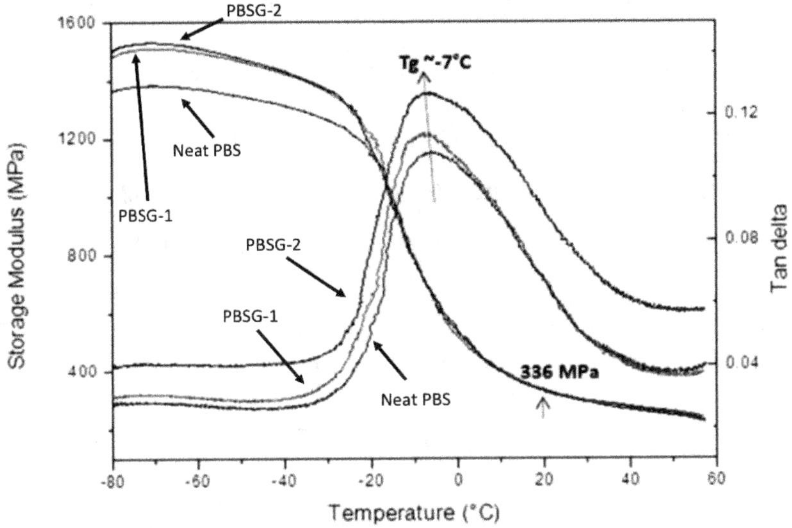

Figure 9.8 Storage modulus and tan δ plots with temperature for neat PBS and its nanocomposites.

to the ability of the material to store energy when an oscillatory force is applied, and tangent delta is related to the dampening ability of the composites. As shown in Figure 9.8, the storage modulus of PBS/graphene nanocomposites exhibited negligible increase with small enhancement in graphene loading up to the glass transition temperature range. Beyond this, the modulus of the nanocomposites was close to that of neat PBS. This implies that graphene nanosheets could enhance the rigidity of the nanocomposites due to their high surface area and good miscibility in the PBS matrix. The glass-transition temperature (T_g) is usually determined by the tan δ peak temperature in DMA. As shown in Figure 9.8, the T_g of ~7 °C for both neat PBS and nanocomposites indicated that the movements of the PBS polymer chains were restricted to a certain degree because of their grafting to the graphene nanosheets.

Nanoindentation tests were performed to measure the modulus and hardness of the PBS/graphene nanocomposites (Table 9.3). The values were averaged across 12 indents ranging over indentation depths of 1200-1500 nm. The incorporation of graphene nanosheets in the composites slightly improved the modulus and hardness of PBS matrix. The presence of homogeneously dispersed graphene nanosheets in the PBS matrix was likely responsible for the observed performance.

Table 9.3 Modulus and hardness measurement using nanoindentation

Sample	Modulus (GPa)	Hardness (GPa)
PBS	1.15 ± 0.39	0.07 ± 0.02
PBS/GNC1	1.02 ± 0.23	0.08 ± 0.02
PBS/GNC2	1.32 ± 0.26	0.09 ± 0.03

Figure 9.9 also shows the storage and loss modulus of nanocomposites as a function of filler fraction and angular frequency. The

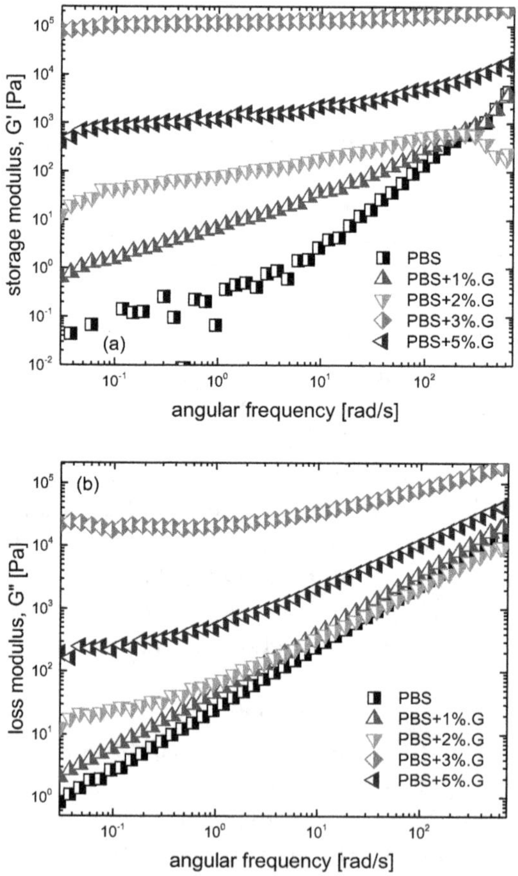

Figure 9.9 (a) Storage and (b) loss modulus of PBS nanocomposites as a function of filler content and frequency. Reproduced from Reference 18.

composites were observed to have significant enhancement in modulus till 3 wt% filler content. In fact, the composite with 3 wt% filler exhibited frequency independent behavior. However, on further increasing the filler loading to 5%, reduction in the modulus was observed probably due to the presence of large filler aggregates. In addition, the polymer exhibited lower storage modulus than loss modulus, due to the dominance of viscous nature. However, the nanocomposites exhibited higher storage modulus than loss modulus, the extent of which also increased with filler content, thus, indicating strong elastic nature due to the hindrance to the chain movement caused by the presence of exfoliated graphene nano-sheets in the polymer matrix.

The enhancement in thermal stability of the composites was also revealed from the viscosity of the composites as a function of angular frequency at 170 °C as well as strain sweep analysis at a temperature of 170 °C, as shown in Figure 9.10. The viscosity of the composites increased significantly with increasing filler fraction, up to 3 wt%. It indicated enhanced ability of the polymer to withstand shear and temperature. Similarly, the strain sweep analysis also indicated enhanced resistance as well as modulus as the filler content was increased till 3 wt%. Further increasing the filler content, however, led to a reduction in these properties.

9.4 Conclusions

Poly(butylene succinate) based nanocomposites with graphene nanosheets as filler were prepared by melt mixing method. Graphene nanosheets were observed to be uniformly dispersed in the PBS matrix. The XRD results showed that the addition of graphene nanosheets had almost no effect on the crystal structure. However, FTIR revealed that the β-form of PBS crystal was formed in the composites, and it was especially significant for 5 wt% graphene loading. The thermal stability of PBS/graphene composites was nearly 35 °C higher than that of neat PBS. The thermal, electrical and mechanical behavior of the nanocomposites with graphene nanosheets showed improvement over the original PBS matrix. The enhancement could be due to the homogeneous dispersion of graphene in the matrix, which also enhanced interfacial interactions. The POM and DSC studies also indicated that the well dispersed graphene nanosheets acted as nucleating agent for crystallization, thus, further compounding the reinforcing effect.

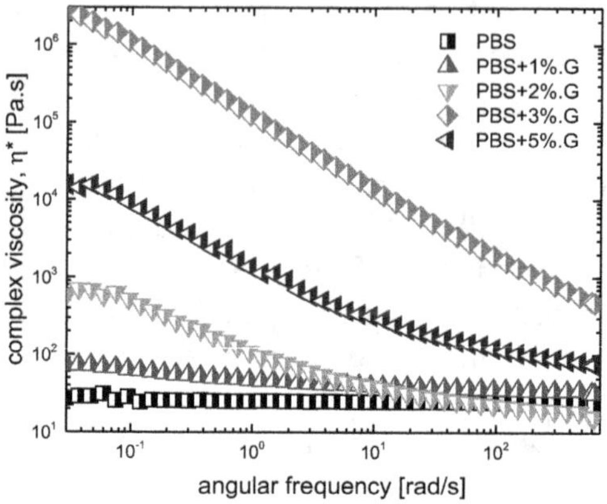

(a)

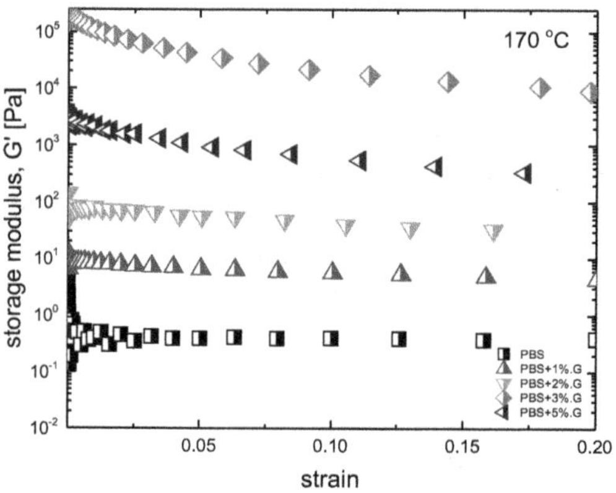

(b)

Figure 9.10 (a) Complex viscosity of PBS/graphene nanocomposites as a function of angular frequency and filler fraction at 170 °C, and (b) strain sweep analysis indicating the storage modulus of the PBS/graphene nanocomposites as a function of strain and filler fraction at 170 °C. Reproduced from Reference 18.

Summary of the Results

Poly(butylene succinate) is a biodegradable aliphatic polyester with semi-crystalline nature. It has a wide range of engineering applications owing to its attractive combination of good processability and mechanical properties. PBS nanocomposites with graphene nanosheets were prepared by melt compounding in the current study. The structure, thermal, electrical and mechanical properties of the composites were studied. Thermal stability of the nanocomposite increased by 35 °C as compared to neat PBS. An interesting phenomenon was also observed from the FTIR investigation. β-form of PBS crystals was formed in the composite samples with 5 wt% graphene loading, which otherwise forms in pure PBS only under stress. Most significantly, the electrical conductivity increased to ~10^5-fold for 5 wt% PBS/graphene nanocomposites.

Acknowledgements

The authors acknowledge the financial support of Agency for Science, Technology and Research (ASTAR), Singapore; Institute of Materials Research and Engineering (IMRE), Singapore and The Petroleum Institute, UAE.

The definite form of this work has also been published in Polymer Composites, 38(51), E42-E48. Reproduced with permission from Wiley.

References

1. *Polymer Nanocomposites: Synthesis, Characterization and Modelling*, Krishnamoorti, R., and Vaia, R. A. (eds.), American Chemical Society, USA (2001).
2. *Nanocomposite Science and Technology*, Ajayan, P. M., Schadler, L. S., and Braun, P. V. (eds.), Wiley-VCH, Germany (2004).
3. Bikiaris, D. (2013) Nanocomposites of aliphatic polyesters: An overview of the effect of different nanofillers on enzymatic hydrolysis and biodegradation of polyesters. *Polymer Degradation and Stability*, **98**, 1908-1928.
4. Ray, S. S., Vaudreuil, S., Maazouz, A., and Bousmina M. (2006) Dispersion of multi-walled carbon nanotubes in biodegradable poly(butylene succinate) matrix. *Journal of Nanoscience and Nanotechnology*, **6**, 2191-2195.
5. Pramoda, K. P., Linh, N. T. T., Zhang, C., and Liu, T. X. (2009) Multiwalled carbon nanotube nucleated crystallization behavior of biodegradable poly(butylene succinate) nanocomposites. *Journal of Applied Polymer Science*, **111**, 2938-2945.
6. Xu, Y., Hong, W., Bai, H., Li, C., and Shi, G. (2009) Strong and ductile poly(vinyl alcohol)/graphene oxide composite films with a layered structure. *Carbon*, **47**, 3538-3543.

7. Miyata, T., and Masuko. T. (1998) Crystallization behaviour of poly(tetramethylene succinate). *Polymer,* **39**, 1399-1404.
8. Papageorgiou, G. Z., and Bikiaris, D. N. (2005) Crystallization and melting behavior of three biodegradable poly (alkylene succinates). A comparative study. *Polymer,* **46**, 12081-12092.
9. Ray, S. S., Okamoto, K., and Okamoto, M. (2003) Structure-property relationship in biodegradable poly(butylenesuccinate)/layered silicate nanocomposites. *Macromolecules,* **36**, 2355-2367.
10. Shih, Y. F. (2009) Thermal degradation and kinetic analysis of biodegradable PBS/multiwalled carbon nanotube nanocomposites. *Journal of Polymer Science, Part B: Polymer Physics,* **47**, 1231-1239.
11. Papageorgiou, G. Z., Achilias, D. S., and Bikiaris, D. N. (2007) Crystallization kinetics of biodegradable poly(butylene succinate) under isothermal and non-isothermal conditions. *Macromolecular Chemistry and Physics,* **208**, 1250-1264.
12. Huang, J., Lu, X., Zhang, N. Yang, L., Yan, M., Liu. H., Zhang, G., and Qu, J. (2014) Study on the properties of nano-TiO_2/PBS composites prepared by vane extruder. *Polymer Composites,* **35**, 53-59.
13. Snowdon, M., Mohanty, A. K., and Misra, M. (2015) Melt processing and characterisation of bionanocomposites made from PBS bioplastic and carbon black. *Macromolecular Materials and Engineering,* **300**, 118-126.
14. Wang, X., Yang, H., Song, L., Hu, Y., Xing, W., and Lu, H. (2011) Morphology, mechanical and thermal properties of graphene-reinforced poly(butylene succinate) nanocomposites. *Composites Science and Technology,* **72**, 1-6.
15. Han, S. I., Lim, J. S., Kim, D. K., Kim, M. N., and Im, S. S. (2008) In situ polymerized poly(butylene succinate)/silica nanocomposites: Physical properties and biodegradation. *Polymer Degradation and Stability,* **93**, 889-895.
16. Bian, J., Han, L., Wang, X., Wen, X., Han, C., Wang, S., and Dong, L. (2010) Nonisothermal crystallization behavior and mechanical properties of poly(butylene succinate)/silica nanocomposites. *Journal of Applied Polymer Science,* **116**, 902-912.
17. Patwary, F., and Mittal, V. (2018) Melt processed PBS/chitosan bio-nanocomposites with silica, alumino-silicate and thermally reduced graphene. In: *Biopolymers Based Advanced Materials,* Mittal, V. (ed.), Central West Publishing, Australia, pp. 1-24.
18. Yassin, A, Al Ghailani, L., and Mittal, V. (2018) Mechanical and thermal properties of poly(butylene succinate)/graphene nanocomposites. In: *Biopolymers Based Advanced Materials,* Mittal, V. (ed.), Central West Publishing, Australia, pp. 269-292.
19. Liang, J., Ding, C., Wei, Z. Sang, L., Song, P., Chen, G., Chang, Y., Xu, J., and Zhang, W. (2015) Mechanical, morphology, and thermal properties of carbon fiber reinforced poly(butylene succinate) compo-

sites. *Polymer Composites*, **36**, 1335-1345.

20. Wang, X., Song, L., Yang, H., Lu, H., and Hu, Y. (2011) Synergistic effect of graphene on antidripping and fire resistance of intumescent flame retardant poly(butylene succinate) composites. *Industrial & Engineering Chemistry Research*, **50**, 5376-5383.
21. Yan, C., Zhang, Y., Hu, Y., Ozaki, Y., Shen, D., Gan, Z., Yan, S., and Takahashi, I. (2008) Melt crystallization and crystal transition of poly(butylene adipate) revealed by infrared spectroscopy. *Journal of Physical Chemistry B*, **112**, 3311-3314.
22. Phua, Y. J., Chow, W. S., and Mohd. Ishak, Z. A. (2013) Reactive processing of maleic anhydride-grafted poly(butylene succinate) and the compatibilizing effect on poly(butylene succinate) nanocomposites. *Express Polymer Letters*, **7**, 340-354.
23. Kim, H. S., Kim, H. J., Lee, J. W., and Choi, I. G. (2006) Biodegradability of bio-flour filled biodegradable poly(butylene succinate) biocomposites in natural and compost soil. *Polymer Degradation and Stability*, **91**, 1117-1127.
24. Socrates, G. (1994) *Infrared and Raman Characteristics Group Frequencies: Tables and Charts*, 3rd edition, John Wiley & Son, USA.
25. Ichikawa, Y., Suzuki, J., Washiyama, J., Moteki, Y., Noguchi, K., and Okuyama, K. (1994) Strain-induced crystal modification in poly(tetramethylene succinate). *Polymer*, **35**, 3338-3339.
26. Ichikawa, Y., Kondo, H., Igarashi, Y., Noguchi, K., Okuyama, K., and Washiyama, J. (2000) Crystal structures of a and b forms of poly(tetramethylene succinate). *Polymer*, **41**, 4719-4727.
27. Adamopoulou, E. (2012) *Poly(butylene succinate): A Promising Biopolymer*, Master's Thesis, University of Piraeus, Greece.
28. Bikiaris, D. (2011) Can nanoparticles really enhance thermal stability of polymers? Part II: An overview on thermal decomposition of polycondensation polymers. *Thermochimica Acta*, **523**, 25-45.
29. Wang, X., Zhou, J., and Li, L. (2007) Multiple melting behavior of poly(butylene succinate). *European Polymer Journal*, **43**, 3163-3170.
30. Hong, M. K., Ko, S. W., Park, J. H., Choi, H. J., and Kim, J. H. (2011) Multi walled carbon nanotube nanocomposites with biodegradable poly(butylene Succinate) and their physical characteristics. *Journal of Nanoscience and Nanotechnology*, **11**, 5352-5357.
31. Ali, F. B., and Raja, M. (2010) Thermal, mechanical, and rheological properties of biodegradable Poly(butylene succinate)/carbon nanotubes nanocomposites. *Polymer Composites*, **31**, 1309-1314.

Index

A

B

C

D

E

Index

F

G

H

I

L

M

R

S

T

U

V